PLANE AND GEODETIC SURVEYING

VOLUME TWO

PLANE AND GEODETIC SURVEYING

BY THE LATE DAVID CLARK M.A., B.Sc.
Professor of Civil Engineering, University of Dublin

VOLUME TWO

HIGHER SURVEYING

SIXTH EDITION REVISED AND LARGELY REWRITTEN BY
J. E. JACKSON, M.A.; F.R.A.S.; F.R.I.C.S.
Lecturer in Geodesy at the University of Cambridge (Retired)

CBS

CBS Publishers & Distributors Pvt. Ltd.

New Delhi • Bengaluru • Chennai • Kochi • Kolkata • Mumbai
Hyderabad • Nagpur • Patna • Pune • Vijayawada

ISBN: 81-239-1173-4

First Indian Edition: 1983
Reprint: 1999, 2004

This edition has been published in India by arrangement with
Constable and Company Limited, London

Published by **Satish Kumar Jain** and produced by **Varun Jain** for
CBS Publishers & Distributors Pvt. Ltd.,
4819/XI Prahlad Street, 24 Ansari Road, Daryaganj, New Delhi - 110002
delhi@cbspd.com, cbspubs@airtelmail.in • www.cbspd.com
Ph.: 23289259, 23266861, 23266867 • Fax: 011-23243014

Corporate Office: 204 FIE, Industrial Area, Patparganj, Delhi - 110 092
Ph: 49344934 • Fax: 011-49344935
E-mail: publishing@cbspd.com • publicity@cbspd.com

Branches:
- ***Bengaluru:*** 2975, 17th Cross, K.R. Road, Bansankari 2nd Stage, Bengaluru - 70 • Ph: +91-80-26771678/79 • Fax: +91-80-26771680 E-mail: cbsbng@gmail.com, bangalore@cbspd.com
- ***Chennai:*** No. 7, Subbaraya Street, Shenoy Nagar, Chennai - 600030 Ph: +91-44-26681266, 26680620 • Fax: +91-44-42032115 E-mail: chennai@cbspd.com
- ***Kochi:*** Ashana House, 39/1904, A.M. Thomas Road, Valanjambalam, Ernakulum, Kochi • Ph: +91-484-4059061-65 Fax: +91-484-4059065 • E-mail: cochin@cbspd.com
- ***Kolkata:*** 6-B, Ground Floor, Rameshwar Shaw Road, Kolkata - 700014 Ph: +91-33-22891126/7/8 • E-mail: kolkata@cbspd.com
- ***Mumbai:*** 83-C, Dr. E. Moses Road, Worli, Mumbai - 400018 Ph: +91-9833017933, 022-24902340/41 • E-mail: mumbai@cbspd.com

Representatives:
- Hyderabad: 0-9885175004
- Patna: 0-9334159340
- Vijayawada: 0-9000660880
- Nagpur: 0-9021734563
- Pune: 0-9623451994

Printed at:
Neekunj Print Process, Delhi (India)

CONTENTS

Section

CHAPTER I

FIELD ASTRONOMY—INTRODUCTORY

Section

CHAPTER II

FIELD ASTRONOMY—OBSERVATIONS

CHAPTER III

GEODETIC SURVEYING—ANGLE MEASUREMENT

Section

CHAPTER VI

GEODETIC COMPUTATION

PREFACE TO THE SIXTH EDITION

Readers acquainted with the fifth edition of this volume will immediately notice that this new edition has far fewer pages. Of course, successive editions of a work such as this generally tend to get bigger, by inclusion of material needed to cover modern developments in the subject, and by the natural reluctance to discard any text that could possibly still be relevant.

But nowadays, even staying the same size means becoming much more expensive, and a considerable reduction of wordage seemed highly desirable. This has been achieved by using more methods than just the removal of obsolete material, and some of the reduction has been possible through developments in the instruments and techniques of the subject itself. Overlapping with subjects treated in Vol. I has been avoided as far as possible, but subjects new in the present Volume are introduced as if new to the reader, so the text may seem very elementary in places.

Much of the matter formerly in Chapter VIII now seems irrelevant or inappropriate in this particular publication: the modern surveyor is likely to do his reconnaissance by examining air photographs or even by inspecting the area from an aircraft! Descriptive detail about particular instruments has been kept to a minimum; either it would be a repetition of old information already available at many sources, or the developments in instrumentation are going on so rapidly that any constructional details now described may be out of date before the words can be printed: the automatic recording of observations is on the way in; new EDM devices seem to be arriving on the market at the rate of two or three a year! Descriptions of purely arithmetic processes have been omitted wherever possible; this applies particularly to the formal solution of sets of simultaneous equations—an operation that can hardly be described as surveying: in any case, computing procedures are largely determined by the type of calculating equipment that is available to the surveyor, and here too the rate of appearance of new models is quite hectic. On the other hand, the modern surveyor may have very little computing to do himself, if his work is so organised.

In this new edition there are no references to photographic surveying, air survey, or photogrammetry. These are subjects that must either be omitted altogether or take up a considerable amount of space if treated adequately. There are now many excellent books devoted solely to these subjects.

In view of the above remarks, it is not surprising that this edition is completely newly written—no text has been directly transferred from the fifth edition. Only the general arrangement has been followed as closely as possible. However, the opportunity has been taken to attempt to get away from the somewhat stereotyped style of presentation that has developed over the years and is particularly noticeable in writings on field astronomy. It is hoped that this change will be acceptable and even pleasant to the student reader, but any comments about it will be noted with interest and no malice.

I am very grateful to numerous friends and colleagues whose opinions I have sought on many technical matters, and especially to Mr. Peter Dale who also has read through some of the typed drafts.

J. E. JACKSON

CAMBRIDGE
October 1973

CHAPTER I

FIELD ASTRONOMY—INTRODUCTORY

1.01 Latitude, Longitude, and Azimuth. Surveying on land is the measurement of angles, distances, and heights, and it results in the description of the relative positions of points either by the angles and lengths in geometrical figures, or as coordinates of positions in a reference system: the coordinate systems in use are spheroid latitude and longitude as described in Chapter VI, and plane rectangular coordinates in linear measure in a defined system of map projection as described in Chapter IX.

In Field Astronomy, the quantities observed, the other data used in the work, and the results of the calculations, are entirely angles. The observer's object is to obtain astronomical latitude, longitude and bearings at points on the Earth. It must be emphasised that astronomical observation cannot directly give any information about linear distances. Astronomical terminology sometimes gives unfamiliar meanings to common words, and one example is the expression 'astronomical position'; this simply means 'astronomical latitude and longitude', and must not be thought of as giving positional information in any of the systems mentioned in the preceding paragraph.

On the other hand, astronomical values are absolute and in no way arbitrary, in the sense that at any point the determinable astronomical quantities are real and unique: any two observers at the same point will get the same results, within the limits of their observational errors, and the results will not have to be referred to any arbitrary system, except that longitudes must be reckoned from some adopted zero meridian, often the meridian of the old Observatory at Greenwich, or perhaps the meridian of one of the points included in the survey.

The navigator at sea or in the air, if he cannot connect his position to land by direct observation or by radio navigation aids, observes latitude and longitude, and describes them as his 'position'. By calling one minute of latitude a 'nautical mile' the men of the sea have added another bit of unfamiliar usage!

Of course it is true that differences of astronomical values give a good idea of relative linear position on the Earth: for instance, two points differing by 1° of latitude and of the same longitude will be about 110 km apart. This quantity comes, however, from our knowledge of the size of the Earth and could not be found from astronomical observations alone. The same two astronomical 'positions' on the Moon would be about 30 km apart!

1.02 Uses of Field Astronomy. Because astronomical observation produces results that are always expressed in angle measure, its function in relation to Surveying and Geodesy must be carefully understood.

(1) In earlier times, explorers travelling in unmapped country often carried simple surveying instruments and made observations of latitude and bearing as they went along; observations of longitude were much more difficult and were liable to large errors. However, the information they obtained enabled them to plot their journeys, necessarily on rather small scale, and they could interpret their astronomical results on the assumption that they were on a simple geometrical surface such as a sphere of radius 6000 km. Salient topographical features could be added to these route surveys. Referring to the astronomical results, the explorer could direct his travel so that he arrived more or less accurately at the place where he intended to terminate his journey.

(2) In a more systematic manner, if astronomical latitudes and longitudes are observed at a number of points distributed over an area, they can be plotted on a network of lines suitably drawn on paper to represent meridians and parallels. The 'positions' so fixed can be used as control for the survey of detail by simple or graphical methods such as compass traversing and plane tabling. A considerable amount of mapping has in the past been done on this basis, and on small scales such as 1/200 000.

(3) The construction of charts of islands, soundings, etc. in oceanic regions is technically the same process as mentioned in the preceding paragraph. Astronomical 'positions' are observed on the islands or by standard navigational astronomy on the ships. Methods of navigational astronomy are introduced in Section 2.27.

(4) Bearings observed by simple methods may be used to check the accumulation of directional errors in ordinary surveys, especially in traverses composed of a large number of short lines.

(5) It is desirable that the plans of any large survey should be correctly oriented in relation to a north line. If the survey cannot be connected to existing reference points such as the national system, one or more astronomical observations will provide the orientation. Moreover, in extensive topographical mapping it is necessary to provide a datum of latitudes and longitudes so that the graticule of meridians and parallels can be drawn on the maps. See Section 8.02.

(6) Precise astronomical observations have essential parts to play in geodetic surveying. In combination with the linear and

angular measurements of triangulation and traverse surveys, the astronomical results are used in calculating the size and shape of the Earth, and in determining the ellipsoidal surfaces to be used as datum surfaces for extended mapping systems. See Section 8.03.

1.03 Celestial Sphere. In field astronomy, use may be made of the Sun, Moon, and Planets, as well as the Stars. However, the stars are so numerous and make such excellent signals on which to direct the telescope, that the most precise results are obtainable from them. Moreover, stellar astronomy is easier to understand and simpler to compute, therefore consideration of other objects mentioned will be deferred till after the use of the stars has been discussed.

That stellar astronomy is comparatively simple is due mostly to the great distances of the stars. One effect of this is that the stars' own movements about in space are almost imperceptible, the familiar patterns of the constellations have remained the same for centuries, and we can identify each star by its position in the pattern. Another effect is that the directions of the stars as seen from the Earth are practically independent of the movements of the Earth or the position of an observer on the Earth: even the nearest star is so far away that the directions to it from two diametrically opposite points on the Earth will differ by less than 0·000 1 second. Because of these circumstances, it is permissible and convenient to regard the stars as point sources of light fixed on the inside of a colossal hollow sphere, the *celestial sphere*, having a radius effectively infinite, and the Earth at its centre.

Because there are so many stars, distributed all over the sky, the observer can choose those that are in the positions to give optimum accuracy of the calculated results.

1.04 Apparent Movement of Stars. Anyone who watches the stars on successive nights will become familiar with the pattern of the constellations and will also note that they pass steadily across the sky, night after night, each star apparently tracing out the same path each time. This phenomenon is of course simply an apparent movement due to the actual steady rotation of the Earth about its axis. An observer in middle northern latitudes, looking towards the northern parts of the sky, will see that the stars move round in circular concentric paths as shown in Fig. 1; the centre P of the tracks is of course a point where the Earth's axis of rotation meets the celestial sphere. For the present we can assume that the axis of rotation maintains a fixed direction in space (see Section 1.11); it is a particular diameter of the celestial sphere and it determines the positions of two *celestial poles*.

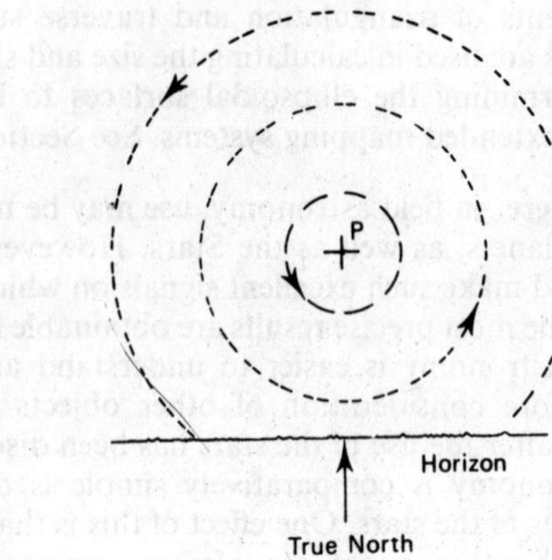

FIG. 1. NORTH POLAR SKY.

1.05 True North: Azimuth: Transit. As seen from a point on Earth in the northern hemisphere, the horizontal direction of the pole P in Fig. 1 is the definition of *true or astronomical north*. If a telescope of a theodolite could be set on point P (in fact there is no star exactly at this point), and then brought down to horizontal, it would be set in the direction of true north; then if the instrument is swung to the right about its vertical axis the angle turned is the *true or astronomical bearing*. Astronomical bearing is usually called *azimuth*, and this word will henceforth be used when the astronomical bearing is intended. It is usually numbered clockwise round to 360°.

If the angle turned is exactly 180° the telescope will be directed at azimuth 180°. The observer (still at a middle northern latitude!) will now note that the stars' tracks in this part of the sky are somewhat as shown in Fig. 2. Each star appears into view at a particular point on the horizon, rises, goes down, then disappears at its

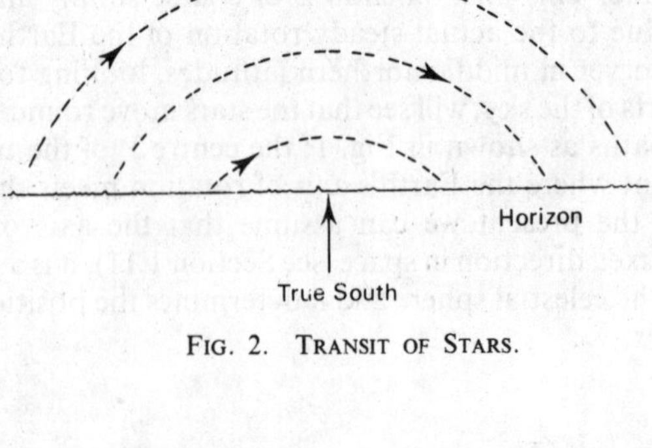

FIG. 2. TRANSIT OF STARS.

appropriate point on the horizon. When at exactly 180° azimuth, or due South, the star is at its greatest altitude and is said then to be at *transit*. For an observer in the southern hemisphere, the figures will be much the same; the arrows will point the other way, and the words 'south' and 'north' will be interchanged.

1.06 Basic Geometry. In astronomical observations for the purposes already mentioned, the observer is concerned with three directions, indicated in Fig. 3. We may suppose point O to be the

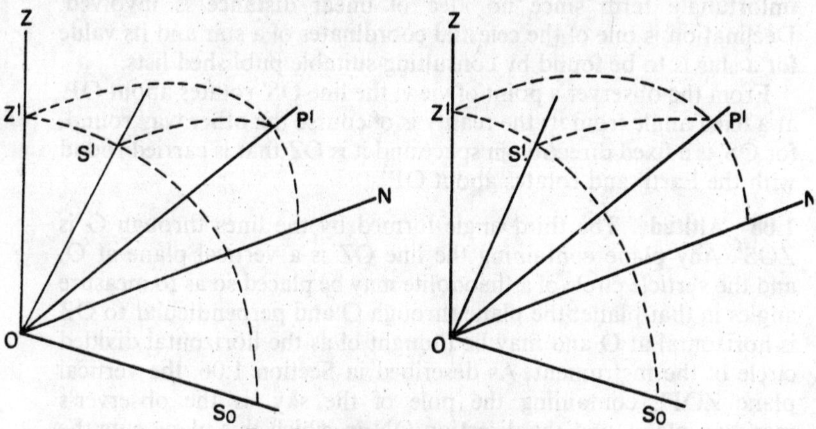

FIG. 3. ASTRONOMICAL GEOMETRY.
(this diagram is a stereoscopic pair)

centre of the theodolite, the point where the telescope axis meets the transit axis. OZ is the direction of the vertical and is the axis of rotation of the instrument, assuming that it has been properly 'levelled'. The point where OZ meets the celestial sphere is the observer's *zenith*. The line OP′ is the direction of the pole of the sky and is therefore parallel to the Earth's axis of rotation; it is the direction from the observer to point P in Fig. 1. These two directions at O are fixed in relation to the Earth, one by the Earth's gravity and the other by the direction of the Earth's axis. The plane defined by the lines OZ and OP′ is the meridian plane of point O; a star at transit is in this plane. The third line OS′ is a typical direction from the instrument to a star; this is not only different for each star, but for any one star it is continuously changing on account of the apparent movement of the stars across the sky. As the celestial sphere apparently rotates, the line OS′ for a particular star will appear to rotate steadily about OP.

1.07 Latitude: Declination. Now the angle ZOP′ is the angle between the axis of the Earth and the vertical at O; this is the definition of the *astronomical co-latitude* of the point. It is more usual to define *latitude* as the angle made by OZ with the equator, but this introduces an unnecessary concept, and the definition of the co-latitude as the angle ZOP′ seems geometrically more direct. Similarly, the angle P′OS′ is determined by the positions of the pole and the star on the celestial sphere, so for a particular star this angle is fixed, and is in fact the definition of the *co-declination* of the star; it is (90°-declination) and is also called *polar distance*, another unfortunate term since no idea of linear distance is involved. Declination is one of the celestial coordinates of a star and its value for a star is to be found by consulting suitable published lists.

From the observer's point of view, the line OS′ rotates about OP′ at a fixed angle from it; the reality is of course the other way round, for OS′ is a fixed direction in space and it is OZ that is carried round with the Earth and rotates about OP′.

1.08 Altitude. The third angle formed by the lines through O is ZOS′. Any plane containing the line OZ is a vertical plane at O, and the verticle circle of a theodolite may be placed so as to measure angles in that plane: the plane through O and perpendicular to OZ is horizontal at O and may be thought of as the horizontal divided circle of the instrument. As described in Section 1.06, the vertical plane ZOP′, containing the pole of the sky, is the observer's meridian plane, and the direction ON in which this plane cuts the plane horizontal at O is the direction of astronomical north (or south in the other hemisphere). The plane of OZ and OS′ is the vertical plane in which the star happens to be situated; the line OS_0 in this plane and perpendicular to OZ is the terrestrial direction of the star. Thus the angle NOS_0 is the astronomical azimuth of the star. Angle S_0OS' is the star's altitude, measurable on the instrument: ZOS′, the complement of the altitude, is called the *zenith distance* (another unfortunate confusion of length with angle!) and many theodolites are so graduated on their vertical circles that the readings on one face or the other are in fact zenith distances.

1.09 Hour Angle. The significance of the three angles formed by pairs of the lines OZ, OP′, and OS′ has been explained. Each pair of these lines also defines a plane, and two intersecting planes make a certain angle with each other. As already shown, the angle between the planes ZOP′ and ZOS′ is the azimuth of the star, illustrated in the angle NOS_0. As for the angle between planes ZOP′ and P′OS′, the rotation of the Earth causes it to change steadily with time. The period occupied by the Earth in making a complete rotation with reference to the stars is called a *sidereal day* (which is slightly shorter than the ordinary solar day of light and darkness), and this sidereal day is divided into the conventional hours, minutes, and seconds. Therefore, the angle we are now considering will change at the rate of 15° in one hour of sidereal time. This angle is called the *hour angle* of the star, a name indicating its close connection with time; indeed, it is customary to express hour angles in time units at the rate mentioned. The student must remember that an hour angle is *really* an angle, measuring the relative orientation of two planes at a particular moment, and is not a measure of a period of time. For computation purposes, it must usually be expressed in angle units.

Once in each sidereal day, the planes P′OS′ and P′OZ will coincide; at these moments, the hour angle of star S is zero and the star is at transit as explained in Section 1.05.

There is not complete agreement among writers as to the sign to be given to an hour angle—whether an hour angle east of the observer's meridian should be positive or negative. There can be no doubt if the descriptions east and west are used.

There is the third angle, that between the planes P′OS′ and S′OZ, which intersect along OS′. This angle is called *parallactic angle;* it does not often enter into the calculations and its significance will not be mentioned until the occasion arises.

1.10 Celestial Coordinates. We are familiar with the concept of a graticule of meridians and parallels for specifying the positions of points on a sphere; a very similar system is used for stars on the celestial sphere. The celestial sphere has two poles naturally located on it by the Earth's axis of rotation as already described, and from these points derives a system of celestial parallels; instead of latitude, the corresponding celestial quantity is called *declination.* For instance, if the angle P′OS′ in Fig. 3 is 20° the star must have declination 70° North. There is of course a *celestial equator* on which are situated all points for which the angle P′OS′ is 90°.

Then there are celestial meridians which may be thought of as semicircles joining the two celestial poles. As on Earth, we have to adopt a particular meridian as zero, but in the celestial case this is not arbitrary: astronomers have always used a special meridian, the position of which is defined in relation to certain celestial dynamical phenomena which need not concern us for the present. This zero meridian crosses the celestial equator at a point which has the curious name of *First Point of Aries*, and is usually denoted by the Greek letter γ. It is located as shown on Fig. 4 near the very conspicuous rectangle of stars in the constellations of Pegasus and Andromeda, to be seen on a clear evening in the autumn.

Suppose that on Earth the Greenwich meridian is taken as zero,

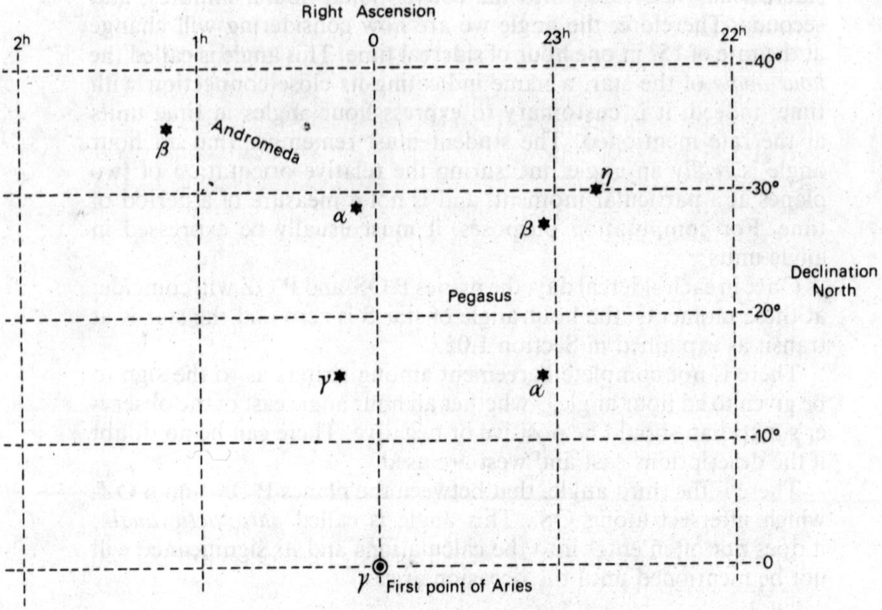

FIG. 4. CELESTIAL MAPPING.

and longitudes are measured eastwards round through 360° back to Greenwich. This is the system adopted for the celestial sphere except that, as mentioned above, the angles are usually expressed in time units of 1 hour instead of 15°, and are called *right ascension*. Thus, right ascensions go from 0 h 0 m 0 s to 24 h 0 m 0 s: they are in fact angles round the polar axis, corresponding exactly with longitude differences in the terrestrial case, but referred to the celestial meridian through the point γ.

1.11 Star Almanac. Declination and right ascension thus provide the mapping system for star positions on the celestial sphere, corresponding with latitude and longitude on the Earth. By use of precise observations at fixed observatories, astronomers 'map' the sky and publish lists of declinations and right ascensions of many thousands of stars. The celestial coordinates for a selection of 650 stars are given in the Star Almanac for Land Surveyors published annually by H.M. Stationery Office. More comprehensive lists are obtainable,* but the Star Almanac is enough for all ordinary purposes. The reader will notice that the coordinates vary slightly through

* The publication "Apparent Places of Fundamental Stars" gives coordinates of over 1500 stars: it is produced at the *Astronomisches Rechen-Insitut* in Heidelberg.

the year; the principal reason for this is the slow change of the direction of the Earth's axis in space, the precession, due to dynamical causes. Since the latitude and longitude system on Earth and the declination and right ascension system on the celestial sphere are by definition referred to the same polar axis, any change in the latter shows up as a change of celestial coordinates, but there will be no changes from this cause in the relative positional pattern of the stars themselves. Periodical variations in the apparent directions of stars are caused by aberration of light, as explained in Section 2.32. To obtain the current celestial coordinates of a star to be used in an astronomical calculation it is sufficient to know the date on which the observations were made, and to refer to a list of 'apparent places'.

1.12 Spherical Geometry. The geographical and celestial systems are not only similar in structure but are centred about the same axis. A consequence of this is that the vertical at a terrestrial place of latitude ϕ will meet the celestial sphere at a point having the same declination; in other words, since the Earth and the celestial sphere are in relative rotation about the same axis, a star on the celestial sphere having declination δ will always be overhead at some point on the Earth where the astronomical latitude is δ.

The three lines OS′, OP′, and OZ, and their associated angles which have been named above, form the basis of many situations in

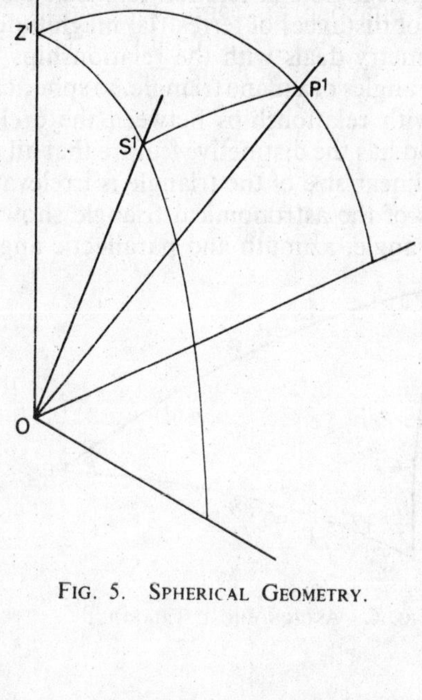

FIG. 5. SPHERICAL GEOMETRY.

field astronomy. This sort of geometry, and its mathematical treatment, are the subject matter of *spherical trigonometry*. If we imagine a sphere centred at O, these lines will cut it at points S′, P′, and Z′ as shown in Fig. 5: further, the three planes on which are situated the angles P′OZ′, Z′OS′, and P′OS′ are diametral planes of the sphere, so each will cut the sphere in a circle, called a *great circle*, which divides the sphere into two equal hemispheres. The curves P′Z′, Z′S′, and S′P′ are portions of these great circles. The figure P′Z′S′ is a *spherical triangle*.

1.13 Astronomical Triangle. It is clear that the spherical triangle embodies the angles that have been discussed as relevant to astronomical geometry. At the vertices of the triangle there are the azimuth at Z′, the hour angle at P′ and the parallactic angle at S′. The three circular arcs P′Z′, Z′S′, and P′S′ are called the *sides* of the triangle but are always expressed in angle measure since they really represent the angles P′OZ′, Z′OS′, and P′OS′ respectively, at the centre of the sphere. Thus, all six 'elements' of the spherical triangle are angles, and it is obvious that the linear size of the sphere is irrelevant: true, a sphere of twice the radius would have produced a triangle of twice the linear dimensions, but the six angles it represents would be the same.

Indeed, we may as well consider the spherical triangle to be situated on the celestial sphere, the point O for all practical purposes as being the centre of the celestial sphere, and the Earth's linear dimensions as being negligible in relation to stellar distances; only directions matter, not distances of terrestrial magnitude.

As plane trigonometry deals with the relationships between the three sides and three angles of a plane triangle, so spherical trigonometry is concerned with relationships between the 6 elements of a spherical triangle and has the distinctive feature that all the elements are angles, and the linear size of the triangle is irrelevant. We have seen that the angles of the astronomical triangle shown as PZS in Fig. 6 are the hour angle, azimuth and parallactic angle: the sides

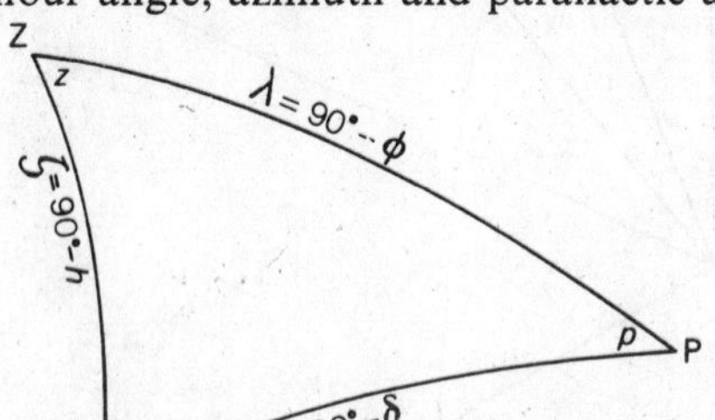

FIG. 6. ASTRONOMICAL TRIANGLE.

are PZ the observer's co-latitude or $(90° - \phi)$ where ϕ is the latitude, PS the co-declination of the star or $(90° - \delta)$ where δ is the declination, and ZS the zenith distance or $(90° - h)$ where h is the angular altitude (sometimes called elevation) which is measurable with a theodolite.

1.14 Spherical Trigonometry. If three elements of a spherical triangle are given, the others can be calculated; this is obvious, for instance, if the three sides are given, because that fixes the relative directions of the lines OP, OZ, and OS, and so fixes the geometry of the situation. To make use of the astronomical triangle, the surveyor must know three of the elements from observation or other sources, and must choose a formula containing the given elements and the one he wishes to compute. Many formulas are available, as can be seen by consulting any elementary treatise on spherical trigonometry; the choice will depend on the problem in hand and on the type of computing facility to be used.

1.15 Looking at the Astronomical Triangle. There are two ways of visualising the role played by astronomical phenomena in enabling the surveyor to determine the quantity he needs—azimuth, latitude, or longitude.

The reader may imagine himself to be a cosmic super-giant looking at the celestial sphere from outside and observing the behaviour of the geometrical figure formed by the three lines OP, OS, and OZ, and the associated spherical triangle PZS. For a particular observer the side PZ is of fixed value equal to his co-latitude: for a particular star the side PS is of fixed value equal to its co-declination. As time goes on, the shape of the spherical triangle is altered by the steady change of the hour-angle as mentioned in Section 1.09. There will be consequential changes in the azimuth, parallactic angle and altitude of the star. It is obvious therefore that there must be exact mathematical relationships between the variable and constant elements of the triangle.

Now the altitude of a star can be measured with a theodolite: this provides the third side (the variable side) of the triangle: ZS is the complement of the altitude. From the three sides, any of the angles of the triangle can be calculated. For instance, if h is the altitude, ϕ is the observer's latitude, and δ is the star's declination, a formula for the angle at Z is

$$\cos Z = (\sin \delta - \sin \phi \, . \, \sin h)/(\cos \phi \, . \, \cos h).$$

1.16 The View from Earth. Another way to visualise the phenomena is quite realistic; it is illustrated in Fig. 7. As a star tracks across the sky, its altitude and azimuth change, and indeed the star's path may be seen as a graph of azimuth against altitude: the vertical

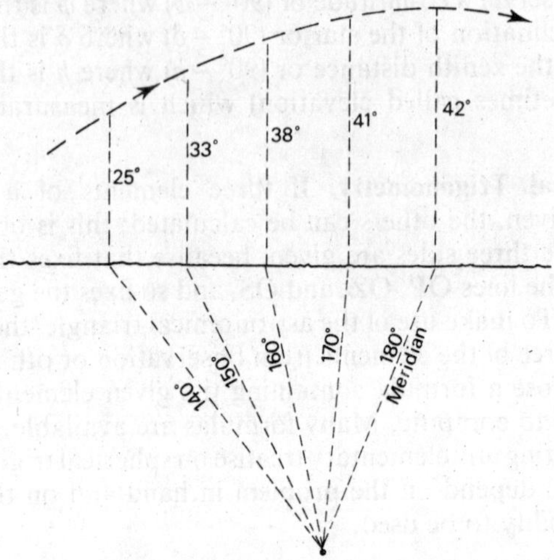

FIG. 7. TRACK OF A STAR.

lines in the figure represent altitudes, and the somewhat panoramic radiating lines are pictures of the various azimuth values of these star positions. The measurement of an altitude will locate a point on the graph and thus determine the azimuth of the star at the moment when the telescope was set on it. The altitudes on Fig. 7 are notional, merely for illustration.

Of course the surveyor is not interested in the ephemeral azimuths of stars at particular moments; he wants the azimuth of some fixed mark or reference object (RO) on the Earth. Thus, to make use of a star observation as just described, he must simultaneously measure both the altitude of the star and the horizontal angle between the star and the ground mark. The practical method is described in Section 2.04.

1.17 Latitude. In its geometrical principle, the determination of latitude is the simplest method of field astronomy. In Fig. 8, the circle represents the section of the celestial sphere by the observer's meridian at a particular moment when the star S_1 is at transit, due south in this case. The dot O at the centre represents the Earth, and OZ is the momentary position of the vertical axis of the observer's theodolite. OP is the direction of the axis of rotation of the Earth, and E is a point on the celestial equator, i.e. 90° from P. By definition, angle ZOE is the observer's astronomical latitude, as is also angle PON, the altitude of the Pole. The line OM, being perpendicular to OZ, is horizontal at the observer's position, so angle S_1OM is the altitude of the star at transit—and S_1OZ is the corresponding

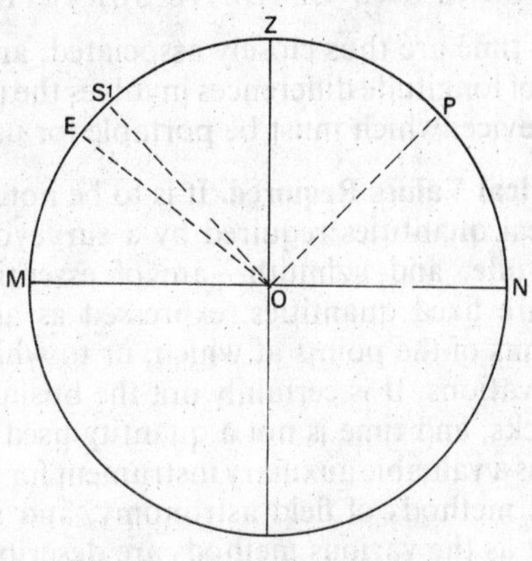

FIG. 8. MERIDIAN GEOMETRY.

zenith distance. But, by definition, the angle S_1OE is the star's declination, which is north in this case. So we get the simple relation $ZOE = ZOS_1 + S_1OE$, or $\phi = (\zeta_0 + \delta)$, where ζ_0 is the star's zenith distance at transit and δ is its declination. Other forms of this relation can be written; for instance EOM is the observer's co-latitude and S_1OM is the star's transit altitude h_0, so $90° - \phi + \delta = h_0$, or $\phi = 90° - h_0 + \delta$.

The direct application of this theory would involve 'catching' the star at exactly the transit moment and there would be no possibility of getting observations on each 'face' of the theodolite; in any case, a single observation is a poor thing, and it is desirable to be able to make several settings so as to minimise the effect of random errors. The practical method is described in Section 2.12.

1.18 Longitude Differences. Any point on the Earth has its astronomical meridian plane, determined by the lines OZ and OP′ in Fig. 3 for instance. The meridian planes of two points cut in a line which will be parallel to the Earth's axis of rotation, and the angle between these planes is the difference of astronomical longitude of the two points. Thus, longitude is essentially a difference property, a quantity associated with two points. When we refer to 'the longitude' of a point, there is always an implied reference to some other meridian—often the meridian of an instrument at the old Greenwich Observatory, but not necessarily so.

Referring to Section 1.09, it is obvious that as the Earth rotates, any star will pass from the meridian plane of one point to the meridian plane of another point in an interval of time directly proportional to the difference of longitude of the points—in fact at the rate of one hour of sidereal time per 15° of longitude difference.

Longitude and time are thus closely associated, and in practice the determination of longitude differences involves the use of an accurate time-keeping device, which must be portable for use in the field.

1.19 Geometrical Values Required. It is to be noted, however, that the astronomical quantities required by a surveyor or geodesist—latitude, longitude, and azimuth—are of essentially geometrical nature. They are fixed quantities, expressed as angles, associated with the positions of the points at which, or to which, the surveyor takes his observations. It is certainly not the business of a surveyor to regulate clocks, and time is not a quantity used in mapping, but a chronometer is a valuable auxiliary instrument for facilitating many of the practical methods of field astronomy, and its functions will become evident as the various methods are described.

In earlier days, accurate chronometers were needed for 'transporting time' for the benefit of travellers, particularly sailors, who could not determine their Greenwich Longitude from astronomical observations unless they also knew the Greenwich Time. Now, Greenwich Time can be received whenever required, at the turn of a switch on a radio equipment.

1.20 Time: Timepieces. Some concepts and definitions are needed, and it may help the reader to note that we customarily use the word *time* in two senses:

(1) as a scale of measurement of the duration of an interval between two events;
(2) as a label for a particular moment, e.g. 11 hours 11 minutes Greenwich Mean Time on November 11th 1918.

For purpose (1) it is necessary to have a device which goes at a constant rate and moves pointers or otherwise indicates the passage of time numerically in terms of a recognised definable unit of time. A *sidereal chronometer* is supposed to be adjusted so as to go at the rate which keeps pace with the rotation of the Earth in space: for instance, such a chronometer would always show the same reading whenever a particular star was at transit at a particular place. By the usual convention, a complete rotation of the Earth corresponds with 24 hours on the chronometer, hence one hour corresponds with 15° rotation and one second corresponds with 15 angle seconds of rotation. As explained in Section 1.09, the hour angle of a particular star in relation to a particular meridian on Earth will change at the above-mentioned rate.

1.21 Chronometer and Chronograph. A chronometer of good quality properly adjusted will keep time over a period of a few

FIG. 9. CHRONOMETER.
(photograph supplied by Thomas Mercer Limited)

hours to an accuracy far better than is needed for ordinary astronomical field observations. Such a chronometer is illustrated in Fig. 9. It is essentially a spring-powered clock mechanism regulated by balance-wheel escapement. The balance wheel makes a complete oscillation in one second and the 'second-hand' moves at half-second intervals: hence it is not possible to read the chronometer to closer than a half second. However, a device actuated by the escapement makes an electrical contact once a second: the contact may be connected through a suitable battery to a magnet which actuates a pen on a chronograph. The chronograph is a mechanism, operated also by clockwork, which moves a strip of paper at a steady rate of the order of a centimetre a second under the pen. When the magnet receives a pulse from the chronometer circuit, the pen is momentarily deflected from its normal position and a trace something like the upper line shown in Fig. 10 is produced on the paper strip. A continuous time scale is thus produced. An alternative to the pen for marking the paper strip is a needle brought down smartly by the electro-magnet: the needle is on a spring-loaded hinge so that it does not drag while in contact with the paper.

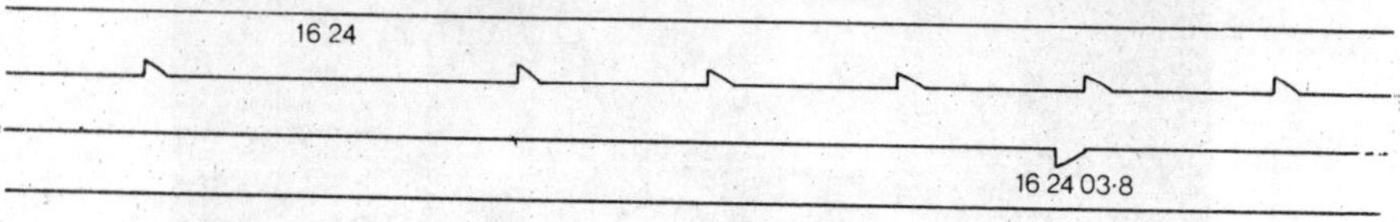

Fig. 10. Chronograph Tape Record.

A second pen on the chronograph is deflected by the observer when he closes a similar circuit by using a contact device held in the hand: this produces a displacement of another trace alongside the time marks as shown in Fig. 10.

In using the equipment just described, the observer must identify one or more of the time marks by writing the chronometer reading against it, and to facilitate this it is usual for the beginning of each minute to be indicated by the omission of the time mark. Thus the reading of the observer's signal shown on Fig. 10 is $16^h\ 24^m\ 03{\cdot}8^s$.

1.22 Local Sidereal Time. Purpose No. (2) mentioned in Section 1.20 refers to the problem of defining what we mean by *the time* at any particular moment: it is equivalent to specifying a zero moment when the clock should read $00^h\ 00^m\ 00^s$. The convention adopted is that at the moment when an observer's meridian intersects the point γ his Local Sidereal Time is $00^h\ 00^m\ 00^s$. Otherwise, the observer's local sidereal time at any moment is the hour angle of γ reckoned from his meridian westwards up to 360° or in time units 24 hours.

It is important to note that the above definition specifies an angle and there is no reference to the duration of a time interval. Time problems are easily illustrated by a diagram such as Fig. 11, which represents a view of the Earth and the celestial sphere from the north pole of the sky. The celestial sphere apparently rotates as shown, and OB is the observer's meridian. By the above definition, the

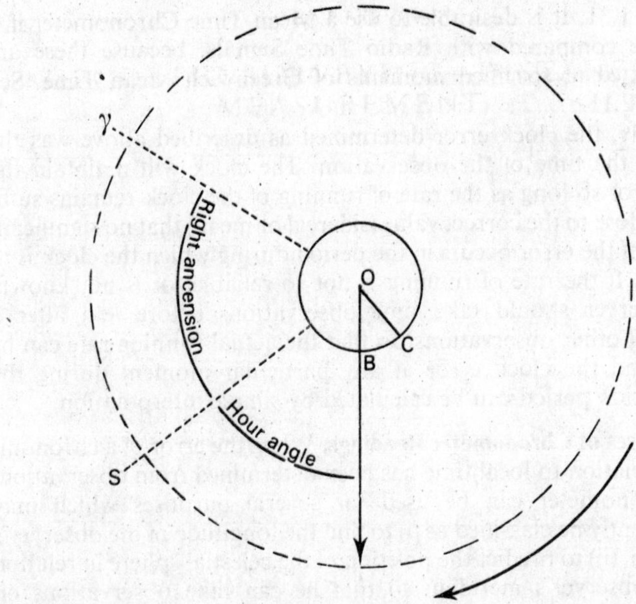

FIG. 11. SIDEREAL TIME.

Local Sidereal Time at meridian OB is the angle γOB: it is about 8 hours in the figure.

1.23 Finding Local Sidereal Time. Of course, there is no visible star at the point γ. To determine local sidereal time the hour angle of a star is found by measuring its altitude and solving the appropriate spherical triangle; in this way the angle BOS in the figure will be found. The angle γOS is the star's right ascension, which can be taken from the Star Almanac, or any other list of *apparent places* of stars. In the case illustrated, the star was observed west of meridian, and so the angle γOB is found by adding the hour angle to the right ascension, and the result is converted to time units: it is the local sidereal time of the moment when the observation was taken. If the observer had noted the clock readings of his observations he can compare them with the calculated sidereal time and determine by how much his clock was fast or slow on local sidereal time. See Section 2.07. The clock 'error' should be noted: it is unwise to try to adjust a chronometer to read the 'correct' time.

It may be remarked here that if, as is usual, a clock rated to go at Mean Time rate (see Section 1.28) is used, the calculated local sidereal time should be converted to local mean time as described in

Section 1.31. It is desirable to use a Mean Time Chronometer if it is to be compared with Radio Time Signals, because these are transmitted at specified moments of Greenwich Mean Time. See Section 2.11.

Strictly, the clock error determined as described above was the error at the time of the observation. The clock will maintain the same error so long as the rate of running of the clock remains sufficiently close to the correct value (sidereal or mean) that no significant change of the error occurs in the period during which the clock is to be used. If the rate of running is not so reliable, or is not known, the observer should take time observations before and after a period of other observations, so that the actual running rate can be found and the clock error at any particular moment during the observation period can be calculated by simple interpolation.

1.24 Uses of Chronometer Readings. When the error of a chronometer in relation to local time has been determined from observation, the chronometer can be used for several purposes which may conveniently be classified as (i) to find the longitude of the observer's meridian, (ii) to predict the position of the celestial sphere in relation to the observer's meridian so that he can take observations on celestial objects when they are in suitable positions, (iii) to calculate accurately the hour angle of a celestial object at the moment when it is observed so that this angle can be used as one of the known elements of the astronomical triangle.

1.25 (i) Difference of Longitude. The difference of longitude between any two places is the difference of their local times at any moment; see Section 1.18. On Fig. 11, the difference of longitude of the two meridians OB and OC is the difference of the two local sidereal times γOB and γOC. A surveyor may find the error of a mean-time clock in relation to his local mean time and compare the corrected clock reading with a radio time signal: the difference is his Greenwich Longitude. See Section 2.08.

1.26 (ii) Observation Programme. An obvious example is the prediction of the moment when a particular star will be at transit, so that observations for latitude can be taken during the proper period, see Section 2.12. A star transits when the local sidereal time is equal to its right ascension. Again, the moment when a star is at elongation, see Section 2.18, may be predicted. If observations with the astrolabe (Section 2.28) are to be made, it is essential to prepare a precise programme.

1.27 (iii) Hour Angle. For calculating the azimuth of a star, angle Z in the astronomical triangle, two of the known elements are the observer's latitude and the star's declination; if the error of the chronometer is accurately determined the third known element can be the hour angle. The observer must measure the horizontal angle between star and a fixed ground mark at the same time as reading the chronometer. See Section 2.17.

1.28 Mean Time. It is appropriate at this point to introduce the Mean Time system. For obvious reasons, it is convenient to use the Sun's apparent rotation around the Earth as the basis of a time unit, but it has long been realised that the Sun's apparent rotation rate

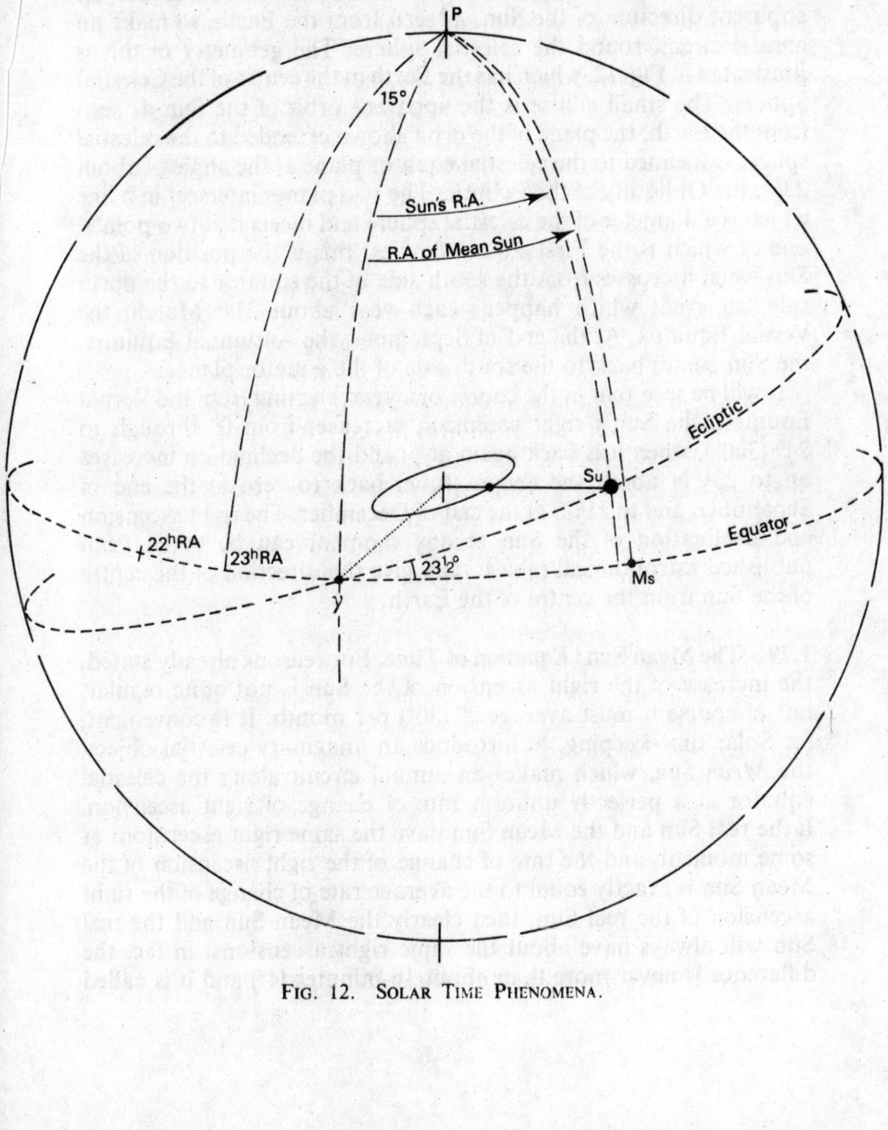

FIG. 12. SOLAR TIME PHENOMENA.

is not quite regular, and so a clock running at a constant rate cannot exactly keep pace with the Sun. It is therefore only possible to make a clock run at a rate equal to the average rotation rate of the Sun, in other words, the mean-time chronometer will go through its 24-hour cycle in the period equal to the average period of rotation of the Earth in relation to the Sun. The irregularity of the Sun's apparent rotation is principally due to the ellipticity of the annual orbit of the Earth, and the fact that the Earth's axis of rotation is not perpendicular to the plane of the orbit.

The movement of the Earth in orbit around the Sun causes the apparent direction of the Sun, as seen from the Earth, to make an annual circuit round the celestial sphere. The geometry of this is illustrated in Fig. 12, which has the Earth in the centre of the Celestial Sphere. The small ellipse is the apparent orbit of the Sun as seen from the Earth; the plane of the orbit shown extended to the celestial sphere is inclined to the celestial equator plane at the angle of about $23\frac{1}{2}°$, the Obliquity of the Ecliptic. The two planes intersect in a line which is a diameter of the celestial sphere and meets it at two points, one of which is the First Point of Aries: this is the position of the Sun when it crosses from the south side of the equator to the north side, an event which happens each year about 21st March, the Vernal Equinox. At the end of September, the Autumnal Equinox, the Sun passes back to the south side of the equator plane.

It will be seen that in the course of a year, starting from the Vernal Equinox, the Sun's right ascension increases from 0^h through to 24^h (360°), when it is back again at γ; and the declination increases up to $23\frac{1}{2}°$N about the end of June, back to zero at the end of September, and to $23\frac{1}{2}°$S at the end of December. The right ascension and declination of the Sun at any moment can be taken from published astronomical tables: they give the direction of the centre of the Sun from the centre of the Earth.

1.29 The Mean Sun: Equation of Time. For reasons already stated, the increase of the right ascension of the Sun is not quite regular, but of course it must average 2^h (30°) per month. It is convenient, for Solar time-keeping, to introduce an imaginary celestial object, the *Mean Sun*, which makes an annual circuit along the celestial equator at a perfectly uniform rate of change of right ascension. If the real Sun and the Mean Sun have the same right ascensions at some moment, and the rate of change of the right ascension of the Mean Sun is exactly equal to the average rate of change of the right ascension of the real Sun, then clearly the Mean Sun and the real Sun will always have about the same right ascensions: in fact the difference is never more than about 16 minutes (4°) and it is called the Equation of Time; it is the angle Su.P.Ms in Fig. 12. The interval between successive transits of the Mean Sun across the same meridian plane is a Mean Solar Day, and the Mean Sun passes over just 15° of terrestrial longitude in one hour of mean time.

1.30 Local Mean Time. The fictitious point Ms is used to define the local mean time, in much the same way as γ is used to define local sidereal time, except that there is a difference arising from our conventional way of reckoning mean time. At any moment the point Ms has a certain hour angle relative to an observer's meridian; if, for instance, the hour angle is 30° west of the observer his local mean time is 2 pm or 14^h in the 24 hour reckoning system. We see therefore that we can define local mean time as the 'western hour angle of Ms *plus* 12 hours'. If Ms is east of the observer's meridian, the hour angle is reckoned negative and the local mean time is less than 12^h: when the Ms is on meridian or due south, it is Local Mean Noon. It is to be noted that an observer's local time refers to the meridian where his instrument is set up, and must not be confused with the conventional time-keeping system of the country or zone in which he is situated. A Mean Solar Day is the duration of the interval between two successive transits of Ms across the same meridian. As stated above, the point Ms moves round the celestial equator at a constant rate of increase of right ascension; this rate is very accurately known.

1.31 Conversion of Time. It is now an easy matter to convert between local mean time and local sidereal time at any moment when either is known and the other is required. In Fig. 13 the outer circle represents the celestial equator and its centre P represents the Earth's axis of rotation seen end on. The small circle round P represents the Earth, which must be vastly exaggerated to show it at all, and PA is a meridian extended to meet the celestial sphere at A′. The Earth rotates anticlockwise on the diagram, carrying the meridian PAA′ with it. By definition the angle γPA′ (expressed in time units) is the local sidereal time for anyone on meridian PA: also by definition, the angle MsPA′ (in time measure) is the (local mean time *minus* 12 hours). Angle γPMs is the right ascension of the Mean Sun. We therefore have the simple relationships:

$$\text{LST} - (\text{RA of Ms}) = \text{LMT} - 12^h, \quad \text{or}$$
$$\text{LST} = \text{LMT} + (\text{RA of Ms} - 12^h) \quad \text{or}$$
$$\text{LMT} = \text{LST} - (\text{RA of Ms} - 12^h).$$

If desired, the LMT itself can be shown as an angle on Fig. 13, by drawing the antipodal meridian PA″, and then the local mean time on meridian PA is the angle A″PMs, that is the angle reckoned

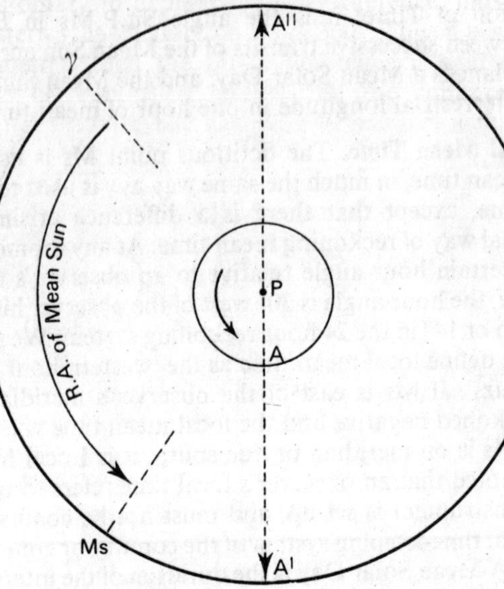

FIG. 13. CONVERSION OF TIMES.

clockwise from PA″ round to PMs—it is greater than 180° in the figure.

Conversion between the local mean time and local sidereal time 'labels' of a particular instant is thus possible if one knows the right ascension of Ms at that moment. As indicated in the equations given above, the direct numerical difference is the quantity (R.A. of Ms − 12 hours): this quantity is tabulated in the Star Almanac in the column headed '*R*' of the Sun tables. As explained in Section 1.30, *R* increases at a uniform rate which is $3^m\ 56{\cdot}56^s$ in a mean solar day or $9{\cdot}86^s$ in an hour. *R* is tabulated at 6-hour intervals of Greenwich Mean Time, so, to get the value of R appropriate to any particular moment, it is necessary to know the GMT of that moment with sufficient accuracy: nowadays, with the aid of a radio receiver, this is no problem. To get *R* to an accuracy of 0·1 second it is only necessary to know the GMT to the nearest half minute.

1.32 Time Proper. The Earth's rotation is no longer a time-keeper. As a basic unit for measurement of time intervals between events, certain oscillatory phenomena associated with the fine structure of atoms are used to control so-called atomic clocks which will maintain a rate constant to within something of the order of one part in 10^{12}. This is Ephemeris Time, the concept of time as it appears in the dynamical equations of Celestial Mechanics.

In comparison with such regularity, the rotation of the Earth has measurable periodic variations and erratic changes.

From January 1st 1972, time signals and standard frequency oscillations broadcast for scientific purposes have been based on a time scale generated by these atomic devices, quite independent of any astronomical phenomena.

On the other hand, concepts such as Greenwich Mean Time and Local Sidereal Time are essentially angles which describe the relative positions of the Earth and the celestial sphere at a particular moment. These positions are related to the independent time scale by means of precise observations at fixed observatories. Greenwich time expresses the relationship of the Greenwich meridian to the celestial sphere, and its value can be found by reference to broadcast time signals. See Section 2.11.

As already mentioned, the surveyor is not a measurer of time. He is interested in measuring certain angles, the values of which are not dependent on any particular definition or concept of time. He simply uses chronometers, radio signals, etc., as aids in making these measurements.

CHAPTER II

FIELD ASTRONOMY—OBSERVATIONS

2.01 Basic Observations. There are three basic observations that can be made in field astronomy, using a theodolite and chronometer:

(i) to measure the altitude of a celestial object,

(ii) to measure the horizontal angle between a celestial object and a fixed reference mark on the Earth,

(iii) to record the reading of a chronometer at the moment when a setting on a celestial object is made, or at the moment of receipt of a radio time signal.

2.02 Altitudes of Stars. In the measurement of the altitudes of stars by theodolite, several special features have to be taken into account:

(i) Observations are done at night, so the bubbles and reading devices of the instrument must be artificially illuminated. Nowadays this is done by small electric lamps attachable, or permanently fitted, to the instrument and powered by dry batteries. The illumination should include a feeble light inside the telescope so that the cross lines can be seen; this light appears quite yellow in comparison with the colour of most stars and does not interfere with the setting of the telescope.

(ii) The stars are apparently moving, often at a rate of the order of 10 minutes of angle in 1 minute of time. To obtain a reading on each face of the theodolite, the observer must be as quick as possible in changing over, and may not be very quick in finding the right star for the second setting! An interval of one minute between the two settings on the star is good going, and the interval should not be more than 3 or 4 minutes. Speed is required because the change of azimuth with altitude is not a linear relationship, so the setting corresponding with the mean of the two altitudes and the mean of the two azimuths is not actually a point on the star's path. A correction for 'curvature of path' can be made, but is best avoided. It is possible to adopt observational procedures that avoid continual changing of 'face' of the theodolite.

(iii) The Earth's atmosphere causes the observed altitude of any celestial object to be greater than its true geometrical altitude. The effect depends on the pressure and temperature of the atmosphere, so the observer's equipment must include a barometer and a thermometer, and their readings must be recorded whenever an altitude is observed. The refraction angle is 58″ × tan (zenith distance) approximately, but the actual value to be used must be calculated from tables such as those included in Chambers Tables or the Star Almanac. See Section 2.03. Refraction is large and erratic at very low altitudes, and it is advisable to avoid taking altitudes lower than about 25°. Though the calculated refraction at reasonably high altitudes will probably be correct within 2″, the actual value at a particular moment depends on the physical condition of the whole thickness of atmosphere at that time, and the observer can never know enough about this to make the calculation of refraction an exact operation. However, it is possible to adopt observational procedures that minimise the effects of any errors in the refraction corrections applied.

2.03 Refraction Angle. To find the refraction angle from the Star Almanac, a basic value is first taken from the MEAN REFRACTION table depending on the observed altitude, and this basic value is multiplied by a factor depending on the temperature and pressure. Suppose the observed altitude is 41° 07′ 53″ at temperature 31°C, pressure 29·3 inches mercury. The MEAN REFRACTION is 67″: the double entry tables give the factor 0·91 for the above temperature and pressure. Thus the required refraction correction is 67″ × 0·91 = 61″ to be subtracted, so the true altitude is 41° 06′ 52″.

2.04 Azimuth by Altitude. One of the standard methods for determining astronomical azimuth can now be described in detail. It does not require use of a chronometer.

The theodolite is set up and levelled at the point at which an azimuth is required, and the artificial lighting equipment is fitted. Thermometer and barometer are available. A luminous signal has been centred at the distant point, the Reference Object. The basic principles of the method are: (i) to measure the altitude of the star and so obtain its corrected zenith distance which is one side of the astronomical spherical triangle: another side is the star's co-declination or *polar distance* which must be taken from the current list of 'apparent positions' of stars (e.g. the Star Almanac), and the third side is the observer's co-latitude which is assumed to be known: (ii) at the same time to measure the horizontal angle between star and RO.

The observer must have a star chart and should select and identify a star which is approximately to east or west of him and at a moderate altitude. A sketch should be made showing approximate directions as in Fig. 14.

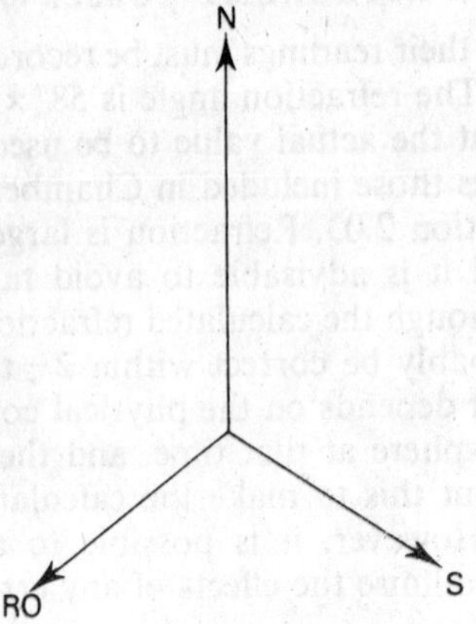

FIG. 14. SKETCH OF DIRECTIONS.

We suppose that the star δ Leonis was seen to be at a suitable position on the east side: the date was 8 March 1972 and the observer's latitude was 52° 12′ 06″N. The practical procedure can be described in the following steps:

(i) Set the diaphragm vertical line on the RO. and read horizontal circle—say 281° 15′ 44″, face left.

(ii) set the telescope exactly on the star i.e. get the image of the star on the intersection of the cross lines: read horizontal circle—say 158° 51′ 02″; having set the vertical arc bubble, read vertical circle—say 53° 27′ 20″. Since the star is apparently moving, setting on the intersection requires some practice; one method is to keep the star on one of the lines by steadily turning the appropriate tangent screw and to stop when the star comes to the intersection.

(iii) change face and as quickly as possible find the same star and set exactly on it as before. Readings, face right, after setting vertical arc bubble are say—vertical 306° 50′ 31″, horizontal 339° 18′ 07″.

(iv) set back on RO. and read horizontal circle—say 101° 15′ 52″. Temperature and pressure were 8°C and 30·1 inches mercury respectively.

Taking into account the graduation system of the theodolite used (Wild T2), the two altitude readings were:

Face Left	36° 32′ 40″
Face Right	36 50 31
Mean	36 41 35

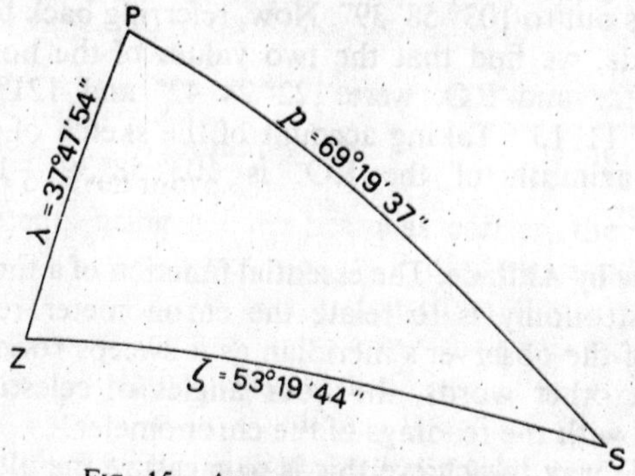

FIG. 15. ASTRONOMICAL TRIANGLE.

The basic refraction for this altitude is 78″ and the temperature/pressure factor obtained from the double-entry table is 1·01, giving the refraction correction 1′ 19″: hence the true altitude was 36° 40′ 16″, or zenith distance 53° 19′ 44″. It may be noted here, that with this graduation system of the vertical circle, the reading on face left is in fact zenith distance, and one side of the astronomical triangle is the zenith distance, so the surveyor may prefer to take the two observed zenith distances 53° 27′ 20″ and 53° 09′ 29″, mean 53° 18′ 25″, and add the refraction correction to give the true zenith distance as above.

The star's declination, from the Star Almanac, was 20° 40′ 23″ North, so the three sides of the astronomical triangle are as shown in Fig. 15.

2.05 Calculation of Azimuth. The angle Z is now calculated: for logarithmic computation the best formula is

$$\tan \tfrac{1}{2}Z = \left(\frac{\sin(s-\lambda) \,.\, \sin(s-\zeta)}{\sin s \,.\, \sin(s-p)}\right)^{\frac{1}{2}}$$

where
$$s = \tfrac{1}{2}(\lambda+\zeta+p),$$

but for computer use there is the cosine formula

$$\cos Z = (\cos p - \cos\lambda \,.\, \cos\zeta)/(\sin\lambda \,.\, \sin\zeta).$$

As explained already, it so happens that the quantities conventionally used—latitude, altitude, and declination—are complements of the actual sides of the spherical triangle, and most books on field astronomy convert the standard spherical formulas for use with these: the cosine formula expressed in the conventional quantities is given in Section 1.15.

In the present example the angle Z, which is the azimuth of the star, works out to 103° 58′ 39″. Now, referring back to the observation records, we find that the two values of the horizontal angle between star and RO. were 122° 24′ 42″ and 121° 57′ 45″, with mean 122° 11′ 13″. Taking account of the sketch of directions, we see that azimuth of the RO. is 103° 58′ 39″ + 122° 11′ 13″ = 226° 09′ 52″.

2.06 Time by Altitude. The essential function of a time observation in field astronomy is to relate the chronometer readings to the position of the observer's meridian as it sweeps round the celestial sphere; in other words, the hour angles of celestial objects are correlated with the readings of the chronometer.

A simple way to achieve this is to measure the altitude of a star and record the chronometer readings at the moments when the settings on the star are made.

As with the azimuth observation (Sections 2.04 and 2.05), the three sides of the spherical triangle are then known, but in this case the angle computed is the hour angle at the pole P. The formulas are obviously

$$\tan \tfrac{1}{2}P = \left(\frac{\sin \tfrac{1}{2}(s-\lambda) \,.\, \sin \tfrac{1}{2}(s-p)}{\sin \tfrac{1}{2}s \,.\, \sin \tfrac{1}{2}(s-\zeta)} \right)^{\frac{1}{2}}$$

or

$$\cos P = (\cos \zeta - \cos \lambda \,.\, \cos p)/(\sin \lambda \,.\, \sin p).$$

Let us suppose that, before taking the observations described in Section 2.04, the observer took time observations on the same star. Since no horizontal angle is to be measured, intersection of the star by the horizontal cross-line only is needed, and in that case the best procedure is to set the cross-line just ahead of the star and record the moment of intersection when it occurs. See Section 2.09.

Without going into full details, we suppose that the readings obtained were:

Face right: vertical circle 304° 07′ 28″: chronometer $20^h\ 22^m\ 29{\cdot}4^s$
Face left: vertical circle 55 38 02 : chronometer 20 23 02·8

The two altitudes were 34° 07′ 28″ and 34° 21′ 58″, mean 34° 14′ 43″. The refraction correction 1′ 26″ gives the true altitude 34° 13′ 17″, zenith distance 55° 46′ 43″. The result of the computation is P = 60° 22′ 38″, or in time units $4^h\ 01^m\ 30{\cdot}5^s$. This was the eastern hour angle of the star when the chronometer reading was $20^h\ 22^m\ 46.1^s$.

Essentially, this result connects the position of the celestial sphere with the chronometer readings. For instance, the observer can predict that the star δ Leonis will transit his meridian $4^h\ 01^m\ 30{\cdot}5^s$ of sidereal time after he made the observation just described. If the chronometer was running at mean-time rate (as most chronometers do, so that they can conveniently be compared with time signals), the interval is $4^h\ 00^m\ 50{\cdot}9^s$ of mean time, so the transit would occur at the chronometer reading $00^h\ 23^m\ 37{\cdot}0^s$.

2.07 Chronometer Error. It is usual practice, however, to express the result of a time observation by relating the chronometer reading to local sidereal time or local mean time.

Reference to the Star Almanac shows that the Right Ascension of δ Leonis was $11^h\ 12^m\ 40{\cdot}2^s$. Hence the point γ, at the mean time of the observation, was $11^h\ 12^m\ 40{\cdot}2^s - 4^h\ 01^m\ 30{\cdot}5^s = 7^h\ 11^m\ 09{\cdot}7^s$ west of the observer's meridian. By definition, this was the local sidereal time. The relative positions round the celestial equator are shown in Fig. 16.

Let us suppose that the observer used a radio receiver and knew that the chronometer was about $5\frac{1}{2}$ minutes fast on GMT. Then the GMT of the observation was about $20^h\ 17^m$. At this time on 8 March 1972, the value of '*R*' was $11^h\ 06^m\ 30{\cdot}5^s$, that is, the right

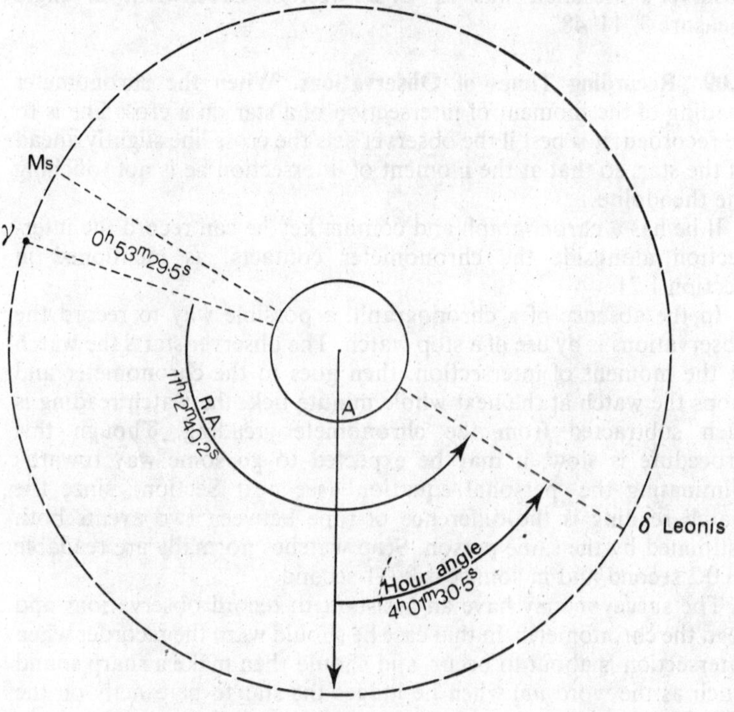

FIG. 16. TIME OBSERVATION.

ascension of the fictitious Mean Sun was $23^h\ 06^m\ 30{\cdot}5^s$. Therefore, the local mean time of the observation was $7^h\ 11^m\ 09{\cdot}7^s - 11^h\ 06^m\ 30{\cdot}5^s + 24^h = 20^h\ 04^m\ 39{\cdot}2^s$, so the chronometer was $18^m\ 06{\cdot}9^s$ fast on local mean time. If the reader is confused about the sudden appearance of 24^h, he is reminded that the quantities we are dealing with are really angles, and the whole process can be worked by sums and differences of angles (expressed in time units) as shown on the figure.

2.08 Longitude. If the time observation is to be used to determine longitude, the chronometer must be compared with precise time signals.

Such signals can now be received from a number of stations, and they generally consist of sharp short sounds emitted at intervals of one second, almost continuously. More details of these signals and their reception are given in Section 2.11.

Meanwhile, let us suppose that by comparison with the signals the chronometer was found to be $5^m\ 19{\cdot}7^s$ fast on GMT. Then the observer's meridian was $12^m\ 47{\cdot}2^s$ west of Greenwich, in angle measure $3°\ 11'\ 48''$.

2.09 Recording Times of Observations. When the chronometer reading of the moment of intersection of a star on a cross line is to be recorded, it is best if the observer sets the cross line slightly ahead of the star, so that at the moment of intersection he is not touching the theodolite.

If he has a chronograph and contact key he can record the intersection alongside the chronometer contacts, as mentioned in Section 1.21.

In the absence of a chronograph, a possible way to record the observations is by use of a stop watch. The observer starts the watch at the moment of intersection, then goes to the chronometer and stops the watch at the next whole minute tick: the watch reading is then subtracted from the chronometer reading. Though this procedure is slow, it may be expected to go some way towards eliminating the 'personal equation' (see next Section), since the watch reading is the difference of time between two events both estimated by the same person. Stop watches normally are readable to 0·2 second and in some cases 0·1 second.

The surveyor may have an assistant to record observations and read the chronometer. In that case he should warn the recorder when intersection is about to occur, and should then make a sharp sound (such as the word *up*) when he judges the star to be exactly on the crossline.

2.10 Personal Equation. The time recording of events involves the observer's use of nerve and muscle action, and almost always there is an unknown difference between the actual event and the recorded reading, usually a delay in the record. This is called the *personal equation* and may be half a second or even more. As a rule, the effects of the personal equation cannot be cancelled out by such devices as changing face or using stars on both sides of the meridian. The only hope of cancellation is that the same observer should take all the observations, including chronometer timing observations before and after each period of work.

Personal-equation effects can be much reduced by use of special equipment, see Section 2.32.

2.11 Time Signals. Several radio stations round the world broadcast, almost continuously and on various wavelengths, precise time signals and frequencies of oscillation for scientific reference purposes. So far as the surveyor is concerned, the useful signals are the short sharp sounds transmitted at intervals of one second. Until recently, these signals were kept very close to Universal Time, which for most practical purposes is the same as Greenwich Mean Time.

However, the Earth's rotation is somewhat irregular and has been slowing down to an extent that is measurable against the very regular time devices now available. This has caused some difficulties, and a new system of precise time signals is now in operation.

As from 1st January 1972, time signals and standard frequencies broadcast from stations like Rugby MSF, etc. have been controlled by atomic oscillators at a fixed rate that is considered to be constant to the order of 1 part in 10^{12}. This time system is completely independent of any astronomical phenomena and the rotation of the Earth.

The signals normally received from MSF are pips at one-second intervals with each whole-minute pip made longer. This signal is called UTC. (Coordinated Universal Time).

Universal time, UT, is defined in relation to the Sun and the rotation of the Earth and is perceptibly changing its rate in comparison with atomic time. In order to keep the readings of the radio signals close to UT, an extra pip will be included in one UTC 'minute' when necessary, but this will be done only at midnight on December 31st or June 30th, and prior notification will be made in appropriate places and publications. Indeed, if the necessity ever arises, one 'second' will be omitted.

To be able to read correct UT from the signals, a system for continuously indicating the differences has been devised.

If UT is in advance of signal time (UTC) by n tenths of a second.

then after each whole-minute pip, n seconds pips will somehow be distinguished from the rest. In this case, the UT of any event = signal reading $+\frac{n}{10}$ second.

If UT is lagging on signal time by n tenths of a second, then n seconds pips *commencing at the ninth pip after the whole minute* will be distinguished. In that case, the UT of any event = signal reading $-\frac{n}{10}$ second.

In the MSF signal, the distinction is done by splitting the seconds pip into two very short pips. In practice, n will never be greater than 7.

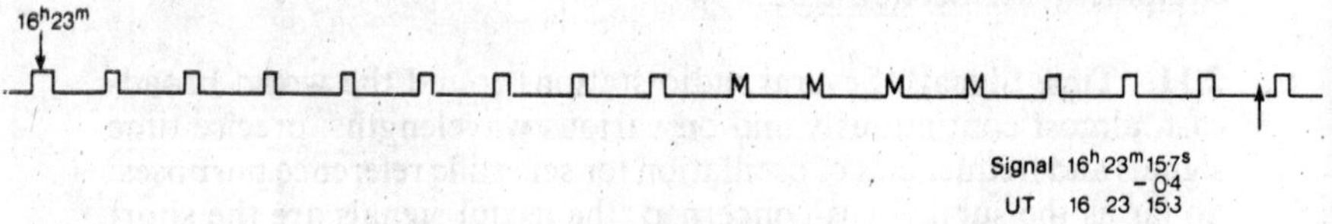

FIG. 17. MSF TIME SIGNAL.

The signal illustrated in Fig. 17 indicates that UT is lagging 0·4 second on UTC, so the UT of the event marked by the arrow was $16^h\ 23^m\ 15{\cdot}3^s$.

Full information on world broadcasts of time signals is to be found in the published Admiralty List of Radio Signals.

2.12 Latitude Observation. The practical method for determination of latitude—by circum-meridian altitudes—is to make settings on a star as rapidly as possible during a period of 10 or 12 minutes each side of transit time, and to read a chronometer at each setting. Change of face of telescope can be done at alternate settings, or in any reasonable manner so as to get equal numbers on each face. (But see Section 2.14.)

For a numerical example we refer to the observer who made time observations as described in Section 2.06, and suppose that he wishes to observe for latitude as soon as a suitable star comes to transit.

The local sidereal time at the time observation was $7^h\ 11^m\ 09{\cdot}7^s$. Stars transit at local sidereal times equal to their right ascensions. Reference to the Star Almanac reveals that the star Procyon (α Canis Minoris) had right ascension $7^h\ 37^m\ 52{\cdot}5^s$ and declination 5° 17′ 47″ North. Transit will therefore occur $26^m\ 42{\cdot}8^s$ of sidereal time after the time observation; in mean time this is $26^m\ 38{\cdot}4^s$, so the mean time chronometer will then read $20^h\ 49^m\ 24{\cdot}5^s$. The observer knows that his latitude is about 52°, so the angle MOE in Fig. 8 will be about 38°; angle EOS_1 is about 5°, so the star will transit at altitude about 43°. Observations should be started when the chronometer reads about $20^h\ 38^m$.

A set of observations might look something like this:

Face	*Vertical Circle*	*Chronometer*
L	46° 56′ 15″	$20^h\ 39^m\ 15^s$
R	313 05 12	41 42
R	313 05 45	43 02·5
L	46 53 47	46 01
L	46 53 31	47 22
R	313 06 48	50 17·5
R	313 06 38	51 59
L	46 54 02	54 07·5
L	46 54 49	56 26·5
R	313 04 17	20 59 03

Chronometer reading at transit $20^h\ 49^m\ 24{\cdot}5^s$
Thermometer 8°C Pressure 30·1 ins. of Hg.

It is clear that for a star transiting as S_1 in Fig. 8, the altitude is a maximum at transit, or the zenith distance is a minimum. For an observation not quite at transit, the altitude will be slightly less, and the principle of the use of this method is that the small differences of altitude, depending on the interval from transit time, are applied to the observed altitudes to convert them to the transit altitude. All the corrected altitudes should then be equal, apart from random errors of observation and any vertical error of collimation of the theodolite (which will cancel out).

2.13 Near Meridian Correction. The theory of this method is based on the narrow spherical triangle shown in Fig. 18. The small hour angle t is the difference between the time of transit and the time of an actual observation. If h_0 is the transit altitude we write $(h_0 - \mathrm{d}h)$ for the observed altitude, and $\mathrm{d}h$ is a small angle. The cosine formula for the triangle then reads:

$$\begin{aligned}\sin(h_0 - \mathrm{d}h) &= \sin\phi \,.\, \sin\delta + \cos\phi \,.\, \cos\delta \,.\, \cos t \\ &= \sin\phi \,.\, \sin\delta + \cos\phi \,.\, \cos\delta(1 - 2\sin^2 \tfrac{1}{2}t).\end{aligned}$$

Using the fact that $\mathrm{d}h$ is small, we get:

$$\sin h_0 - \mathrm{d}h \,.\, \cos h_0 = \cos(\phi - \delta) - 2 \,.\, \cos\phi \,.\, \cos\delta \,.\, \sin^2 \tfrac{1}{2}t.$$

But $(\phi - \delta)$ is $90° - h_0$, and we can write $\tfrac{1}{2}t$ for $\sin \tfrac{1}{2}t$, hence;

$$\sin h_0 - \mathrm{d}h \,.\, \cos h_0 = \sin h_0 - \cos\phi \,.\, \cos\delta \,.\, \tfrac{1}{2}t^2,$$

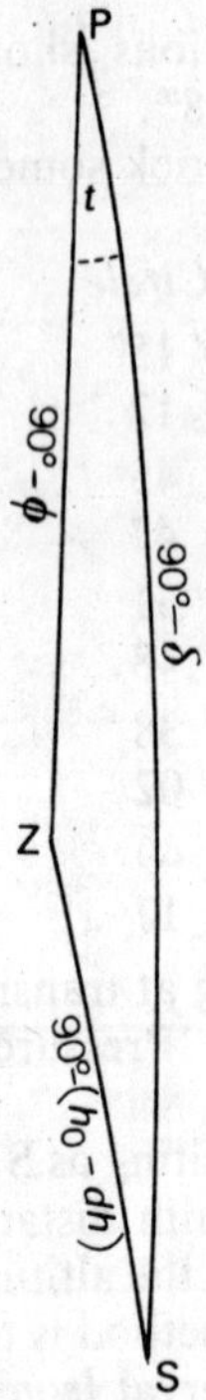

FIG. 18. STAR NEAR TRANSIT.

and the required correction term dh is given by:

$$\mathrm{d}h = \frac{\cos\phi \,.\, \cos\delta}{\cos h_0} \,.\, \tfrac{1}{2}t^2.$$

This formula is of course valid for dh and t expressed in radian, but it is convenient to be able to use t in time seconds and get dh in angle seconds, so we substitute $t = t_s \,.\, 15/206\ 265$ and $\mathrm{d}h = \mathrm{d}h''/206\ 265$, and the practical formula is:

$$\mathrm{d}h'' = \frac{\cos\phi \,.\, \cos\delta}{\cos h_0}\left(\frac{15 \,.\, t^2}{27\ 502}\right).$$

For instance, if t is 10 minutes = 600 seconds, we get

$$\mathrm{d}h'' = \frac{\cos\phi \,.\, \cos\delta}{\cos h_0} \,.\, 196''$$

The trigonometrical factor depends on the circumstances of the particular observation: values of the numerical coefficient depending only on t are given in the Star Almanac TABLE FOR CIRCUM-MERIDIAN OBSERVATIONS.

To find the trigonometrical factor, one may take the maximum altitude observed, or the altitude nearest to transit, and use this to get an estimate of the latitude. In the set quoted, the zenith distance at transit was evidently about 46° 53′ 12″, altitude 43° 06′ 48″, to which a refraction correction of 61″ is applicable; hence h_0 is about 43° 05′ 47″.

With the declination of the star, 5° 17′ 47″N, we get $\phi \sim$ 52° 12′ 00″, and the value of $\cos\phi \, . \cos\delta/\cos h_0$ works out at 0·835 7. For the 10 observations, the intervals t, converted to sidereal time, are given below, with the corrections dh and the corrected altitudes.

t	*Correction:* *m* × 0·835 7	*Corrected Altitudes*
10ᵐ 11·2ˢ	170″	43° 06′ 35″
7 43·8	98	50
6 23·0	67	52
3 24·1	19	32
2 02·8	7	36
0 53·1	2	50
2 34·9	11	49
4 43·8	37	35
7 03·2	82	33
9 40·1	154	51
	Mean	43 06 42
	Refraction	1 01
		43 05 41
	Declination	5 17 47
	Co-latitude	37 57 54
	Latitude	52 12 06

The vertical collimation error shows up as a systematic difference between face-right and face-left results.

It is to be noted that each single observation should be corrected: face-right and face-left pairs should not be averaged. Considering the geometry of the path of the star, it is obvious that the mean of two consecutive settings must be below the true path, so a systematic error would be introduced by taking means.

An alternative is to calculate the mean of all the observed altitudes and add the mean of the corrections; this is not recommended, since a faulty observation would probably remain undetected.

2.14 Balancing Observations. A star in such a position as S_2 in Fig. 19 is said to be at *lower transit*: its altitude will then be minimum, so the corrections to observed circum-meridian altitudes will have to be subtracted. It is to be noted that the right ascension of S_2 will be 12 hours different from that of S_1. It is desirable to balance latitude observations by using stars on both sides of the zenith if

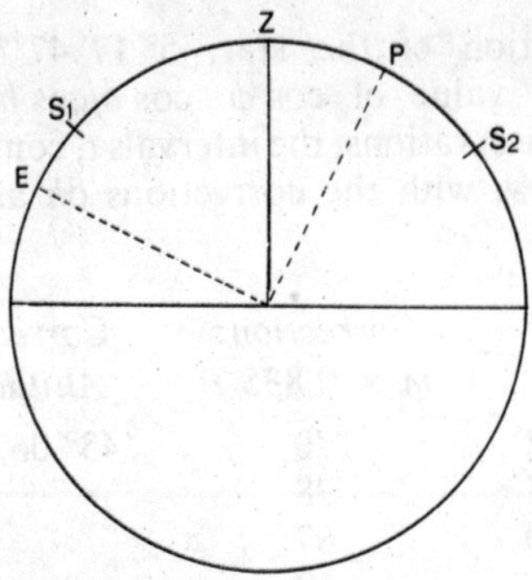

FIG. 19. BALANCED POSITIONS.

possible. If stars at about equal altitudes can be found, so much the better, as any error in the calculated refraction will also balance out. For instance, to balance the Procyon observations described above one would need a star of right ascension about 20 hr and polar distance about $52° - 43° = 9°$; κ Cephei (No. 557 in the Almanac) is available though it is smaller than 4th magnitude; β Cephei would come along about an hour later. At latitude 52° the field of choice is very restricted.

The spherical triangle for lower transit is shown in Fig. 20 where the angle t is the time from transit, and the basic formula is $\sin(h_0 + dh) = \sin\phi \,.\, \sin\delta - \cos\phi \,.\, \cos\delta \,.\, \cos t$. The relationship on meridian is $\phi = 90° + h_0 - \delta$.

Another kind of balancing can profitably be adopted with the modern quick-reading theodolite. Using two stars, one on each side of the zenith, *all* the altitudes are taken on *one* face of the instrument. It is easily seen that the zero error of the vertical circle will cancel out in this case, and it is possible to get a great many more settings in a given period if time does not have to be spent in changing face, re-finding the star and re-setting the bubble.

In low latitudes it is possible to get a balanced pair of stars both at upper transit, as illustrated in Fig. 21.

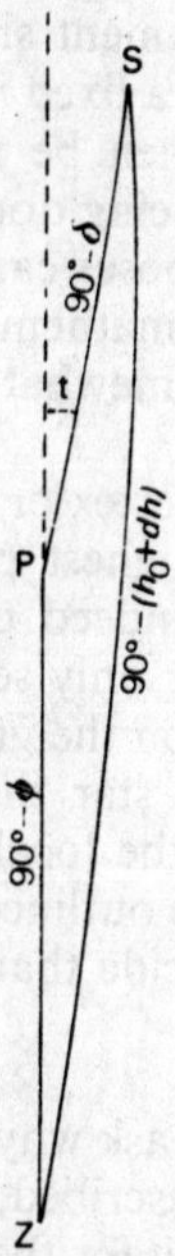

FIG. 20. LOWER TRANSIT.

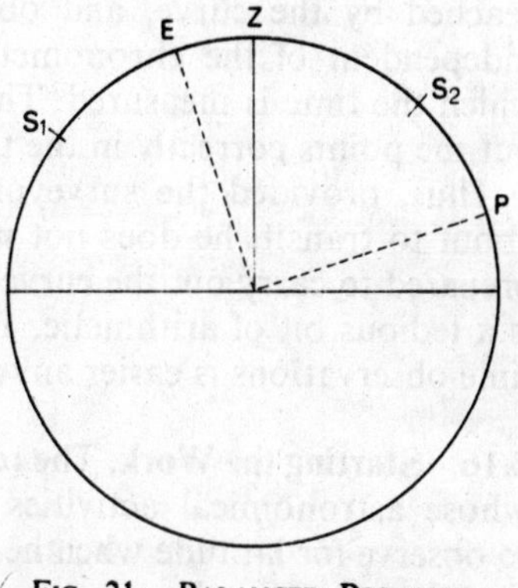

FIG. 21. BALANCED POSITIONS.

2.15 Latitude by Geometry. It is worth noting that the method of finding latitude by circum-meridian altitudes is essentially geometrical, and an exact knowledge of the time of transit is not really necessary. Consider the observations to be plotted on a graph of altitude against chronometer reading, at any suitable scales, as shown in Fig. 22.

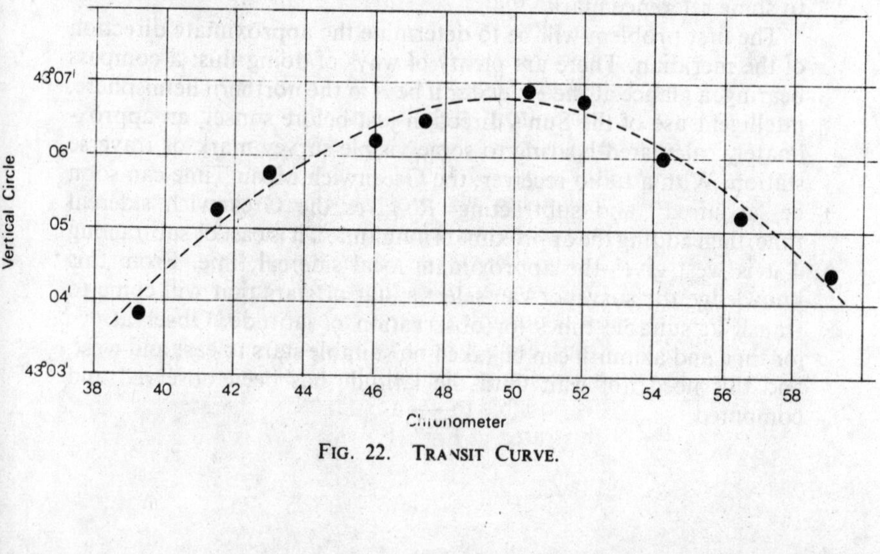

FIG. 22. TRANSIT CURVE.

The vertical collimation error of the instrument shows up: this error could easily be found by observations to a fixed mark on both faces, and the latitude observations could then be moved up or down to the positions of true altitudes. This being done, the points should lie on a curve which for practical purposes can be taken to be a parabola. It is then possible by a standard mathematical process to find a parabola of best fit to the points, somewhat as shown on the figure.

Now the surveyor is only interested in the vertex or highest point reached by the curve, and obviously this highest point is quite independent of the chronometer error and indeed of the unit in which the time is measured. The chronometer only serves to space out the points correctly in the time direction on the graph.

Thus, provided the surveyor can locate a star that is actually about to transit, he does not need to know the local time if he is prepared to carry out the curve-fitting process outlined above. This is a tedious bit of arithmetic, and he may decide that taking some time observations is easier anyway!

2.16 Starting the Work. The reader may well ask why the surveyor, whose astronomical activities have been described, was obliged to observe for latitude when he already knew it for the azimuth and time computations!

The short answer is that the worked examples were given in order to illustrate standard observational methods, and the fact that they all refer to the same place on the same evening is incidental.

However, there remains the question of what a surveyor would do if he arrived at a place of which he did not know the latitude or longitude, and had to find either of these quantities, or an azimuth to some reference mark.

The first problem will be to determine the approximate direction of the meridian. There are plenty of ways of doing this: a compass bearing, a glance at the Pole Star if he is in the northern hemisphere, intelligent use of the Sun's direction just before sunset, an approximately calculated bearing to some visible survey mark or traverse station. With a radio receiver, the Greenwich Mean Time can soon be 'acquired', and subtracting '*R*' gives the Greenwich sidereal time, then adding the approximate longitude if it is east or subtracting if it is west gives the approximate local sidereal time. From this knowledge the surveyor can select a star or stars that will come to transit at suitable times for observation of latitude. Observations for time and azimuth can be taken on suitable stars to east and west, and this need not wait until the latitude has been observed and computed.

If longitude is not required, the radio and chronometer are not really necessary; the surveyor can follow a star up to transit with the theodolite and measure its maximum (or minimum) altitude, and this will give the latitude within a few seconds, which may be accurate enough in the circumstances. Then azimuth can be found by the purely geometrical method (Section 2.04), and by observing east and west stars the effect of any small error in the latitude used in the computations is eliminated. It is even possible to dispense with the latitude observation, as explained in Section 2.30.

If the direction of a rising star at a particular altitude is recorded, and the theodolite allowed to stand until the direction of the same star when it comes down to the same altitude before setting is recorded, the mean of the two directions gives the meridian.

However, this sort of exploratory situation will nowadays be quite exceptional. Many modern surveyors do their work in the frame of an existing survey control system, and occasions for astronomical observation will be infrequent. On such an occasion, the surveyor will know his position quite closely, to a few minutes, perhaps within a few seconds, before work is started. He could even make up a programme before the stars appear.

It must not be forgotten, however, that latitudes and longitudes read from a map are coordinates in a map projection system and are not astronomical values. The only way to find the true astronomical quantities at a point is to go there and measure them!

2.17 Azimuth by Hour Angle. This is an important method for finding azimuth (see Section 8.03) and has some practical advantages over the method described in Section 2.04. In this method, the third known element of the spherical triangle is not the star's altitude but its hour angle, obtained from chronometer readings of the exact moments of setting on the star. The chronometer is thus an essential instrument in this method and its error on local sidereal time must be accurately known from time observations.

The stages of the observation procedure are:

(i) set telescope on RO and read horizontal circle,

(ii) set vertical cross line on star, record simultaneous chronometer reading, and read horizontal circle,

(iii) change face and repeat (ii),

(iv) set back on RO and read horizontal circle.

The mean of the two chronometer readings, corrected for the known error of the chronometer, gives the local sidereal time of the observation. Conversion from mean to sidereal will usually be necessary. The hour angle *P* of the star can then be found, being the difference between the star's right ascension and the local sidereal time. Now, two sides and the included angle of the astronomical triangle are known—the star's polar distance ($90° - \delta$), the observer's co-latitude ($90° - \phi$) and the hour angle P which must be converted to angle measure for the computation. The star's azimuth Z is then obtained from the formula:

$$\cot Z = (\cos \phi \,.\, \tan \delta - \sin \phi \,.\, \cos P)/\sin P.$$

As before, horizontal angles are taken from the records and the azimuth of the RO is calculated.

This method is somewhat easier and quicker than the method by altitude observation, because only the vertical cross line has to be set on the star; also there is no refraction correction.

2.18 Elongation. Twice in each revolution of the celestial sphere, a star of sufficiently high declination will appear, to an observer at a particular place, to be momentarily moving vertically. This occurs at either side of the meridian when the angle at S in the spherical triangle, Fig. 23, is 90°: clearly, the star's declination must be

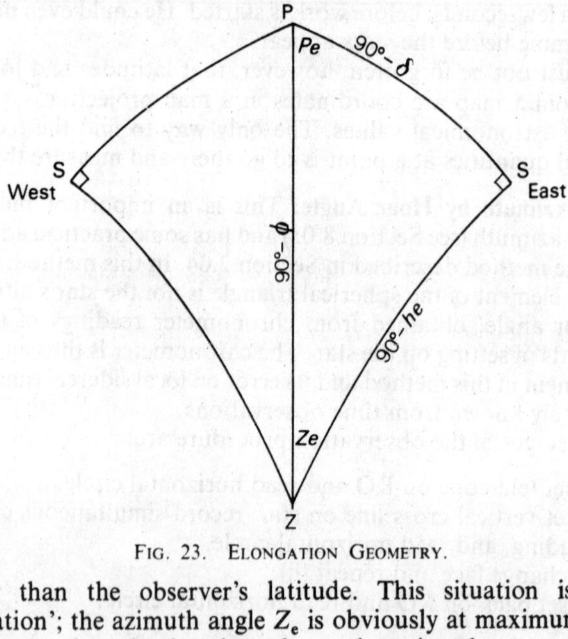

FIG. 23. ELONGATION GEOMETRY.

greater than the observer's latitude. This situation is called 'elongation'; the azimuth angle Z_e is obviously at maximum value. The hour angle and azimuth angle are then given by:

$$\cos P_e = \tan \phi \,.\, \cot \delta,$$
$$\sin Z_e = \sec \phi \,.\, \cos \delta.$$

Having a chronometer of known error, and the calculated value of P_e, an observer can predict the moment when a suitable star will be at elongation. A measurement of the horizontal angle from star to RO, combined with the value of Z_e, will give the azimuth of the RO.

In practice, of course, it is not possible to get two settings on the star exactly at the moment of elongation, but if stars near the elongation are used for finding azimuth by hour angle it is obvious that any small error, such as personal equation, in timing the intersections, will have a very small effect on the calculated azimuth.

From inspection of the spherical triangle, it is clear that at elongation a star will be at an altitude greater than the observer's latitude: in fact $\sin h_e = \sin \phi \cdot \operatorname{cosec} \delta$. The use of stars near elongation is therefore inconvenient in high latitudes: see also Section 2.29.

2.19 Observing on the Sun. The Sun is a star which is so near to us that its right ascension and declination change rapidly, and the size of the Earth is not negligible. In principle, the methods that have

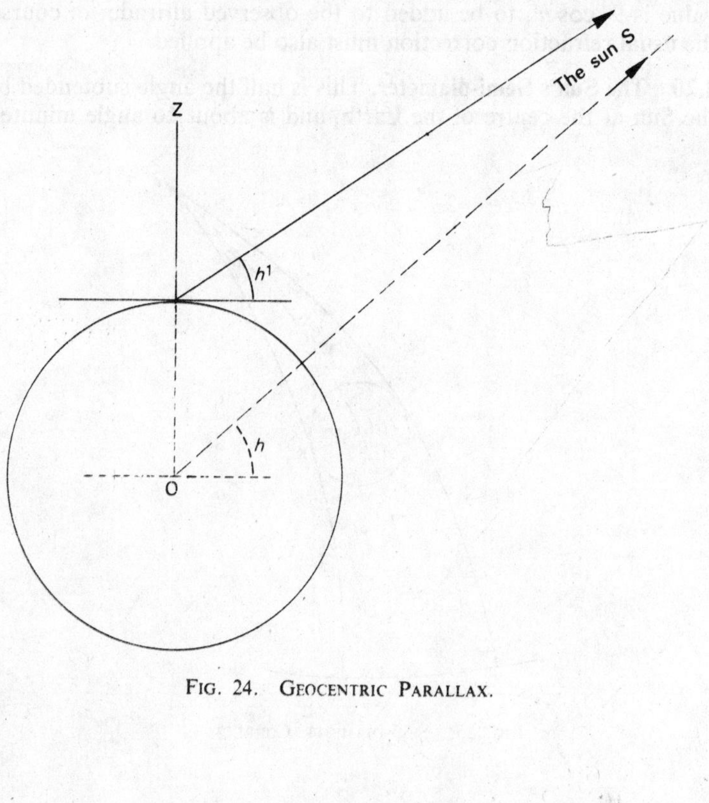

FIG. 24. GEOCENTRIC PARALLAX.

been described can be applied also to observations on the Sun; the practical peculiarities are:

(i) since a dark glass must be fitted to the eyepiece of the theodolite, the cross lines cannot be seen except when they fall across the circular image of the Sun.

(ii) the centre of the Sun cannot be estimated, so the cross line must be set on opposite edges of the disc at the two settings of the telescope: alternatively, both settings made on the same edge must be corrected by a suitable allowance for the angle subtended by the Sun's radius—about 16′ angle as given in the Star Almanac. It is possible to obtain an attachment to the telescope so that four images of the Sun are seen not quite overlapping but leaving a small dark central space which is easily set on the cross lines.

(iii) the published positional information about the Sun is all referred to the centre of the Earth. As illustrated in Fig. 24, an observer on the surface will measure altitude h' from his own horizontal, or zenith distance $90° - h'$ from his vertical line, but the zenith distance that must go into the computations is ZOS or $90° - h$. The difference $dh = h - h'$ is the *geocentric parallax* and its value is $9''.\cos h'$, to be added to the observed altitude: of course, the usual refraction correction must also be applied.

2.20 The Sun's Semi-diameter. This is half the angle subtended by the Sun at the centre of the Earth, and is about 16 angle minutes,

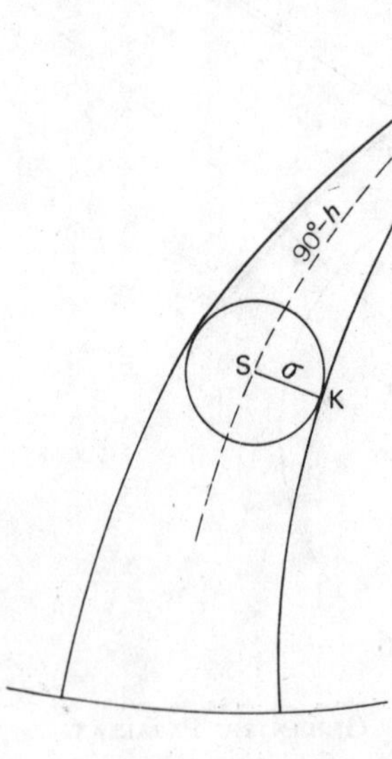

FIG. 25. SEMI-DIAMETER CORRECTION

varying slightly through the year because of the elliptic shape of the Earth's orbit.

When attempting to set a crossline of the telescope to be tangent to the Sun's disc seen through the dark glass, it is easier to set on the trailing edge, because then the crossline is seen on the disc right up to the moment of tangency. If altitude observations are made on one edge, or 'limb' as it is called, the semi-diameter must be added to or subtracted from the corrected altitude, as the case may be. When the vertical line is set on one limb, as in measuring a horizontal angle between the Sun and a reference mark, a small calculation is needed. In Fig. 25, if the setting is made on the limb at K, the horizontal circle reading must be increased by the angle θ to get the reading to the Sun's centre. The relevant spherical triangle SKZ is right-angled at K, the side SK is the semi-diameter σ, and the required formula is $\sin\theta = \sin\sigma \,.\, \sec h$. The altitude h required for this calculation must be observed for this purpose if it cannot be calculated from available data.

2.21 The Sun's Right Ascension. This is not tabulated in the Star Almanac, but can be found in the Astronomical Ephemeris if needed. However, when observations are made on the Sun there need be no reference to the point γ or to sidereal time.

A 'time' observation on the Sun results, as with a star, in the calculation of the hour angle of the observed object. The next step is to get to the hour angle of the time-marking point on the celestial sphere, in this case the fictitious Mean Sun, Ms, and this requires knowledge of the Equation of Time, as explained in Section 1.29.

When Ms is on our meridian it is Local Mean Noon, or $12^h\,00^m$ Local Mean Time. The actual Sun may transit before or after Noon, according to the time of year, so the Equation of Time q has to be given plus or minus sign. To provide for direct conversion from the hour angle of the Sun to the desired local mean time, the quantity 'E' is tabulated in the Star Almanac.

Suppose that the observation has shown that the Sun's hour angle was 45° west: (this can be called 3 pm *Apparent Solar Time* in traditional terminology). Suppose also that Ms was 1° (4 minutes in time measure) ahead of the Sun. The situation at the time of the observation is shown geometrically in Fig. 26. The observer's meridian is OAA′; the angle A′OS is 45° and the angle A′OMs is 46°. This means that the local mean time was $15^h\,04^m$, which is the angle reckoned clockwise round from OA″ to OMs. Defining the equation of time q as 'the angle from the Sun to Ms, positive if Ms is west of the Sun', the quantity 'E' is $(12^h - q)$.

From the diagram we see that:

$$\text{Local Mean Time} + `E' = 24^h\,(360°) + \text{Hour Angle West},$$

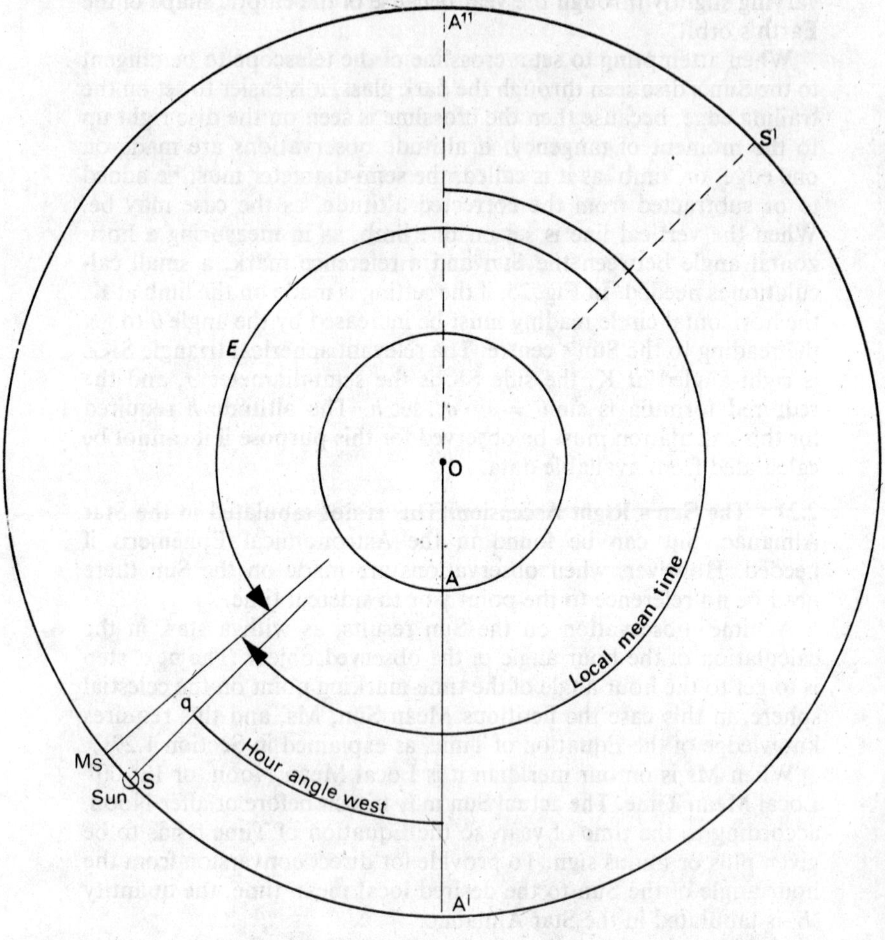

FIG. 26. SOLAR TIME GEOMETRY.

that is:

$$\text{Local Mean Time} = \text{Hour Angle West} + 24^h - \text{'}E\text{'}.$$

Of course the hour angle is to be taken as negative if it is east.

In working the above equation, all quantities are expressed in time units.

2.22 Accuracy of Sun Observations. Results obtained from Sun observations are less accurate than those taken to stars: not only is there some difficulty in setting the telescope on the Sun, but there is also very little scope for balancing out errors by observational arrangements. Nowadays, there is not much occasion for Sun observation by the land surveyor. Bearing checks on detail traverses with numerous lines may be got from Sun azimuth observations if an accuracy of 10″ is useful: the astronomical bearings must of course be converted to the bearing system in which the survey is to be computed. A Sun azimuth can provide a good reference bearing for orienting an isolated survey.

Altitudes of the Sun taken with a sextant on ships are used for navigation in circumstances where radio navigation aids are not available: see Section 2.27.

2.23 Position Lines. Having a chronometer accurately checked against radio time signals, an observer anywhere in the world can determine the Greenwich Mean Time at any moment. By using the appropriate value of 'R', the GMT is converted to Greenwich Sidereal Time, which is the same as the Greenwich West Longitude of the meridian of the point γ. This gives the relationship between the celestial and terrestrial meridian systems at the moment in consideration. At that moment, each star is transiting on some terrestrial meridian, the longitude of which is obtainable by applying the star's right ascension reckoned eastwards from the longitude where γ is.

Further, any particular star at any particular moment is not only transiting but actually overhead at some point on the Earth, and the latitude of that point is the same as the star's declination. This point is called the 'sub-stellar point'; the significant thing about it is that the terrestrial vertical at the point is the direction to the star. In Fig. 27 where O is the centre of the celestial sphere, if OP is

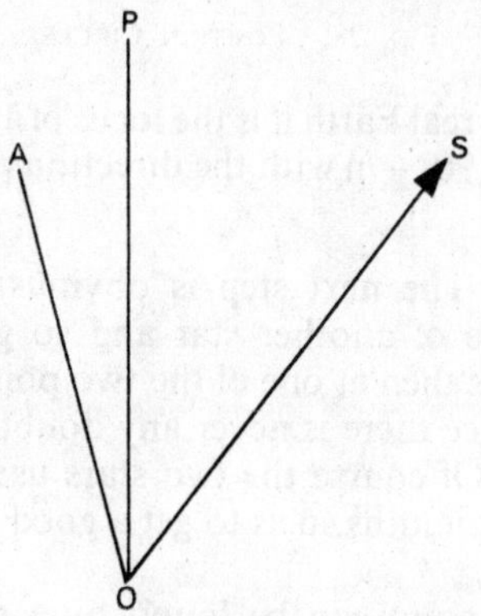

FIG. 27. CELESTIAL DIRECTIONS.

the direction of the Earth's axis and OS is the direction of the star, the plane POS is the celestial meridian of star S, and the terrestrial meridian plane which momentarily coincides with OPS has Greenwich longitude calculated by the procedure explained above. Also, angle POS is the star's co-declination.

Now, suppose that an observer is at a point on the Earth where the direction of his vertical is OA and his meridian plane is therefore POA. If he measures the altitude h of the star he has determined the angle AOS which is $(90° - h)$. In terms of spherical trigonometry with S as the star's position, the observation determines that the observer was at some point on the spherical 'small circle' of angular radius $(90° - h)$ centred at S, as shown on Fig. 28. This circle is a

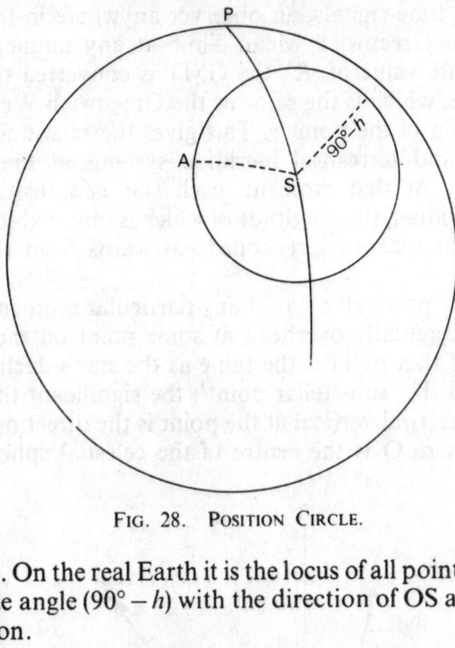

FIG. 28. POSITION CIRCLE.

position line. On the real Earth it is the locus of all points at which the vertical made angle $(90° - h)$ with the direction of OS at the moment of observation.

2.24 Position Fix. The next step is obvious: the observer must measure the altitude of another star and so get another position circle. His position is then at one of the two points of intersection of the circles. In practice there is never any doubt which is the appropriate intersection. Of course the two stars used should be in substantially different azimuths so as to get a good angle of intersection of the position lines.

The intersection point can be found by a purely mathematical solution, but a semi-graphic method should be used because in practice there should always be more than two position lines, to give a check and perhaps some indication of the accuracy of the result.

One may imagine having a large sphere with parallels and meridians marked on it: sub-stellar points could be plotted on it, and with compasses the arcs of angular radii equal to the observed zenith distances could be drawn, to intersect at the observer's position. In fact, a similar procedure is followed, but on a flat map. Nowadays, a surveyor always knows his position fairly accurately before any observations are made.

Let us return to the surveyor on the 8th March 1972, and now suppose that he *does not* know his latitude and longitude, but he can find the Greenwich Mean Time (UT) at any moment, having listened to a precise time signal and noted the chronometer reading. We now use the observational data of Section 2.06 as follows:

Mean altitude of star δ Leonis corrected for refraction	34° 13′ 17″
Mean chronometer reading of observation	$20^h\ 22^m\ 46{\cdot}1^s$
Chronometer correction to GMT	− 5 19·7
∴ GMT of observation . . .	20 17 26·4
'*R*'	11 06 30·6
∴ Greenwich Sidereal Time of observation	7 23 57·0
Right Ascension of δ Leonis	11 12 40·2
∴ East longitude of δ Leonis	3 48 43·2
	≡57° 10′ 47″

The position of the substellar point was therefore:

Latitude	20° 40′ 23″N
Longitude	57 10 47 E.

The position line is a circle of spherical radius 55° 46′ 43″ centred at the substellar point.

The surveyor knows that his latitude is about 52° 10′N and he decides to find the longitude of the point where the position circle cuts this parallel. The spherical geometry and quantities are shown on Fig. 29.

By standard spherical formula the angle at P is found to be 60° 23′ 16″: this means that the point A is that much west of the SSP, hence the longitude of A is 3° 12′ 29″ West. The point is plotted on the chart in Fig. 30.

There are two ways to get the direction of the position line through A. The bearing of the star can be computed, angle PAS in the spherical triangle, and it is found to be 100° 21′; one-minute accuracy is quite sufficient. Then the bearing of the position line is perpendicular

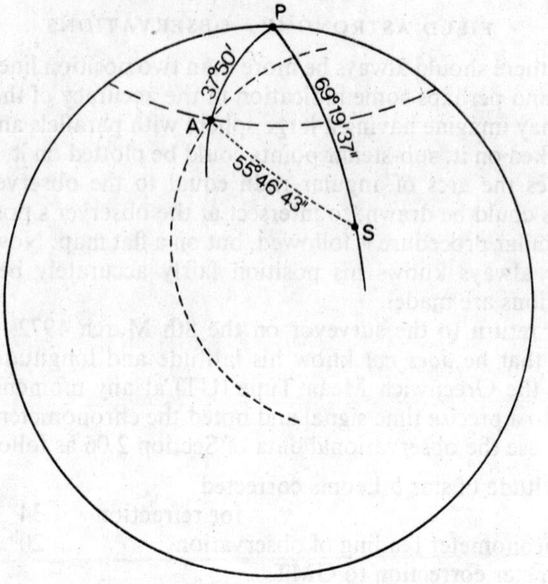

Fig. 29. Position Line Geometry.

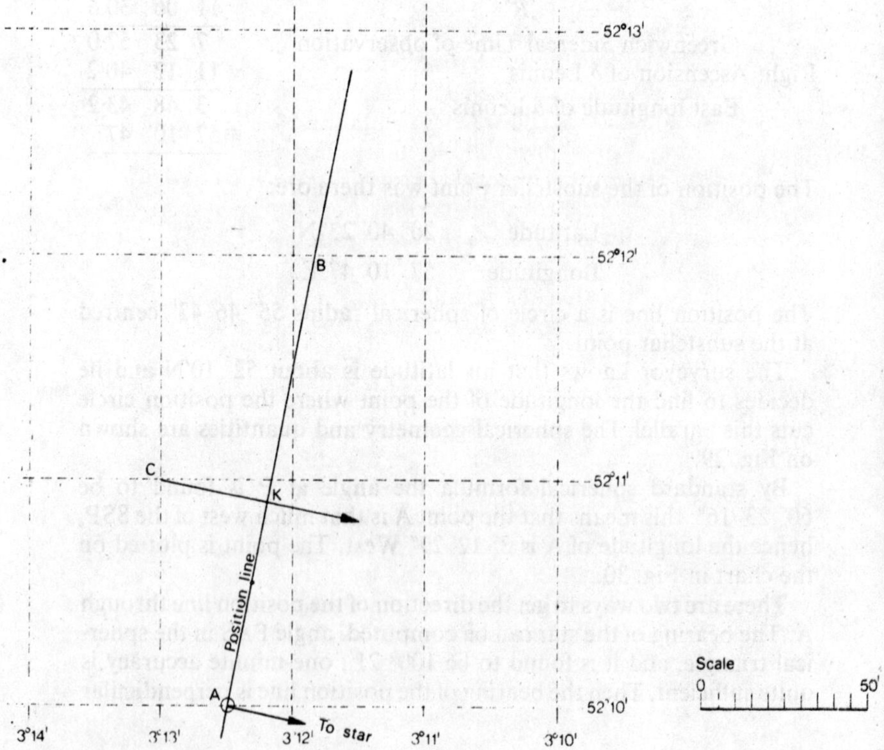

Fig. 30. Position Line.

to this. i.e. 10° 21′, and the line is drawn through A with this bearing. It is important to note that the meridian and parallel grid on Fig. 30 must be an orthomorphic (angle-true) construction (Section 9.02), otherwise the position line could not be placed by this simple method. The ratio of the sides of the rectangles is cos (latitude).

An alternative method for placing the position line is to calculate its intersection with another parallel, e.g. 52° 12′. To do this, the computation would be carried out again, using 37° 48′ for the co-latitude: the result would be the point B in Fig. 30.

2.25 Intercept Method. A position line fixation may also be computed by taking an assumed position of observation and calculating the altitudes that would be measured from that position at the moments when the actual observations were taken.

Suppose the surveyor assumes he was at the point C on Fig. 30. At the moment when δ Leonis was observed, it was over the meridian of longitude 57° 10′ 47″E, so its hour angle from point C was 60° 23′ 47″ eastwards. In the spherical triangle CPS we now have two sides and included angle known. The side CS works out as 55° 47′ 12″, and this is the zenith distance that would have been measured at C when the zenith distance actually measured was 55° 46′ 43″. Thus, the actual position of the observer must have been nearer to the star by 29″ in angle measure. To plot the position line, a line is drawn from C in the azimuth of the star, 100° 21′, the distance CK representing 29″ at the scale of the plot is marked off, and the position line is drawn through K at its bearing of 10° 21′.

A strong fixation will be obtained if a star in each quadrant of bearing is observed. Probably the four position lines will not be concurrent, and the surveyor will have to select a best position on the plot.

It is to be noted that the position line on Fig. 30 does go very close to the point representing the observer's latitude and longitude as given in the earlier examples.

2.26 Offset Position Lines. If three or more stars are observed, the observational procedure can be made simpler and easier by taking all the observations on one face of the theodolite. This means that say 4 or 5 settings on each star can be made in a very short time, just as fast as the settings can be made and the readings recorded; no time is used in re-finding a star after changing face. The effect of this procedure is that each altitude will be incorrect by the amount of the vertical collimation error of the instrument. Suppose, for instance, that each altitude is too great by 26″; then the radius of each position circle will be too small by 26″, and the position lines drawn on the map will 'miss' the true position by the distance

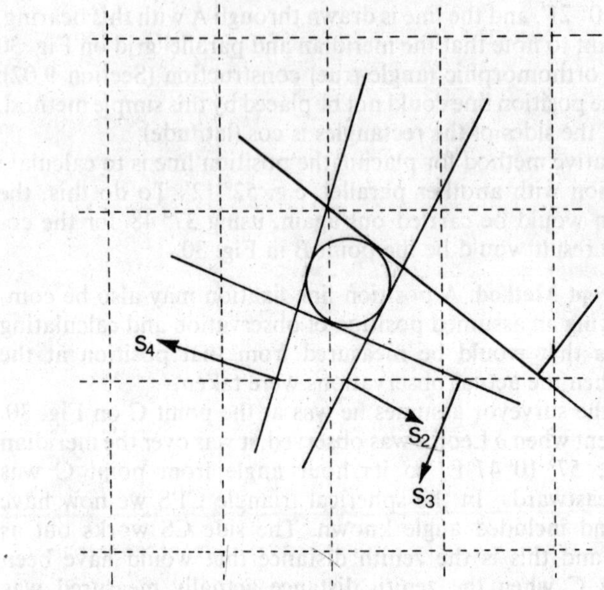

FIG. 31. OFFSET POSITION LINES.

representing 26″ at map scale. They might fall somewhat as shown in Fig. 31. Obviously, the position lines will be tangents to a circle of map radius 26″ and the centre of the circle will be the correct position. In practice, with random errors also affecting the results, a 'circle of best fit' must be selected by the surveyor.

2.27 Navigation. In position finding at sea, the sextant is used to measure the vertical angle between the visible horizon and the lower edge of the Sun's disc. This is illustrated in Fig. 32. The measured angle obviously requires correction as follows:

(i) subtract the dip: this depends on the height of the observer above sea surface. Values can be obtained from tables such as Chambers, and are usually a few minutes of angle.

(ii) add the Sun's angular semi-diameter σ, about 16′ as given in the Almanac.

(iii) subtract refraction; usually about a minute.

(iv) add geocentric parallax (Section 2.19). As this is not more than 9″ it may be considered negligible, or a round value of 0·1′ used.

At the moment of taking the observation, the navigator reads the ship's chronometer, which has been checked against time

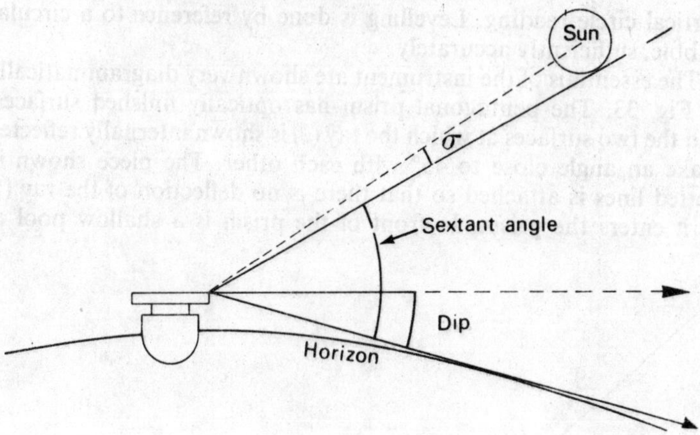

FIG. 32. SEXTANT OBSERVATION.

signals, and thereby obtains the Greenwich Mean Time of the observation, T_m. This means that the Greenwich western longitude of the Mean Sun was $T_m - 12^h$. Then (see Section 2.21) the Greenwich western longitude of the real Sun was $(T_m - 12^h - q)$, which may also be written $(T_m + (12^h - q) - 24^h)$ or $(T_m + E - 24^h)$; this will be negative if the Sun was actually east of Greenwich at the time of the observation. Converted to angle measure, this gives the Greenwich longitude of the sub-solar point, and the latitude was of course the Sun's declination, taken from the Almanac.

A position line is calculated and drawn on the chart as already described. A second position line is not obtainable immediately, but the ship moves on. Thus the sailor's practice is to get a position line in the morning as soon as the Sun is at a reasonable altitude; then at Solar Noon, Sun at maximum altitude, a position line running due east–west is obtained and the morning position line is shifted to the position estimated from dead reckoning. The intersection of the two lines was the ship's position at noon. Later, an evening position line is obtained and the noon line is shifted to give the position at the evening observation. Some of the brighter stars or planets can be used in clear weather when the horizon is also visible during twilight. Astronomical information about positions of planets is to be found in the Astronomical Ephemeris.

2.28 Equal Altitudes Method: The 45° Astrolabe. With this special instrument the position-line method is applied to stars which are observed all at exactly the same altitude—about 45°. This altitude is determined by the fixed shape of a glass prism, and there is no vertical circle reading. Levelling is done by reference to a circular bubble, sufficiently accurately.

The essentials of the instrument are shown very diagrammatically in Fig. 33. The pentagonal prism has optically finished surfaces, and the two surfaces at which the ray (*2*) is shown internally reflected make an angle close to 45° with each other. The piece shown in dotted lines is attached so that there is no deflection of the ray (*1*) as it enters the prism. In front of the prism is a shallow pool of

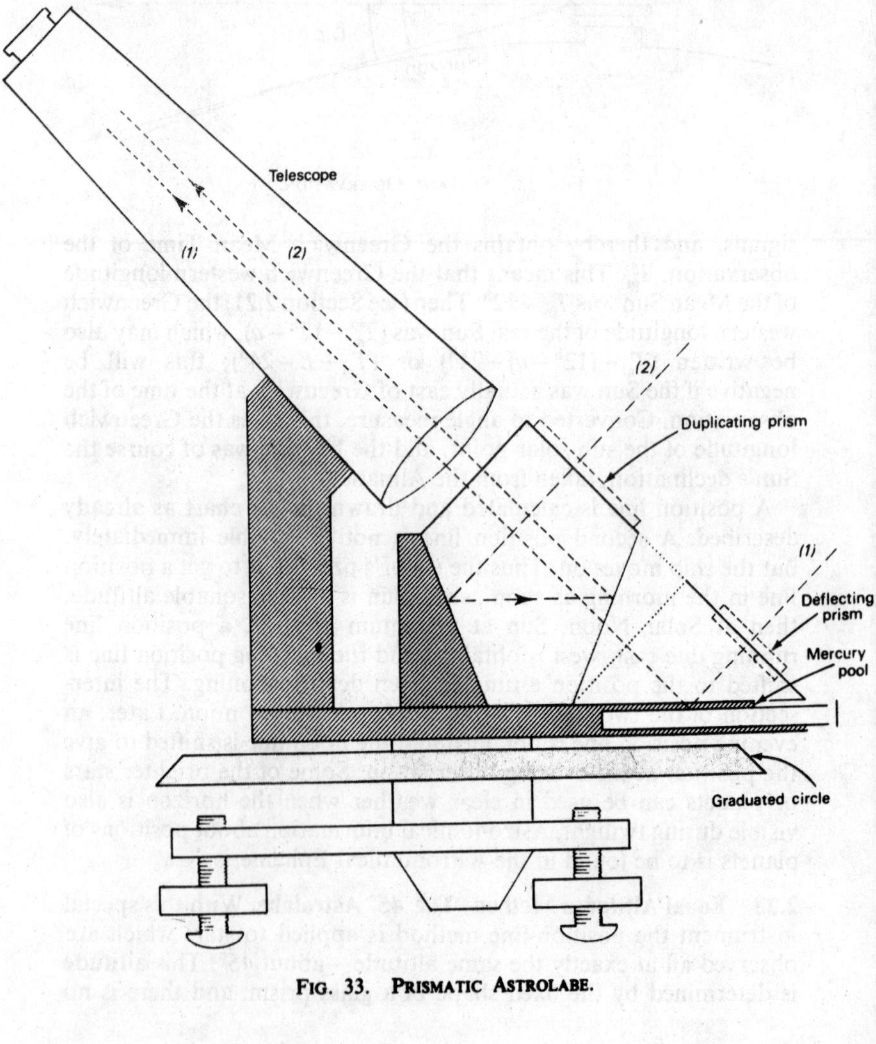

FIG. 33. PRISMATIC ASTROLABE.

mercury on a tray which can be taken out, and this pool reflects the ray (*1*) as shown. It is easily demonstrated that when a star is at the altitude equal to the angle mentioned above, the two images of the star formed in the telescope by the two numbered beams, are coincident. Since these two images appear to the observer to be moving in opposite directions, one route containing one reflection and the other two, the moment of coincidence of the images can be estimated quite sharply.

It is usual to observe stars in quadrantal sets of four, and of course it is essential to make a programme for astrolabe observations, as each star comes to the necessary position for a suitable programme only once per sidereal day. The selected stars should come to altitude 45° within an interval of half an hour or so, but each two consecutive events must not be too close in time. An approximate direction for finding each star must be calculated: usually the magnetic bearing is estimated from appropriate calculations of the true bearings. The astrolabe has a magnetic compass by which the instrument can be set in magnetic meridian, then a floating graduaated circle can be set to zero and clamped. The astrolabe is then set to the appropriate magnetic bearing ready for the recording of each star in turn. The only quantity to be recorded is the chronometer reading at each coincidence.

To improve accuracy, two devices are available. A prism of small deflection is placed over one half of the main prism face entered by beam (*2*), and the effect is to double horizontally this image of the star: the observer then sees a pair of images approaching the single image produced by the other beam, as indicated in Fig. 34. Thus the coincidence is more accurately estimated as the moment when the single image lines up with the pair. The other device is a set of three small deflecting prisms that are placed in succession in beam (*1*)

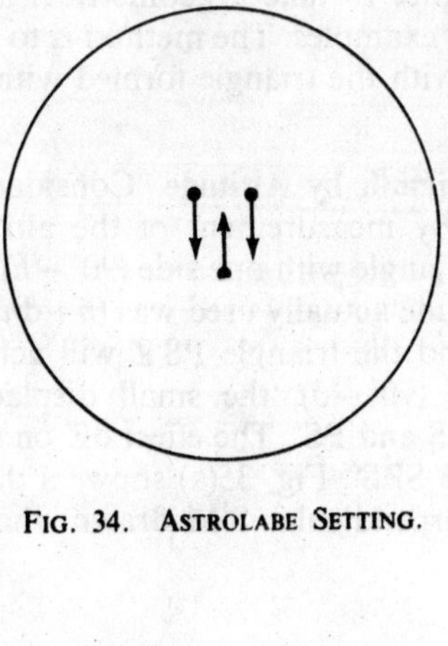

FIG. 34. ASTROLABE SETTING.

in a prescribed order: as each prism is put in place after a coincidence is recorded, it separates the beams again, and six coincidences can be timed and averaged.

The computations are done as for position lines at an assumed altitude of 45°, or such altitude as is found from experience to be more nearly the value determined by the prism of the particular instrument. In any case, all the position lines will be offset by the inaccuracy of the assumed altitude as well as the refraction which of course is taken to be the same for all the stars, so need not be calculated.

2.29 Effects of Errors in Astronomical Work. The sources of errors likely to affect the results of the observation procedures that have been described are:

(i) Errors of observed altitudes: both random errors of setting on the star and reading the theodolite, and constant errors in a set of observations due to the refraction not actually having the value calculated;

(ii) If the observer takes the latitude from a map it may differ from the astronomical latitude that should be used in the spherical triangle (see Section 2.16);

(iii) Personal equation in recording the chronometer readings of settings: this tends to be systematic in effect.

It is assumed that there are no errors in the declinations and right ascensions taken from the Almanacs.

In general, the required astronomical value will be calculated from a spherical triangle formed from an observed quantity and other data: the relationship between a small error in one of the elements and the consequent small error of the calculated quantity, can be obtained by standard calculus differentiation. As a rule, however, it is simpler to take a geometrical approach, as will be illustrated by a few examples. The method is to compare the correct spherical triangle with the triangle formed with the erroneous item of data.

2.30 Error in Azimuth by Altitude. Consider the calculation of azimuth angle Z by measurement of the altitude of a star. The correct spherical triangle with one side $(90° - h)$ is PSZ, as shown in Fig. 35. If the altitude actually used was $(h + \mathrm{d}h)$, the side SZ will be too small by $\mathrm{d}h$ and the triangle PS′Z will actually be calculated. Since PS = PS′ = $(90° - \delta)$, the small displacement SS′ will be perpendicular to PS and PS′. The effect $\mathrm{d}Z$ on the azimuth angle is shown as the angle SZS′. Fig. 35(a) shows a detail at the points S and S′ with the perpendicular S′M drawn: the difference between

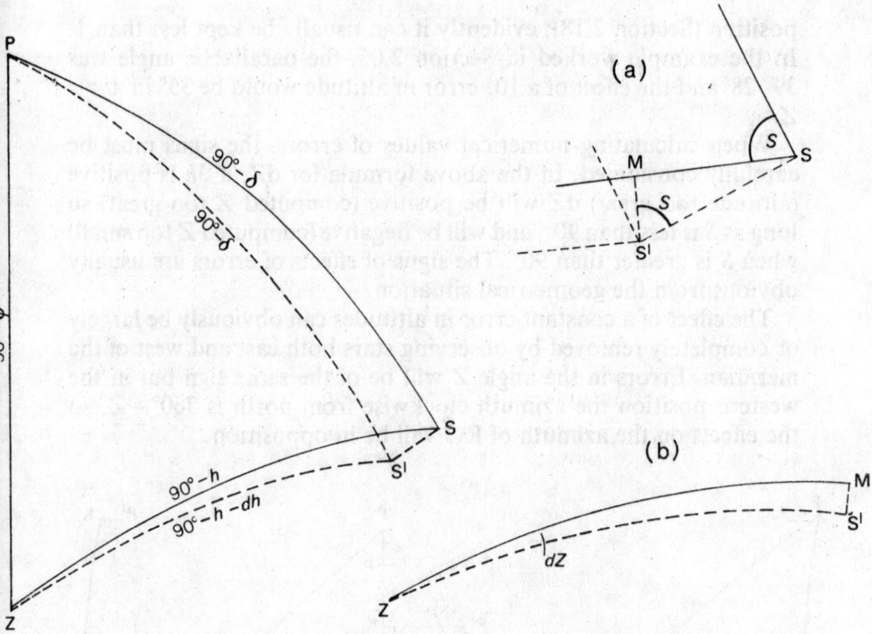

FIG. 35. ERROR IN AZIMUTH CALCULATION.

ZS and ZS′ is dh which is therefore the angular value of SM. Obviously S′M is dh . cot S where S is the parallactic angle of the triangle (Section 1.09). We now consider the long thin triangle MZS′, Fig. 35(b). The angles at M and S′ are practically right angles, and the sine formula reads:

$$\frac{\sin(dZ)}{\sin(dh \cdot \cot S)} = \frac{\sin 90°}{\cos h},$$

but the angles in brackets are small and we can write

$$\frac{dZ}{dh \cdot \cot S} = \sec h, \quad \text{or} \quad dZ = dh \cdot \cot S \cdot \sec h.$$

Relationships like this can be used to guide the choice of stars in positions that will minimise the effects of expected errors. The factor is least for $h = 0°$ but this conflicts with the obvious requirement to observe a star at a reasonable altitude to avoid uncertainty in the calculated refraction; sec h does not increase rapidly through moderate altitudes and is, for instance, only 1·22 at $h = 35°$. The factor cot S is zero at $S = 90°$, which of course is the elongation position (Section 2.18): evidently it can usually be kept less than 1. In the example worked in Section 2.05, the parallactic angle was 39° 28′ and the effect of a 10″ error in altitude would be 35″ in angle Z.

When calculating numerical values of errors, the signs must be carefully considered. In the above formula for dZ, if dh is positive (altitude too great) dZ will be positive (computed Z too great) so long as S is less than 90°, and will be negative (computed Z too small) when S is greater than 90°. The signs of effects of errors are usually obvious from the geometrical situation.

The effect of a constant error in altitudes can obviously be largely or completely removed by observing stars both east and west of the meridian. Errors in the angle Z will be of the same sign but in the western position the azimuth clockwise from north is 360° − Z, so the effects on the azimuth of RO will be in opposition.

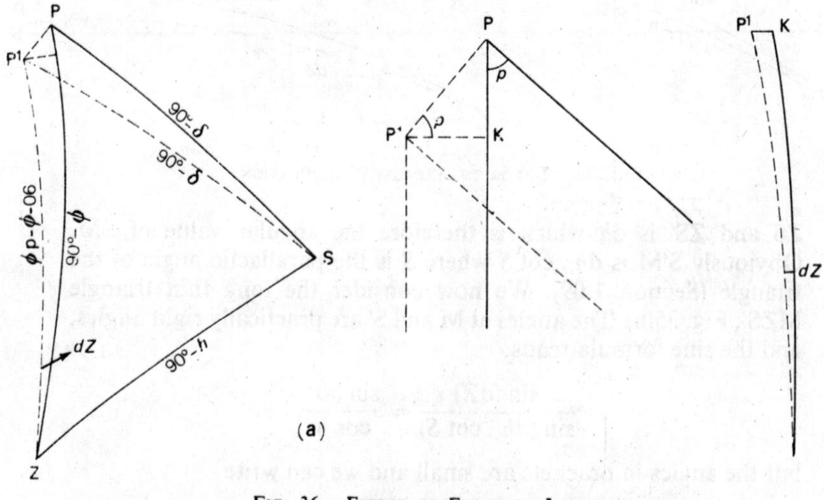

Fig. 36. Effect of Error in Latitude.

To find the effect of using a wrong value of the latitude, the correct and erroneous triangles are drawn in Fig. 36. The angle value of PK is the error dϕ, so P′K is dϕ . cot P and we get dZ = dϕ . cot P . sec ϕ. Balancing out by taking east and west stars is again possible. Moreover, when there is an obvious systematic difference between an azimuth obtained from east stars and the same azimuth obtained from west stars, it will be possible to work out a more correct latitude, such that the mean azimuths from the two sets of stars are equal.

2.31 Effect of Dislevelment. There are possible sources of error in astronomical observations, arising specially from the fact that the observed celestial objects are usually at considerable altitudes. If the telescope of a theodolite when elevated by rotation about the transit axis does not move in a vertical plane, any horizontal circle reading to an elevated mark will differ from what it would be if the instrument was in perfect mechanical adjustment and perfectly levelled. The two conditions just mentioned are quite independent, and must not be confused: the expression often seen—'dislevelment of the transit axis'—is not always clearly defined, and the reader is advised to avoid using it.

The transit axis may be out of level because it is not perpendicular to the vertical axis of rotation of the theodolite; this is a mechanical maladjustment of the instrument, and its effects are completely eliminated by the device of changing face. This source of error therefore will not be considered further here. The other reason for 'dislevelment' of transit axis is simply that the theodolite has not been accurately levelled, that is its axis of rotation is not truly vertical. Effects of this state of affairs cannot be eliminated by any observational procedure.

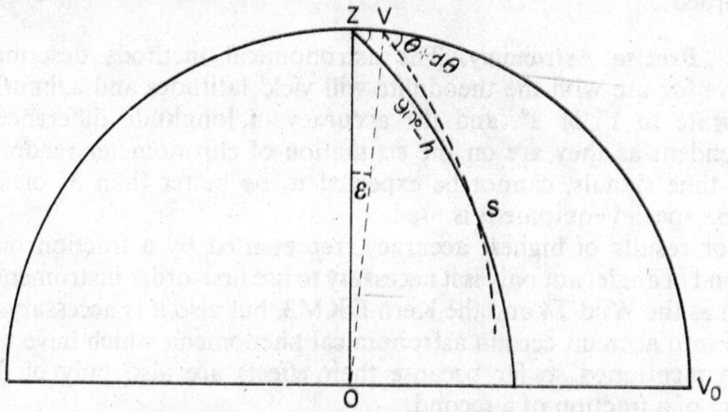

FIG. 37. DISLEVELMENT ERROR.

In Fig. 37 let OZ be the true vertical through the theodolite, and let OV be the direction of the actual axis of rotation: OV_0 is the horizontal direction of the dislevelment. Let OS be the direction of a star at elevation h, so that in the spherical triangle ZVS the side ZS is $(90° - h)$. The horizontal angle θ, angle VZS, is the direction of the star referred to the direction OV_0: with the dislevelled instrument, this direction would be measured as the angle SVV_0, or $(\theta + d\theta)$. Let the dislevelment angle ZOV be ε. Four elements of the spherical triangle are named, the angle at V being $180° - (\theta + d\theta)$.

The relation between them, by a standard spherical formula, is

$$\cos \varepsilon, \cos \theta = \sin \varepsilon \,.\, \tan h + \sin \theta \,.\, \cot (\theta + d\theta).$$

Since ε and $d\theta$ are very small, we can write;

$$1 \,.\, \cos \theta = \varepsilon \,.\, \tan h + \sin \theta \,(\cot \theta - d\theta \,.\, \mathrm{cosec}^2\, \theta),$$

which reduces to $0 = \varepsilon \,.\, \tan h - d\theta \,.\, \mathrm{cosec}\, \theta$, or $d\theta = \varepsilon \,.\, \sin \theta \,.\, \tan h$. Obviously, the factor $\varepsilon \,.\, \sin \theta$ is the component of ε perpendicular to the direction of the star. Thus, if the horizontal angle between two marks of altitudes h_1 and h_2 in directions θ_1 and θ_2 is measured it will be in error by $\varepsilon \,.\, (\sin \theta_2 \,.\, \tan h_2 - \sin \theta_1 \,.\, \tan h_1)$. In practice, one of the sighted objects is usually the RO which will be a terrestrial signal at very small altitude, and then the term relating to the RO will be negligible.

To cope with possible errors due to non-verticality of the rotation axis, the theodolite used for astronomical observations should have a sensitive bubble placed perpendicular to the transit plane of the telescope. The readings of this bubble will then indicate values of the factor $\varepsilon \sin \theta$, and the necessary corrections to horizontal circle readings can be made, providing that an approximate value of h is recorded.

2.32 Precise Astronomy. The astronomical methods described above for use with the theodolite will yield latitudes and azimuths accurate to 1″ or 2″, and the accuracy of longitude differences, dependent as they are on the estimation of chronometer readings and time signals, cannot be expected to be better than 3″ or so, unless special equipment is used.

For results of highest accuracy, represented by a fraction of a second of angle, not only is it necessary to use first-order instruments, such as the Wild T4 and the Kern DKM3, but also it is necessary to take into account certain astronomical phenomena which have not been mentioned so far because their effects are also only of the order of a fraction of a second.

There is not space here for detailed descriptions of first-order methods, and the reader may refer to special treatises on this subject. The following notes briefly mention some of the problems.

Instruments. Apart from the use of first-order instruments, the most obvious requirement in instrumental precision is the removal of personal equation in recording the timing of events. This is achieved by use of an *impersonal eyepiece*. In this device, the cross line (or close parallel pair of cross lines) that is to be set on the star is carried on a slide which is traversed across the field of view by the observer, who turns a fine-pitch screw. The star is kept on the moving cross line (or between the pair of cross lines) and the slide makes a number (e.g. 5) of electrical contacts as it is moved along. The contacts operate the chronograph.

The recording of time signals and impersonal micrometer contacts can be done very precisely by use of the 'Chronocord', an instrument which prints out readings of electric impulses applied to it, to $0{\cdot}001^s$ on a time scale determined by internal crystal-controlled electronic circuits.

Star Positions. Celestial coordinates must be taken from a catalogue such as the Apparent Places of Fundamental Stars which give positions to higher precision than in the Star Almanac. More comprehensive catalogues are also available.

Nutation etc. The direction of the Earth's axis of rotation is not invariable, either in relation to the celestial sphere, or within the body of the Earth itself. In space, there is the steady slow movement, the precession, which is taken into account in the published 'apparent places' of stars. In addition to the precession there are 'nutations' of shorter period mainly due to the influence of the Moon and therefore having a period of a lunar month. Variations of such short period obviously cannot be included in values that are tabulated at intervals of a month, or, as in the Apparent Places of Fundamental Stars, at intervals of 10 days. Factors and instructions for applying the nutation corrections are given in the catalogues and certain other astronomical publications.

Diurnal Aberration. The Earth travels round the Sun in its orbit at the speed of about 18 km per second. This causes apparent shifts in positions of stars, analogous to the apparent change of direction of a cross wind when the observer is moving. This aberration effect may amount to about 20″, the radian value of which is the ratio of the orbital speed to the speed of light, but since its period is a year its effects are included in the tabulated apparent positions. An observer on the Earth has an additional speed due to the rotation of the Earth and directed towards the east point of his horizon. On the equator the speed is nearly $\frac{1}{2}$ km/second, and the apparent shift, the 'diurnal aberration' may be as much as 0·3″. Of course this effect cannot be included in published tables and its value must be calculated for each observation if it is needed.

Tidal Effects, etc. The direction of the vertical at a point on the Earth undergoes small oscillatory changes caused by the varying of positions of the Sun and Moon: these changes are too small to be detected by field surveying instruments. However, the direction of the Earth's axis of rotation is not quite fixed in relation to the physical body of the Earth, but slowly moves about in a somewhat irregular circuit with amplitude of a few tenths of a second and period about a year. A strict definition of latitude must be referred to an adopted mean direction of the axis of rotation, and the information required for applying a correction to the results of precise astronomical observations is derived from work done at the stations of the International Latitude Service.

What all this amounts to is that, to the order of small fractions of a second, no astronomical quantity at any point is absolutely constant: they have mean values averaged over suitably lengthy periods.

2.33 Astronomical Resection. Several recent contributions to the literature of field astronomy have described methods for finding astronomical quantities by recording horizontal directions and chronometer readings of settings on stars, without any measurement of vertical angles. Obvious advantages of any such procedure are the absence of errors due to uncertain values of refraction, and the freedom to use stars at rather low altitudes. Most of these methods are essentially resections.

If the sub-stellar positions of three stars are known, and the horizontal directions to them are measured at a ground station, the position of the station can be calculated. This is resection on a sphere. The mathematics of the problem is very cumbersome, but in any case, as in the plane situation, graphical aids should be used, because then it is quite easy to take account of settings on any number of stars.

A practical procedure would go as follows. It can be assumed that the surveyor knows his astronomical 'position' quite closely and he can calculate the approximate azimuth of a convenient star that will appear at a convenient time for beginning the observation programme. He sets up the theodolite, and at the calculated time he sets the vertical cross line on the star and then adjusts the micrometer and horizontal plate to read the calculated azimuth. Then, any reading of the circle will be close to the true azimuth, and all readings will be in error by a small unknown but constant amount.

The surveyor takes readings on a fixed RO and sets the vertical cross line on a number of stars in different directions, recording the circle and chronometer readings for each. Settings should be made on both faces to avoid trouble with instrumental errors; bubble readings for dislevellment correction should also be taken, though errors from this source will be small if stars at low altitudes are used.

In Fig. 38(a) let S be the sub-stellar position of one of the stars, and let D be the direction recorded on the theodolite. The computation problem is then to plot on a chart some points at which the

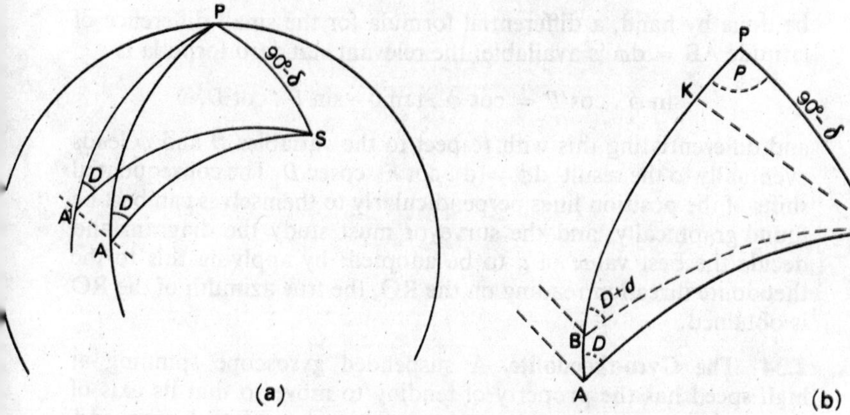

FIG. 38. AZIMUTH POSITION LINES.

azimuth to point S would be exactly *D*. The surveyor chooses a meridian close to his estimated position and solves the spherical triangle: the angle at P is the difference of longitude of this meridian and the SSP, and the other two known elements of the triangle are of course $(90° - \delta)$ of the star and the direction *D*. The side PA, the co-latitude of A, is required. There is no simple direct formula for this: one method is to calculate the altitude of the star by the sine formula $\cos h = (\sin P \,.\, \cos \delta)/\sin D$, then, SK being perpendicular to PA, the two portions of PA are given by $\tan(PK) = \cot \delta \,.\, \cos P$ and $\tan(KA) = \cot h \,.\, \cos D$. Taking another meridian a minute or two away from PA, a similar computation gives point A′ at which the azimuth to S would also be exactly *D*.

The broken line AA′ is thus part of the locus of all points at which the azimuth to S is *D*: it is not a small circle but of course it can be drawn as a straight line for the short distance needed in practice, just as was the altitude position line. A 'direction position line' can be drawn for each star observed. The locations of these lines on the diagram are affected not only by normal random errors of observation but also by the fact that all the used values of *D* differ from true azimuths by a constant small amount *d*. The next stage of the process is therefore to find the best value of *d* that would cause all the position lines to gather together and pass more or less through a common point. It is necessary to calculate how much each line would move on the diagram for a given value of *d*, say 10″. Carrying out the same computation with a direction $(D + 10'')$ will give the latitude of a point B as shown on Fig. 38(b). The electronic computer can be instructed to do this as part of the programme. The position line through B can be drawn parallel to AA′. If the calculation is to be done by hand, a differential formula for the small difference of latitude AB = $d\phi$ is available: the relevant standard formula is

$$\sin \phi \,.\, \cos P = \cos \phi \,.\, \tan \delta - \sin P \,.\, \cot D,$$

and differentiating this with respect to the variables D and ϕ leads eventually to the result: $d\phi = d \,.\, \cot h \,.\, \text{cosec}\, D$. The consequential shifts of the position lines perpendicularly to themselves can best be found graphically, and the surveyor must study the diagram and decide the best value of d to be adopted: by applying this to the theodolite direction reading on the RO, the true azimuth of the RO is obtained.

2.34 The Gyro-theodolite. A suspended gyroscope spinning at high speed has the property of tending to move so that its axis of spin is in the true meridian plane. Thus a combined gyro and theodolite provides a means of finding true bearings of terrestrial objects, and such instruments are available. The accuracy obtainable is not equal to that of good astronomical observations. Though it is not an astronomical method, the use of the gyro-theodolite is mentioned here because it is a means of getting a true bearing, and has the great advantage that it is available in conditions of poor visibility, and in enclosed situations such as mine workings and tunnels.

EXAMPLES

1. A surveyor observes that a star of declination 68° 58′ 53″ South transits at a maximum altitude of 33° 22′ 34″. Taking the refraction to be 1′ 21″, what was the surveyor's latitude?

2. What will be the approximate local sidereal time at sunset at a place in the Tropics early in November?

3. At a place in latitude 36° 16′ North, what will be the hour angle, azimuth, and altitude of a star of declination 56° 06′ North, when it is at elongation?

4. Prove that the rate of change of the altitude of a star is greatest when the star is due East or due West.

5. At a place in latitude 1° 24′ 17″ South the true altitude of a setting star of declination 9° 44′ 58″ North was found to be 41° 07′ 39″ when the chronometer reading was $20^h\ 16^m\ 48{\cdot}4^s$. The star's right ascension was $21^h\ 42^m\ 49{\cdot}6^s$. What was the local sidereal time of the observation?

The clock was running at mean time rate. At what altitude and at what reading on the chronometer will a star of right ascension $1^h\ 58^m\ 02{\cdot}7^s$ and declination 61° 41′ 56″ South be at transit?

6. A surveyor receives the Greenwich signal for 16^h GMT when his mean-time chronometer reads $21^h\ 17^m\ 04{\cdot}5^s$. He observes a star approximately east of his position, and finds that its true altitude is 31° 42′ 09″ when the chronometer reads $21^h\ 38^m\ 12{\cdot}2^s$. The star has declination 19° 24′ 58″ North, right ascension $19^h\ 57^m\ 36{\cdot}6^s$. The surveyor estimates his latitude to be 29° 15′ 11″

North. At 12^h GMT the right ascension of the Mean Sun was $6^h\ 22^m\ 46{\cdot}6^s$ increasing at the rate of $9{\cdot}86^s$ per hour. What was the surveyor's longitude?

7. A navigator takes a morning sight on the Sun, finding that the apparent altitude of the Sun's lower limb above the visible horizon is 48° 19′ 50″ when the GMT is $12^h\ 21^m\ 56^s$. His estimated position is 6° 53′ North, 44° 52′ West.

Using the information given below, calculate data for plotting a position line:

Sun's declination	17° 55′ 40″N
semi-diameter	15 50
Dip of horizon	4 10
Refraction	50

The equation of time was $6^m\ 45^s$ in the sense that the right ascension of the Mean Sun was less than that of the actual Sun.

8. At a place in latitude 28° 16′ 54″ North a surveyor observes a star in the NW quadrant, obtaining the undermentioned readings:

	Hor. Circle	*Ver. Circle*
Face Left on RO	96° 15′ 23″	
Face Left on Star	217 33 18	42° 00′ 11″
Face Right on Star	37 31 03	138 10 59
Face Right on RO	276 15 38	

The theodolite reads altitudes on face left. The declination of the star was 53° 50′ 49″ North, and the refraction was 61″. Calculate the azimuth of the RO.

What error would be caused by an error of +10″ in (i) the latitude, (ii) the altitude?

REFERENCES

ADAMS L. P. 'Astronomical position and azimuth by horizontal directions'. *Survey Review* No. 148, April 1968.

ADAMS L. P. 'Astronomical position and azimuth by horizontal directions: a rigorous solution'. *Survey Review* No. 159, January 1971.

Admiralty List of Radio Signals, Vol. V.

BARRELL H. AND ESSEN L. 'Atomic standards of length and time'. *Science Progress* No. 186, April 1959.

BEBB G. 'Personal equation in field astronomy'. *Australian Surveyor*, December 1965.

BENNETT G. G. 'Precise astronomical longitude determinations in Australia'. *Survey Review* No. 129, July 1963.

BIDDLE C. A. *Textbook of field astronomy*. H.M. Stationery Office, 1958.

BOWIE I. G. 'Latitude determination without vertical circle readings'. *Empire Survey Review* No. 102, Oct. 1956.

CHIAT B. 'Azimuth and longitude from near-meridian star observations by theodolite'. *Canadian Surveyor*, June 1970.

CHICHOWICZ L. *Methods of geodetic astronomy for the intertropical zone*. Warsaw, 1969.

GLEINSVIK P. 'Astronomical determination of latitude by horizontal angles'. *Bulletin Géodésique* No. 90, 1968.

INTERNATIONAL ASTRONOMICAL UNION. *Apparent Places of Fundamental Stars.* (Published annually.)

KOCHLE R. 'Simultaneous determination of latitude and azimuth from observations of the Sun'. *Survey Review* No. 145, July 1967.

LAUF G. B. 'Precise determinations of azimuth using gyroscopic methods'. *Australian Surveyor*, January 1968.

MACKIE J. B. *Elements of Astronomy for Surveyors* (7th Edn). Griffin, London, 1971.

MOREAU R. L. 'Photoelectric observations in geodetic astronomy'. *Canadian Surveyor*, September 1966.

MUELLER I. I. *Spherical and Practical Astronomy as Applied to Geodesy.* New York, 1968.

MUNSEY D. F. AND PUGH K. T. 'Observations with PIM'. *Survey Review* No. 161, July 1971.

PRESCOTT N. J. D. 'A photographic method of determining azimuth, latitude and longitude'. *Survey Review* No. 134, October 1964.

OPIE B. P. 'The accuracy of circum-elongation observations for azimuth'. *Survey Review* No. 137, July 1965.

ROBBINS A. R. 'Personal equation in the determination of geodetic azimuth'. *Bulletin Géodésique* No. 57, 1960.

—— 'Time in geodetic astronomy', *Survey Review* No. 143 January 1967.

—— 'The Chronocord Mk. III: a portable recording crystal chronometer'. *Bulletin Géodésique* No. 87, 1968.

—— 'Azimuth control in Canadian latitudes'. *Canadian Surveyor*, September 1969.

ROELOFS R. 'A new device for Solar observations'. *Bulletin Géodésique* No. 6, 1947.

—— *Astronomy applied to land surveying*. Amsterdam, 1950.

RUSHWORTH W. D. 'Use of a photo-electric cell for impersonal longitude observations'. *Empire Survey Review* No. 114, October 1959.

SPENCER JONES, H. *General Astronomy* (4th Edn). Edward Arnold, London, 1961.

STEPEC W. A. 'Visible record of time signals'. *Canadian Surveyor*, December 1969.

Star Almanac for Land Surveyors. H.M. Stationery Office. (Published Annually).

STRASSER G. J. AND SCHWENDENER H. R. 'A north-seeking gyro attachment for the theodolite'. *Bulletin Géodésique* No. 79, 1966.

TAIT G. B. 'A proposed method of solution of the position line problem'. *Empire Survey Review* No. 107, January 1958.

WHITE L. A. 'General theory for horizontal angle observations in astronomy'. *Survey Review* No. 141, July 1966.

WILLIAMS H. S. AND BELLING G. E. 'The reduction of gyrotheodolite observations'. *Survey Review* No. 146, October 1967.

ORDNANCE SURVEY *The History of the Retriangulation of Great Britain 1935–62*. H.M. Stationery Office, 1967.

CHAPTER III

GEODETIC SURVEYING—ANGLE MEASUREMENT

3.01 Geodetic Mapping Systems. For a period of about two centuries, say 1730–1930 if numbers must be quoted, geodetic surveying and topographical mapping were works conducted in close co-operation for the purposes of measuring the Earth and accurately describing the shape of its surface. Great national surveying organisations carried out these operations and in course of time most of the land areas of the world have been covered by systems of triangulation controlled in scale by a few laboriously measured base lines; some traversing, levelling and astronomical observations were also made in connection with these surveys.

From these essentially geometrical measurements, various Figures of the Earth were derived: these were ellipsoid-shaped surfaces calculated to be closely fitting the limited areas covered by the separate survey systems.

Of course the survey systems consisted physically of points marked on the Earth's surface. To describe the relative horizontal positions of trigonometrical points, traverse stations, etc., they were represented on the ellipsoidal reference surface by calculating geographical coordinates, i.e. latitudes and longitudes. This is described in detail in Chapter VI.

The next stage in the process was to represent the curved surface of the ellipsoid on to a plane coordinate system for the purpose of constructing maps. This stage made use of the exact geometry and mathematical formulas of a chosen map-projection. Topographical details were measured on the ground—by chain survey, plane table, or other method—and represented on the maps in correct positions by suitable calculations and graphical procedures. Map construction is described in Chapter IX.

So much for the horizontal geometry. The third dimension, height, was treated separately. For reasons which need no explanations here, the definition of height as usually stated was not strictly compatible with any of the methods used for measuring that quantity. Maps showing horizontal detail but no indication of heights are useful, even desirable, for some purposes. However, a complete survey includes the third dimension. In most triangulation systems, vertical angles were observed: otherwise, heights could be obtained by independent methods, very roughly by use of barometers, much more accurately by levelling. Contours were put on the maps by surveyors with their plane tables. Height determination is described in Chapter VII.

3.02 New Methods and Means. Mention of the year 1930, and use of the past tense above, are not to be taken as implying that the methods of surveying described are no longer in use. The traditional methods are employed where appropriate, and we may describe them as traditional only because the surveyor and the geodesist now have at their disposal some new tools which have altered practices and wrought changes of considerable economic and scientific importance. Five significant developments during the past few decades have been:

I *Air Survey*. After the first world war the use of an air-borne camera as an accurate observational instrument was rapidly developed, so that it is now the usual method of gathering information about topographical detail on the Earth's surface. In the drawing of the maps, choice of features to be shown is made by the photogrammetric operator at his machine and not by the surveyor on the ground. But the surveyor on the ground is required for measuring the control points for the photogrammetry. Moreover, there are types of country, and types of task, in which air survey is not the most suitable tool.

II *Electromagnetic Distance Measurement*. Instruments for measuring distance by timing the passage of electromagnetic waves along the lines have revolutionised survey procedures in the field. A line can be measured in a few minutes, once the instruments have been set up, with an accuracy of a few parts per million. Almost any intervisible line can be measured and it is now possible to construct a survey framework by measuring distances only: systems in which all angles and all distances have been measured actually exist. Another effect of the introduction of EDM has been to increase the use of traverses for establishing geodetic framework.

III *Gravimetry*. The gravimeter has made it possible to measure the acceleration of gravity at the rate of tens of stations per day. Moreover, gravity can be measured at sea. Gravity observations give information about the shape of the Earth, and they make possible a precise definition of the third dimension, in a manner described in Chapter VII.

IV *Artificial Satellites*. From the surveyor's point of view, these are survey beacons situated at some hundreds of kilometres above the Earth's surface, and although they are rapidly moving they can be sighted or photographed simultaneously from stations on the Earth that are hundreds of kilometres apart. This means that, for the first time, it has become possible to make geometrical measurements, by a kind of solid or spatial triangulation, to connect together all the land areas of the Earth's surface. Furthermore, observations of the dynamical behaviour of satellites have provided new and accurate estimates of the overall shape of the Earth.

V *Computers*. The advent of the large electronic computer has made it possible to undertake computation tasks of such magnitude as would previously have been quite beyond the capacity of any human operator even with the aid of mechanical calculators. No longer need the surveyor divide his computing into little bits, or make simplifying assumptions, or contrive special formulas, for the purpose of keeping the computing time within reasonable bounds.

3.03 The Gravity Field. These recent developments have emphasised the fact that geodetic surveying is very much concerned with the phenomenon of gravitation. A satellite moves under the influence of the forces of gravitation along its path; the surveyor on the ground sets up theodolites and levels with reference to the direction of gravity at each station.

The Earth is a piece of matter which exerts gravitational influence on other objects in accordance with Newton's formula. This influence can be treated as a mechanical force. Now force is a vector, having strength as well as direction, so we can think of the Earth as being surrounded by a *field of gravitation*, meaning that at any point there is a direction of gravitation, and in some suitable unit of measurement the strength of the vector can be given a numerical value.

The situation can be illustrated by the idea of a network of *lines of force* somewhat as shown in Fig. 39. At any point the direction of gravitation is the tangent to the line of force through that

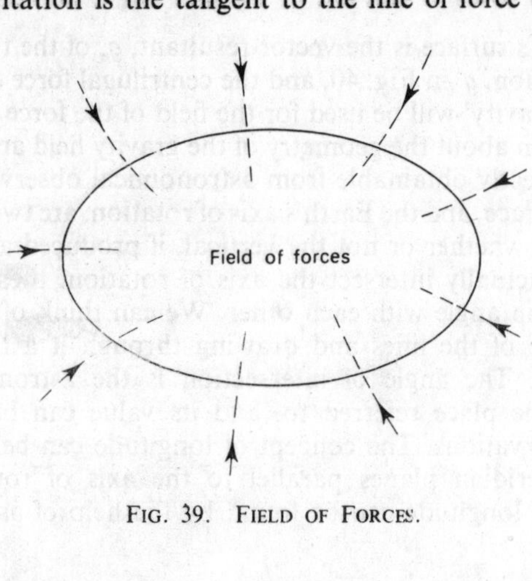

FIG. 39. FIELD OF FORCES.

point. The shapes of the lines are determined by the distribution of the mass of material of the whole Earth: the strength of the vector will obviously decrease with increasing distance from the Earth. A satellite moves in this gravitation field: the word 'gravitation' has been deliberately used in distinction from 'gravity' as described below.

3.04 Terrestrial Gravity. The Earth is also rotating. Any object constrained to partake of this rotation, by being situated on the Earth's surface for instance, behaves as if it is affected by an additional influence, the so-called centrifugal force. This effect is proportional to the distance from the axis of rotation, and is perpendicular to that axis. At any place on the Earth's surface it is much smaller than gravitation. The force of 'gravity' which we actually experience

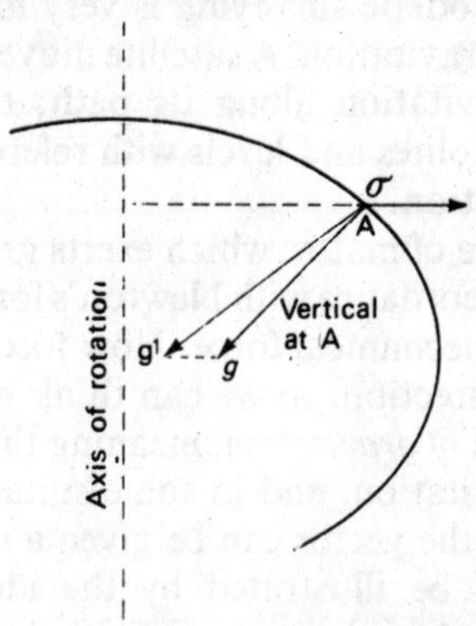

FIG. 40. GRAVITY ON THE EARTH.

on the Earth's surface is the vector resultant, g, of the true gravitational attraction, g' in Fig. 40, and the centrifugal force σ as shown. The word 'gravity' will be used for the field of the force g.

Information about the geometry of the gravity field at the Earth's surface is directly obtainable from astronomical observations. The vertical at a place, and the Earth's axis of rotation, are two directions in the Earth: whether or not the vertical, if produced as a straight line, would actually intersect the axis of rotation, these two lines make a certain angle with each other. We can think of taking any point on one of the lines and drawing through it a line parallel to the other. The angle of intersection is the astronomical co-latitude of the place referred to, and its value can be found by suitable observations. The concept of longitude can be defined by means of meridian planes parallel to the axis of rotation, and differences of longitude can be found by the help of astronomical observations. The principles and methods of astronomical observations have been described in some detail in Chapters I and II. Essentially, the results of latitude and longitude determinations are the angles between vertical directions at points on the Earth's surface, and their relations to the direction of the axis of rotation.

3.05 Geometrical Surveying. Survey measurements on the Earth's surface are conducted in close relationship to the gravity field. At any point, the vertical is the direction of g, and an instrument such as a theodolite is levelled by placing its axis of rotation in coincidence with the vertical, usually by the aid of a bubble tube. The plumb line is a direct indicator of the vertical.

Any other direction through the centre of the theodolite will make a certain angle with the vertical direction. Such angles, or zenith distances, are measurable on the vertical circle of the theodolite, but it is more usual practice to quote them as elevations (or depressions) which are angles referred to the plane through the centre of the theodolite and perpendicular to the vertical.

A level is an instrument for determining distances of points perpendicularly below the plane just mentioned. Such a plane is often described as a horizontal plane but this is confusing terminology because a plane, in the strict geometrical sense, can only be horizontal at one point, if we accept that horizontal means perpendicular to the vertical.

A horizontal angle is measured round the vertical axis of a theodolite; it is the angle between two planes, each of which is defined by the vertical axis and one of the observed points.

3.06 Level Surfaces: Potential. A horizontal surface is everywhere perpendicular to the lines of force. In the Earth's gravity field these surfaces can be visualised as surfaces of liquid at rest. They are also called 'level' surfaces and 'equipotentials'. The idea of potential is a general one in mechanics and is connected with the measurement of energy or work. Mean sea level is one such surface which it is supposed that the oceans of the world would take up if they could be freed from the tides, waves and other oscillatory disturbing influences.

In Fig. 41 the broken lines represent lines of gravity and the full

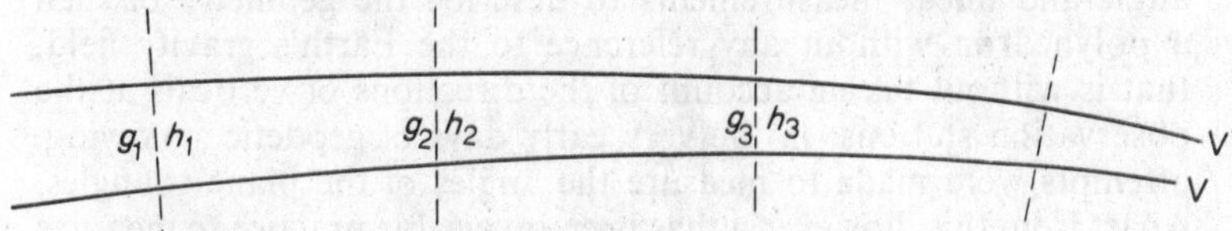

FIG. 41. LEVEL SURFACES.

lines represent sections of level surfaces cutting the verticals everywhere orthogonally. The two level surfaces $\mathfrak{v}$ and $\mathfrak{v}'$ are separated by linear distances h_1, h_2, and h_3 measured along the vertical lines as shown. g_1, g_2, and g_3 are measures of the strength of gravity at the places indicated. Then the property of level surfaces is that $g_1h_1 = g_2h_2 = g_3h_3 = \ldots$, and these products are differences of *potential*. In mechanical terms they measure the work done in taking a unit mass from any point on surface $\mathfrak{v}$ to any point on surface $\mathfrak{v}'$. Thus, level surfaces can be given numerical values in a suitable scale of potential units. This idea is further developed in Chapter VI.

3.07 Triangulation. Surveying is essentially a process of measuring and describing the geometry of the pattern formed by points permanently and visibly marked on the Earth's surface. This is done by measurements of angles and distances, followed by suitable calculations. Geodetic surveying is done to the highest practicable standard of accuracy, usually covering large areas with a basic network of fixed points. Most of the geodetic surveying already done in the world has used the familiar and long-established method of *triangulation*. In the typical small triangulation system illustrated in Fig. 42, the horizontal angles at all the points would be measured (94 angles in this case) and the scale would be fixed by finding the lengths of a few lines such as AB and CD by base-measurement. In a primary triangulation system, the distances between adjacent points would mostly be in the range 20–70 km. As a rule, the lines AB and CD themselves would be unsuitably situated for direct measurement by the very precise but laborious method of suspended invar tapes, and short lines more suitably placed would be measured; special pieces of triangulation called *base extension figures* would have to be observed, to connect these bases to the main system.

A triangulation network is really a geometrical figure in three dimensions. In any such survey system, adjacent points must be intervisible. If we imagine adjacent points joined up by straight lines as indicated in Fig. 43, we get a many-faceted structure formed of plane triangles. In theory, it would be possible to make sufficient angle and linear measurements to describe the geometry of such a polyhedron without any reference to the Earth's gravity field, that is without taking account of the directions of verticals at the observation stations: in the very early days of geodetic surveying, attempts were made to measure the angles of the plane triangles. Apart from this, however, it has been invariable practice to measure horizontal angles at the points of the system. This introduces the vertical line at each observation point, and the measured angles are

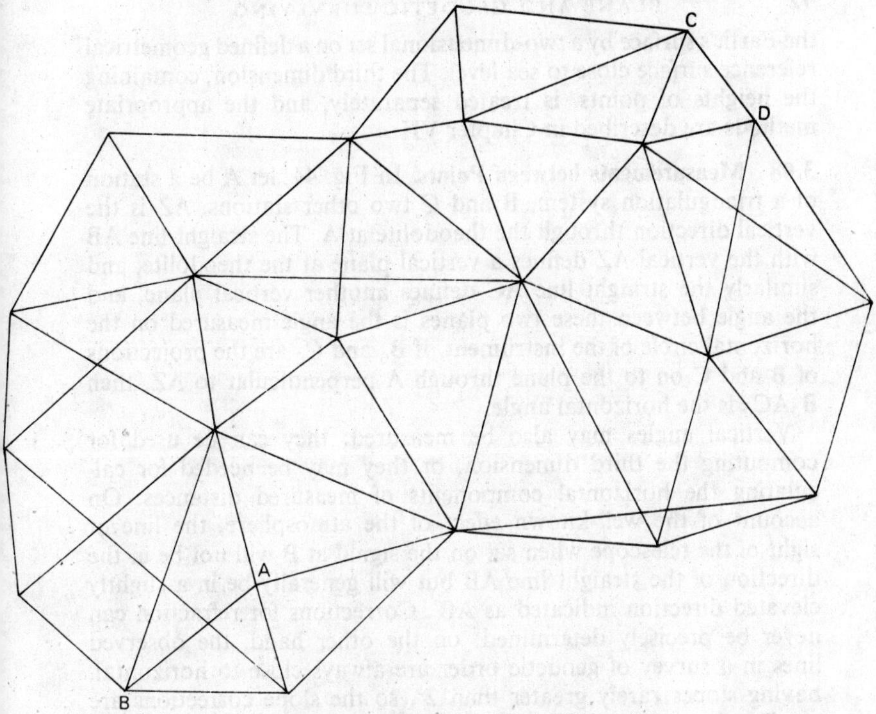

FIG. 42. TRIANGULATION SYSTEM.

the projections of the plane angles on to horizontal surfaces. As for linear measurements, it is also normal practice to convert the measured distances between points to their horizontal components, and further to reduce these to sea level.

Thus, the customary treatment of survey measurements has the effect of replacing the three-dimensional set of real points on

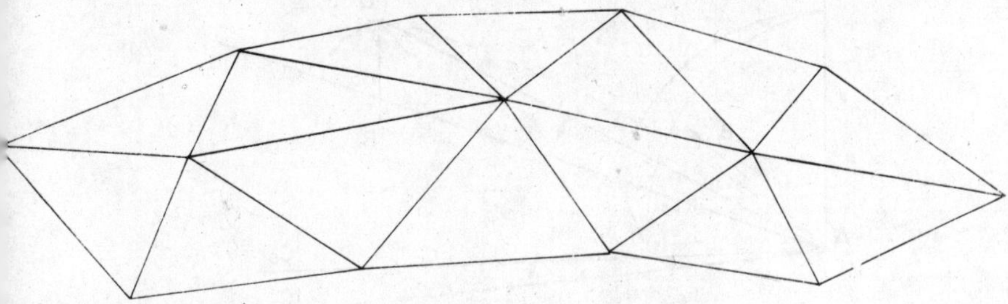

FIG. 43. 3D TRIANGULATION FACETS

the Earth's surface by a two-dimensional set on a defined geometrical reference surface close to sea level. The third dimension, containing the heights of points, is treated separately, and the appropriate methods are described in Chapter VII.

3.08 Measurements between Points. In Fig. 44, let A be a station of a triangulation system, B and C two other stations. AZ is the vertical direction through the theodolite at A. The straight line AB with the vertical AZ defines a vertical plane at the theodolite, and similarly the straight line AC defines another vertical plane, and the angle between these two planes is the angle measured on the horizontal circle of the instrument. If B_0 and C_0 are the projections of B and C on to the plane through A perpendicular to AZ, then B_0AC_0 is the horizontal angle.

Vertical angles may also be measured: they can be used for computing the third dimension, or they may be needed for calculating the horizontal components of measured distances. On account of the well-known effect of the atmosphere, the line of sight of the telescope when set on the signal at B will not be in the direction of the straight line AB but will generally be in a slightly elevated direction indicated as AB′. Corrections for refraction can never be precisely determined: on the other hand, the observed lines in a survey of geodetic order are always close to horizontal, having slopes rarely greater than 2°, so the slope corrections are small and an error of a few seconds in the supposed slope angle will have small effect on the horizontal distance.

It is to be noted that 'horizontal distance' really means the distance along a horizontal surface between A and B: any such line

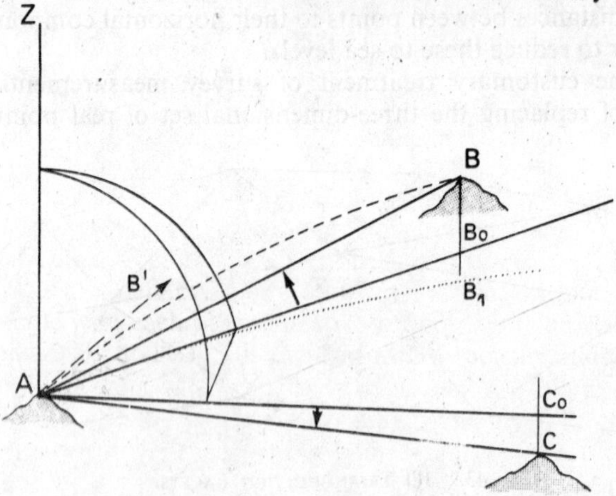

FIG. 44. HORIZONTAL AND VERTICAL ANGLES.

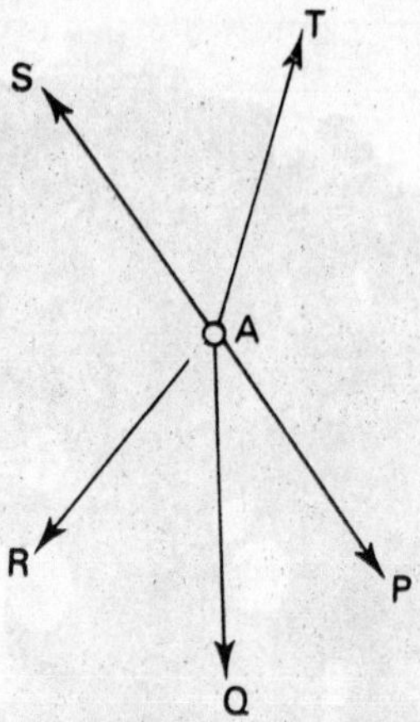

FIG. 45. SKETCH OF DIRECTIONS.

is curved (Section 3.06) as indicated by the dotted line AB_1. This is discussed in more detail in Chapter IV.

A feature of modern geodetic surveying is that distance measurement presents no more difficulty than angle measurement. Many trigonometrical systems nowadays include a large number of measured distances that are treated as observational data along with the angles; indeed it would clearly be possible to determine the geometry of a system of points by measuring distances only, though an angle-measuring instrument would still be necessary for finding the slopes. For the long distances in geodetic systems, the Tellurometer has been widely used: the Geodimeter will also measure fairly long lines, and there are more recently developed instruments such as the Distomat DI.160. With such instruments, the lengths of lines AB and AC can be found, to an accuracy comparable with that of the angle measurement. By the effect of atmospheric refraction, the distances actually measured are also along curves, not necessarily the same as those followed by optical rays, so the curve lengths must first be reduced to chord lengths. Length measurement methods are described in Chapter IV.

3.09 Theodolite Observations. Modern theodolites for first-order work are similar in design to those described in Vol. I. They are merely somewhat larger, with wider object glasses for better resolution, and their micrometer scale graduations are less than one second: it is customary to book readings to 0·1″. Of course this is not the accuracy of a single measure, but by taking the mean of some 12 or 16 rounds the attainable accuracy seems to be a few tenths of a second.

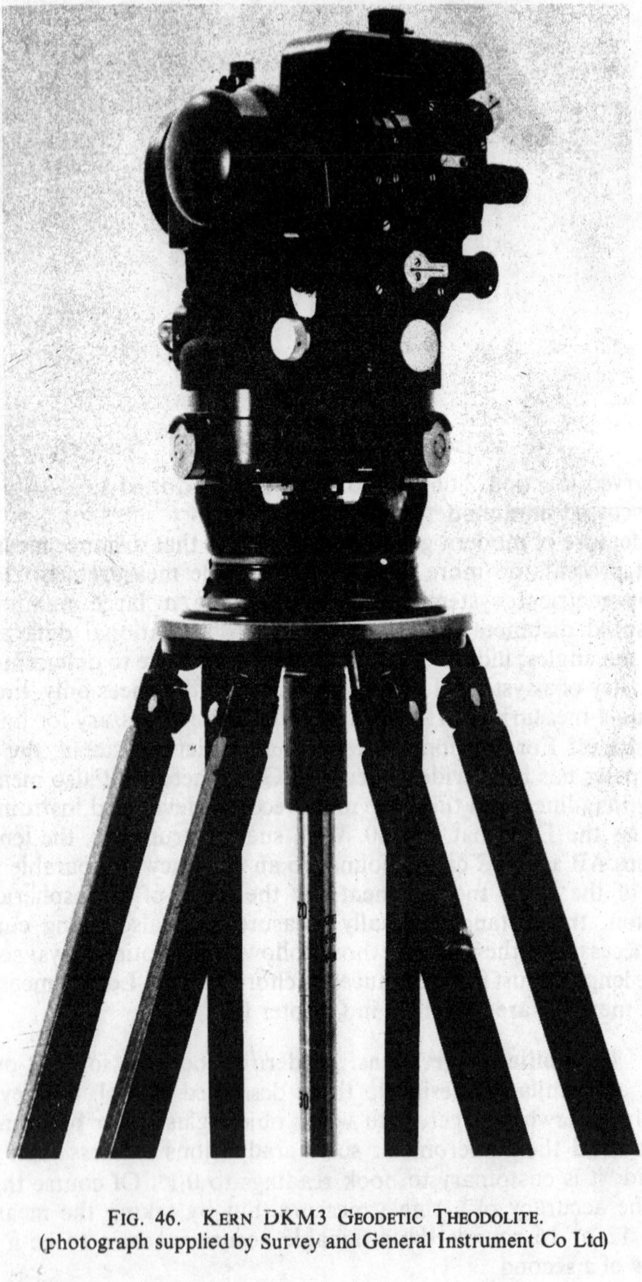

FIG. 46. KERN DKM3 GEODETIC THEODOLITE.
(photograph supplied by Survey and General Instrument Co Ltd)

Typical instruments used for geodetic work are the Geodetic Tavistock theodolite as used by the Ordnance Survey, the Wild T3 and the Kern DKM3.

Fig. 46 is a photograph of a DKM3. The aperture of the telescope is 72 mm and the design is kept compact by use of mirrors, which bring the image to the eyepiece at one end of the transit axis.

At an observation station in a triangulation system, there will be a number of surrounding visible points to which observations have to be made, rarely as few as three, possibly as many as ten or twelve.

For first-order accuracy, each point must be sighted a considerable number of times and a systematic procedure of observation must be adopted. Several such procedures have been advocated. In any case, of course, there must be equal numbers of measures on each face of the theodolite and there must be systematic changes of zero on the horizontal circle so that graduation errors will largely cancel out.

Whatever the number of pointings made to the signals, the observations at a station must finally be presented as a list or abstract of some sort of mean values, which will be used by the computer in his further processing of the triangulation. An important consideration is whether the results of the field work are to be presented as angles or directions.

Fig. 45 represents the observation programme at station A to the five stations P, Q, R, S, and T. Three distinctive systems for observing are described below:

(I) *Single Closed Rounds*. Select a sharp clear signal as Reference Object (RO). Set on this point first, then on all the others in succession, going either clockwise round the diagram i.e. 'swing right', or anticlockwise i.e. 'swing left', ending with a second setting on the RO. The record might look something like columns (a) and (b) below:

Forepoints	*Face*	(a) *Readings*	(b) *Included Angles*
Q(RO)	L	87° 33′ 17·2″	
			35° 17′ 28·9″
R		122 50 46·1	
			106 46 30·6
S		229 37 16·7	
			56 51 07·6
T		286 28 24·3	
			126 15 04·7
P		52 43 29·0	
			34 49 48·9
Q(RO)		87 33 17·9	
			360 00 00·7

A balancing round would be observed face right, and the points could be taken in the opposite order, swing left. The setting of the horizontal plate could remain for this round, or a change could be made in accordance with a suitable rule to spread the zero settings round the circle. There are two matters for discussion about this method:

(i) When a closing setting is taken on RO there is the problem of what to do about the two, usually different, readings. If included angles are abstracted, as in the example, their total will not generally be exactly 360° and the question shifts to (ii) below. Another procedure is to take a mean RO reading, say 17·5 in this case; this unloads the misclose on to two of the observed angles, one each side of the RO. Another procedure is to make the second RO reading the same as the first and apply proportional adjustments to the other settings, so that in this case the new column would read 17·2, 46·0, 16·4, 23·9, 28·4, 17·2: arithmetically this is the same as applying equal adjustments, totalling −0·7″, to the five included angles.

(ii) When observations at a point are abstracted in such a way as to lead to a set of angles having a total differing from 360° 00′ 00·0″, there is the question of whether to adjust immediately to the correct total. This is really a problem for the computation stage. If the adjustment of the whole system is to be applied to observed angles, there is no problem: any discrepancies from 360° round a point will be removed in the process: the observer may consider that his job is done when he has produced a usable and reliable set of measures at each station. What is important, on the part of the observer, is to comment on any circumstances that might affect accuracy, such as a particular signal appearing indistinct or unsteady.

(II) *Double or Complete Rounds for Directions.* In this method, used by the Ordnance Survey for example, the round is not closed up to 360°. Observations at the station A of Fig. 45 would be done as follows: choose a suitable signal as RO, say point Q, and after taking a reading on it set on R, S, T, P, change face, set on P, T, S, R, then back to Q(RO). The observations might be recorded as below:

Forepoint	*Face Left* ↓	*Face Right* ↑	*Mean*	*Direction*
Q(RO)	87° 33′ 17·2″	267° 33′ 18·5″	17·8″	00° 00′ 00·0″
R	122 50 46·1	302 50 46·4	46·2	35 17 28·4
S	229 37 16·7	49 37 17·1	16·9	142 03 59·1
T	286 28 24·3	106 28 22·9	23·6	198 55 05·8
P	52 43 29·0	232 43 28·7	28·8	325 10 11·0

The readings are booked in the order indicated by the arrows, and on the return part of the round the degrees and minutes need not be booked but should be read by the observer to check gross errors. In the last column the directions are angles referred to the RO reading, and there is no question of misclosure on 360° to be dealt with. The differences of the two settings on each point, including the RO, will contain any effects of instrumental errors, collimation or transit axis, but the mean readings will be free from these errors. In practice, the observer will be instructed to re-observe a round if any difference exceeds a certain amount.

After the required number of rounds has been observed, a list of mean directions can be prepared.

An advantage of observing for directions is that it is easy to divide the work into groups of say 5 or 6 directions if the number of signals to be observed at a point is large. Also, in conditions of poor visibility or temporary obscuration, a missed signal can be picked up and referred to RO whenever it comes into view.

(III) *Observed Angles*. It is possible to measure the angles separately, one by one. For instance, 12 observations on each face could be made on angle QAR only, with systematic changes of zero; then angle RAS could be observed similarly and so on for all five angles. The mean values would not generally add to exactly 360°. More elaborate procedures by which every possible angle in the set is separately measured a number of times, have been suggested but they are not favoured in English practice.

3.10 Multiple Readings. It is now quite common practice, and should certainly improve accuracy, to take several micrometer readings at each setting on a signal. The observer, after setting on a signal, sets the micrometer and reads it, then displaces it and makes another reading, and does this a few times. In the Ordnance Survey primary triangulation three settings were made. In this system the first line of the bookings given above might look like this:

		17·2″		18·6″	
Q(RO)	87° 33′	17·6	267° 33′	18·5	17.9″
		18·1		17·4	

where the 17·9 is the mean of all 6 readings.

3.11 Station Adjustments. For computation purposes, one value of each angle or direction at each station is required. As indicated in II, these can be obtained by taking simple means of the numerous values recorded in the field. However, some surveyors or surveying organisations prefer to apply a more elaborate treatment of observations, in accordance with the theory of errors, to decide the values to go into the next stage of the work. Consider the method II and suppose that 8 rounds are taken to 5 signals: this will lead to a list of 40 'observed' directions, 8 to each point. All these, including the zero values written for the RO, are assumed to be in error, and 40 adjustments $\theta_1, \theta_2, \ldots, \theta_{40}$ are to be applied to them. The adjustments are calculated to satisfy the requirements:

(a) the 8 directions to each signal calculated from the adjusted directions must agree exactly,

(b) the 40 adjustments satisfy some minimum condition such as the principle of least squares derived from the theory of accidental errors.

It is debatable whether such elaborate treatments are really necessary: a great amount of arithmetic is involved, but this is no problem if a suitable programme and an electronic computer are available.

The observation method III may require station adjustment. Again, each mean observed angle receives an adjustment and the adjusted values must be in perfect agreement: for instance, if angle QAS is measured as well as the single angles QAR and RAS, the adjusted values of QAR and RAS must exactly equal the adjusted value of QAS; and so on.

Closure Checks. In the course of the observations in a system of triangulation, the misclosures of triangles provide simple and immediate checks on the quality of the work. The allowed limit may be as small as 2″ in first-order triangulation. Of course the spherical excess of each triangle must be included in the correct total of the three angles (Section 6.02).

3.12 Signals. Distances between stations in geodetic triangulation will generally be in the range 20–50 km, sometimes considerably longer. At such distances, luminous signals must be placed on the points to which observations have to be made. In sunny climates it is possible to use heliographs which are set up to shine the rays of the Sun in the direction of the theodolite station; they need to be continually adjusted as the Sun moves across the sky. Otherwise, electric lamps are nowadays widely used. These are constructed so that the lamp filament is at the focus of a parabolic mirror and also over the centre of the tribrach on which the lamp is supported. The lamp casing should be fitted with a sighting telescope for setting on to the theodolite station. When observing to lamps, the work will usually be done in the early hours of the evening: in clear air, lamps may become visible before the end of twilight, and even in daylight if the weather is overcast and dull.

For each observer in an extensive triangulation scheme, there will have to be 6 or 8 signal parties, and a major organisational problem is that of getting all the signals to the required points at the right times. The observer should be able to communicate with the parties by Morse Code, or by portable radio apparatus.

3.13 Station Marks. In the past, insufficient attention was paid to the proper marking of the important stations of national surveys, and to the protection of marks for future use. In modern extensive surveys it is usual to construct station marks to a standard design, where the station is at ground level. Most of the points established by the Ordnance Survey are now marked by concrete pillars standing about 1·2 m above ground, and they have three radiating grooves in bronze cast into the top surface, so that the theodolite, or the signal lamp, is automatically centred when placed on its levelling screws in the grooves.

In addition to the pillar, there should also be an underground mark on rock or on a block of concrete, over which the pillar is carefully constructed, but not structurally connected: even if the pillar is overturned the underground mark remains. It is also advisable to establish three or four witness-marks on rock or on buried concrete blocks round an important station at distances of 0 to 20 m measured with a tape and noted on the description sheet of the station.

Other kinds of centre mark are needed in certain situations, for instance a metal bolt fixed in the roof of a flat-topped building or a water tower.

In earlier times it was common practice to fix points by intersection only, that is, using directions observed at other stations. Many of these points, such as church spires and factory chimneys or even isolated trees, would be inaccessible as observation stations, but they were very useful to a surveyor with a plane table. In these days of aerial surveying, such points are not so valuable.

3.14 Observation Towers. In flat country, particularly if it is wooded, stations may not be intervisible at ground level. The theodolite will have to be placed on a tower temporarily erected over the ground mark. The well-known Bilby Towers have been extensively used: they are double structures of light metal girders or tubes, the inner tower supporting the theodolite, and the outer tower, nowhere touching the inner one, supporting a platform for the observer. The towers are guyed to stakes driven into the surrounding ground. They can be erected in a few hours to heights up to about 30 m.

In modern practice, when distances can easily be measured by EDM instruments (Chapter IV), the tendency is to use traversing for providing survey control points in flat country. There is more flexibility in choosing the route of a traverse, so that the requirement in observation towers can be reduced or even eliminated, but low towers will probably be needed in wooded country or on long lines, since an unobstructed line of sight is needed for EDM operations.

REFERENCES

ADLER R. K. AND SHMUTTER B. 'Precise traverse in major geodetic networks'. *Canadian Surveyor* Vol. 25 No. 4, 1971.

BANCROFT A. D. 'A new type of lightweight surveying tower'. *Survey Review* No. 137, July 1965.

BURNETT W. H. 'The encoder for automatic angle readout'. *Surveying and Mapping*, December 1967.

HAGGETT B. C. 'The rejection of theodolite readings in the field: a proposed method'. *Survey Review* No. 142, October 1966.

JACKSON J. E. 'Gravity and the Land Surveyor'. *Chartered Surveyor* No. 103, 1970.

ORDNANCE SURVEY. *The History of the Retriangulation of Great Britain 1935–62*. H.M. Stationery Office, 1967.

RAINSFORD H. F. 'Triangulation: the reduction of angles and station adjustments'. *Empire Survey Review* No. 67, January 1948.

RAINSFORD H. F. 'Angles versus directions'. *Survey Review* No. 146, October 1967.

CHAPTER IV

GEODETIC SURVEYING—DISTANCE MEASUREMENT

4.01 Modern Distance Measurement. The invention and practical application of the technique known as *electromagnetic distance measurement*, or EDM, have made it possible to measure directly, to an acceptable accuracy, the distance between any two intervisible points in a survey system. Base-lines of the traditional type—short, few in number, measured to extreme accuracy with great expenditure of time and manpower—are things of the past. It is now customary to measure many, or even all, of the lines in a triangulation, and it is possible, though not common practice, to determine the geometry of a survey system by measurement of distances only.

4.02 Speed of Light. The speed of light was measured by Fizeau in 1849, by finding the time taken for the light to travel a known distance from his apparatus to a mirror and back. He passed the beam through the edge of a rapidly rotating toothed wheel; this switched the light on and off at regular intervals, and he was able to make the wheel go so fast that a 'switched on' portion of the beam was not reflected back to the wheel soon enough to avoid hitting the next oncoming tooth of the wheel, so the light seemed to be cut off entirely. When the wheel was driven faster still, the light re-appeared, having come through the next gap to the one by which it went out. In short, a signal was transmitted and the time it took to do the double journey was equal to the time taken by the edge of the wheel to move the distance between consecutive teeth. Fizeau calculated this time interval by finding the rate of rotation of some part of his apparatus in terms of true clock time.

In theory, this procedure could be reversed, by measuring the time interval and calculating the distance, but it was just about a century after Fizeau's experiment that the idea was realised in a practical method that could be used in surveying and geodesy. The solution of the problem of measuring the very short time interval to acceptable accuracy had to await the development of suitable electronic apparatus. Over ordinary geodetic distances the time interval is less than a thousandth of a second, so an accuracy of something like a thousand-millionth of a second (a nanosecond) is required to determine such a distance to one part in a million!

Light is just one type of electromagnetic radiation: other types are the radio waves used for transmitting messages and television programmes. In the radar system, pulses or flashes of radiation are sent out from a transmitter and are detected when they return after reflection by a physical object. Further development produced the Shoran system by which distances of several hundred kilometres can be measured to an accuracy of the order of 1/20 000: transmitting and recording equipment are carried in an aircraft and the signals received at a ground station are re-transmitted to the source. The aircraft is flown across the line joining two ground stations and the minimum total time of transit of the signals to the two stations is used in calculating the distance between them. An extensive trilateration framework has been surveyed in northern Canada by this method, and several long connections over sea crossings have been measured.

The absolute accuracy of pulse-timing methods is not good enough for ordinary ground survey measurement. The application of a somewhat different technique to the idea of using the known speed of electromagnetic waves to measure the distances normally required in geodetic and topographical surveying really began with the Geodimeter in 1950 and the Tellurometer in 1957. There are now many varieties of EDM equipment in use. It would not be possible, in the limits of a chapter, to give details of the modes of operation of many instruments: moreover, the rate of appearance of new instruments is such that some descriptions would be obsolete by the time of publication. However, these instruments have a common basic principle of working, independent of electronic, optical and mechanical design features, and it is possible to describe that basis without reference to any particular instrument.

4.03 Modulated Beams. In Fig. 47 there is a transmitter Tr at point A, which sends out electromagnetic waves, and to fix ideas we suppose that it sends a beam of light from an ordinary incandescent electric lamp. The light is focused into a parallel beam by a lens and is directed to point B where there is a reflector F1 which returns the light back along almost the same path, to the Receiver Re. The light takes a finite, but very short, time to do the journey ABA, but a simple beam like this cannot be used for measurement, because it is not possible to distinguish a particular point on the beam which could be recognised or detected on its return to Re.

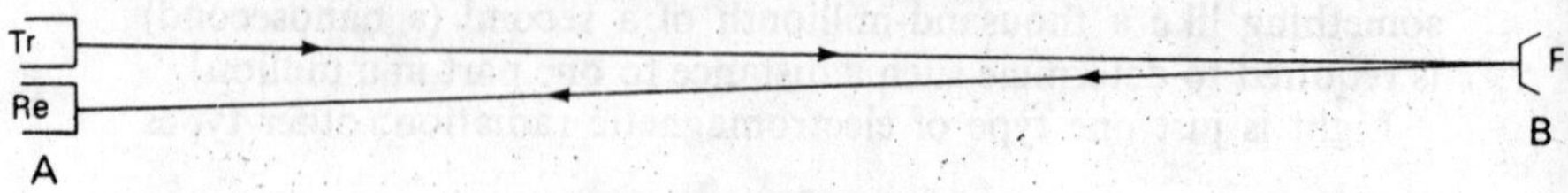

FIG. 47. CARRIER BEAM.

It is desirable at this point to get some idea of the numbers and quantities involved. The speed of light and other electromagnetic radiations is about 300 000 km per second. More accurately, the speed is 299 792·5 km/sec when travelling in a vacuum. In going through a transparent material the speed is reduced; for instance, in air at normal temperature and pressure the speed is reduced by about 90 km/sec. For purposes of illustration, however, the simple number of 300 000 000 metres per second will be used.

Now suppose that the light could be switched off momentarily at the rate of 60 million times a second: this would put 'marks' on the beam at 5 metre lengths along it. However, no mechanical switch could possibly operate at the rate of 60 million times a second; in any case it is much simpler to use electronic methods. An electrical circuit containing inductance (a coil of wire) and capacitance (a capacitor) has a natural resonant frequency, that is it can be caused to generate strong oscillations of voltage at a particular rate depending on the sizes of the components used. The circuit only needs to be kept supplied with sufficient electric power to make up for the inevitable wastage of energy in operation.

The rate of oscillation is usually described as frequency—number of times per second. In earlier writings, 60 million times a second would perhaps be described as 60 megacycles; now, it is more customary to say 60 million hertz, or 60 megahertz, or 60 MHz.

Now it is possible to use oscillations produced, as mentioned above, to cause cyclic modifications of the beam of light coming from Tr. The beam will be impressed with a repeating pattern which, at 60 MHz, will recur at 5 metre intervals along the beam. In technical terms, the carrier, in this case the light beam, has been modulated at the frequency of 60 MHz. The beam is no longer undistinguished from point to point, but carries the repeated pattern along with it, and when it returns into the receiver Re the regular repetition of the electromagnetic pattern can be made to cause oscillations at the same rate in suitable electronic components.

Fig. 48 represents the beam divided up into modulation patterns of length p. It shows a situation where the length of the line AB happens to be an exact number of patterns, in this case nine. Between Tr and Re there are 18 patterns. The figure has been drawn for a particular moment when one of the pattern marks is

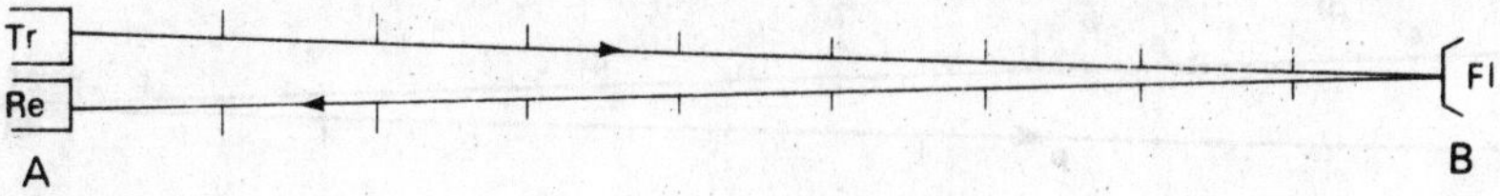

FIG. 48. MODULATED BEAM.

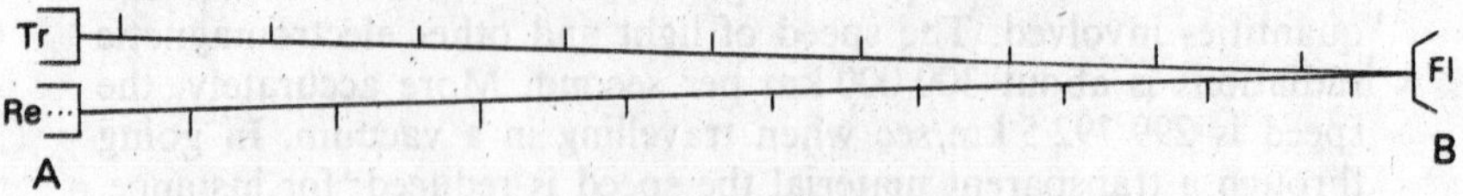

FIG. 49. IN PHASE.

just emerging from Tr therefore another pattern mark, the one 18 patterns ahead on the beam, is just entering Re. The situation a little later is represented in Fig. 49. A portion of a pattern has emerged from Tr and a portion of the 18th pattern has been absorbed into Re and is indicated by dots. A little consideration will show that at all times the two portions mentioned will be of equal length. Whatever the nature of the electromagnetic variations that make up a pattern, the variations entering Re will always be similar to those just coming from Tr. This situation is described by saying that the transmitted and received patterns are in phase, and this will occur when the length AB is a whole number of pattern lengths.

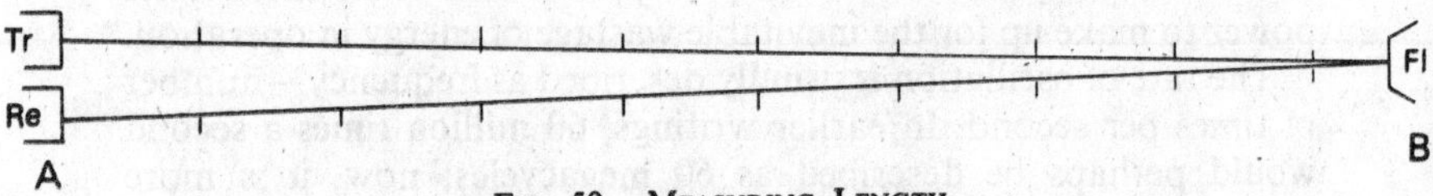

FIG. 50. MEASURING LENGTH.

Now consider Fig. 50. This shows the distance AB as $9\frac{1}{2}$ pattern lengths. It is obvious that when the distance is an integral number of half patterns, the return and outgoing signals will also be in phase. Thus the transmitted and received signals will be in phase whenever the distance AB is a multiple of $\frac{1}{2}p$. The effective unit of measurement in an instrument operating this principle is half the pattern length, which would be $2\frac{1}{2}$ metres in the numerical example used above.

By a suitable electronic device it is easy to detect when the signals are in phase, but of course the line to be measured will generally not be an exact multiple of $\frac{1}{2}p$, that is the transmitted and received signals will not generally be in phase.

Fig. 51 illustrates the general situation. The drawing shows the moment when a pattern mark is arriving at the receiver, and a later pattern mark has proceeded a distance s out of the transmitter,

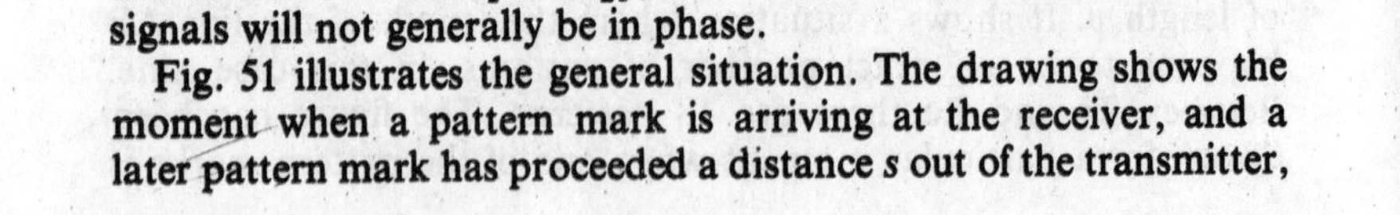

FIG. 51. PHASE COMPARISON.

the length of s being less than one pattern. Then the double distance ABA is $(N.p+s)$ where N is an integer, and the length of the line is $(N.\frac{1}{2}p+\frac{1}{2}s)$. Obviously, an essential part of any measurement will be the determination of the length s to the desired accuracy.

4.04 Phase Comparison. One way of doing this is indicated in Fig. 52. By means of two mirrors α and α' a small part of the outgoing beam is diverted and sent into Re, to be compared with the signal coming in from the distant reflector. Now if the length of the path $A\alpha\alpha'A'$ is made equal to s the diverted part of the pattern mark (1) will arrive at Re along with a pattern mark from F1, and so the two signals entering Re will be in phase. The two mirrors are adjustable along a sliding graduated scale which will read the value of s directly.

This device was used in early forms of the Geodimeter, but has now been generally superseded by electronic devices for the same purpose. Note that the effect of the mirrors is to delay the comparison

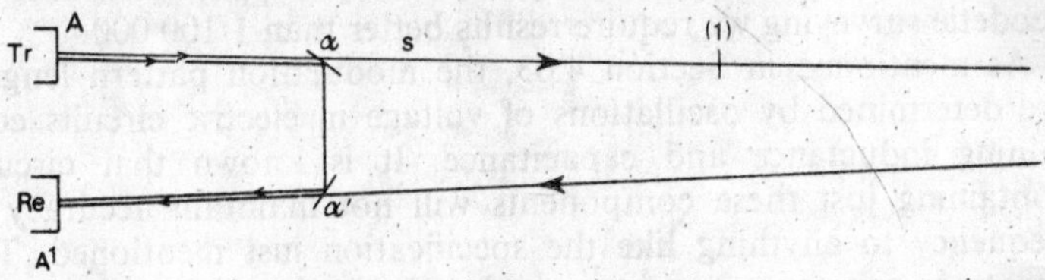

FIG. 52. VARIABLE LIGHT PATH.

of the outgoing and returning signals by the time taken to travel the distance from A to (1).

When an alternating electric signal is sent into an inductance (coil) the effect at the other end is also alternating but there is a time lag the duration of which can be adjusted by varying the amount of the inductance. This effect can be used to produce the delay referred to above, and the adjustment control can be graduated and calibrated to give a direct reading of the distance s.

The ultimate precision of a measurement will depend on the sensitivity of the electronic delay device in detecting the coincidence of phase: with some arrangements, this sensitivity can be as good as one thousandth of a phase cycle. For instance, if the effective measuring length is $2\frac{1}{2}$ metres, as mentioned in Section 4.03, the delay device should then give a reading to $2\frac{1}{2}$ millimetres. This practical limit of fine reading largely determines the pattern length that can be used: if reading to 1 centimetre is acceptable the measuring unit could be 10 m and so the modulation pattern could be 20 m long, which would be produced by a modulation frequency of 15 MHz.

It is customary to express a phase difference, illustrated by the partial pattern s, as the fraction s/p of 360°. Thus if p is 5 metres and s is 3·17 metres, then the phase difference is said to be 228·2°. This mode of description arises from the common method of illustrating repeated cyclic change by means of a pointer moving round a circle.

4.05 Modulation Frequency. Methods of finding the partial pattern length s have been mentioned: the measuring of the length of the whole line will then be achieved if the number of whole pattern lengths in the double distance is counted. Since there will be hundreds, or perhaps thousands, of pattern lengths in a distance to be measured in surveying, it is obvious that the pattern length must be kept constant to a high degree of precision, and its actual length must be known to the accuracy required of the measurement. The strict maintenance of modulation frequency is obviously one of the key problems in the design of EDM instruments for field use, since their scale accuracy ultimately depends on this. In geodetic surveying we require results better than 1/100 000.

As mentioned in Section 4.03, the modulation pattern lengths are determined by oscillations of voltage in electric circuits containing inductance and capacitance. It is known that circuits containing just these components will not maintain accuracy of frequency to anything like the specification just mentioned. The necessary constancy can however be obtained by including in the circuit a piece of quartz crystal. This is a thin slab of quartz with opposite sides ground to be accurately plane and parallel. The two sides are coated with electrically conducting material. When oscillatory changes of voltage are applied to the two sides at a certain frequency, depending largely on the thickness of the quartz, internal oscillations of crystal structure are set up, and it is found that these maintain a frequency that is constant to the accuracy required for EDM applications in surveying. When the crystal is included in an electrical circuit that has the same natural period of oscillation, the crystal will take charge and maintain constant frequency, so long as the voltage oscillations are kept going by the action of the electric components and power supply. To maintain the frequency at a known value, the crystal must be kept at a certain temperature, so it is the practice to encase the crystals in an oven which is warmed up to the correct temperature when the instrument is switched on.

Another device for maintaining precise frequency of modulation is the resonant cavity. Its frequency depends on its geometrical dimensions so that, in effect, it constitutes a linear unit in terms of which the desired ground distance is measured. The cavity contains dry air at the ambient temperature and pressure, thus automatically compensating for most of the meteorological effects. An instrument using this modulation method is the Mekometer, developed originally at the National Physical Laboratory by Froome and Bradsell. The carrier beam is light and the measuring length is 0·3 metre: the range in daylight is up to 3 km.

4.06 Counting the Patterns. To determine the number of whole pattern lengths in a distance, an obvious method is to use a series of longer and shorter pattern lengths. Suppose that the length of a line is 283 52·76 m. If a measuring length ($\frac{1}{2}p$) of 10 000 m is applied the instrument will show a phase fraction (s/p) of about 0·835: this means that the length is about 8350, or 18 350, or 28 350, or 38 350 . . . metres. We suppose that it is known that the correct choice is 28 350: in modern circumstances, the surveyor will know this much about the distance, by reference to a map or a reconnaissance sketch. Now the measuring length is reduced to 1000 m and the delay device will read about 0·353, so the length is now known to be about 28 353 m ($28 \times 1000 + 0{\cdot}353 \times 1000$). The measuring length is reduced to 100 m, then finally to 10 m when the 'fine reading' will give the fraction 0·276 and the full reading to 1 centimetre. This method is widely used in EDM systems.

Another method which has recently been developed is suitable for instruments that measure the shorter distances, and operates by continuously changing the modulation frequency between fixed limits. The instrument has a rotating component (resolver) on which one revolution indicates one complete cycle of phase relationship. Suppose the distance to be measured is 647·38 m. Starting with an effective measuring length of 11·111 m, that is 100/9 metres, there will be $647{\cdot}4 \times 9/100 = 58{\cdot}26$ measuring lengths in the line. As the measuring length is continuously reduced there will come a time when there are 59·26 measuring lengths, and the resolver will have done just one revolution to keep up with the change. When the measuring length comes down to just 10 m there will be 64·738 measuring lengths in the line and the resolver will have done $(64{\cdot}74 - 58{\cdot}26) = 6{\cdot}48$ revolutions. A little consideration of the arithmetical basis of this method shows that the last figure, the count of the resolver revolutions, is an estimate of 1 % of the length of the line. Thus, the estimate of 648 m is got from the revolution counter and this is refined to 647·38 by the final setting indicating the phase fraction of 0·738 of 10 m in the usual way. Briefly, the modulation frequency is increased continuously by 10 % and the number of measuring intervals in the distance will consequently increase by 10%. This number is counted by noting the number of times the patterns go in and out of phase during the change. Multiplying by 10 gives the approximate number of measuring lengths.

It is to be noted that EDM instruments do not actually measure distance; they are basically devices for counting the number of measuring lengths in the distance to be measured. The modulation frequency is precisely known and the number of measuring intervals is determined to an accuracy of the order of 1/1000 of an interval. With m intervals at the frequency of f intervals per second, the time taken by the signal to travel the length of the line is the fraction m/f of a second, and this is the quantity that is accurately found by using the instrument. The distance is then $(m/f)V$, where V is the actual speed of the signal through the air at the time of measurement. It is the practical difficulty of knowing V that is one of the limitations on the accuracy of EDM for measuring medium and long distances. At short distances, 1 km or so, other causes of error are likely to be more prominent.

4.07 Optical and Radio Waves. EDM instruments for surveying fall into three classes:

(i) those in which a beam of light is the carrier and is reflected back from some sort of mirror. An obvious advantage of this method is that only one active instrument with power supply is needed and the equipment at the other end of the line is comparatively inexpensive. Most of these instruments are suitable for measuring the shorter distances required in surveying, since they are limited by the ability to get enough power back from the reflector to operate the receiving components. An attached telescope for aiming the beam at the mirror is necessary. Longer distances can be measured, however, by using a laser as carrier, because this has a very small angle of spread.

(ii) those which emit short radio waves (microwaves) as carrier: there is an active instrument and operator at each end of the line, and the signal is re-transmitted on another carrier wave back to the originating instrument. The wave lengths of the carriers are of the order of centimetres, in some recent instruments about 8 millimetres. Because the beam has a spread of some 5°–10° these instruments do not require precise directional setting. They are generally used for the long distances required in first-order triangulation and traversing.

(iii) radio systems widely used for navigation are also essentially EDM. These use longer waves, require heavier equipment, and are less accurate than the types of instrument mentioned above. As surveying instruments their principal use is for fixing position at sea in relation to fixed points on land.

Perhaps the greatest advance in techniques available to the geodetic surveyor, since the introduction of invar tapes, has been the invention of the Tellurometer by T. L. Wadley, about 1955. To measure a line, two instruments are required, and are usually referred to as 'master' and 'remote' as indicated in Fig. 53. The two

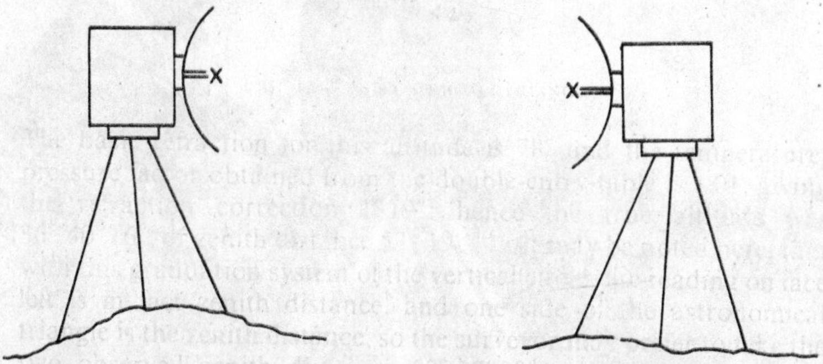

FIG. 53. MICRO-WAVE LONG DISTANCE INSTRUMENTS.

operators can speak to each other, using telephone sets connected to the instruments. A recent model of this instrument is illustrated in Fig. 54.

Transmission and reception are by means of aerials at the focus of parabolic cylinder reflectors which in standard instruments are about 300 mm diameter. The outward and return carriers are at slightly different frequencies so that they can be distinguished by the electronic circuits, and the receipt of the modulation pattern at the remote instrument controls the emission of a corresponding pattern back to the master. Pattern frequencies are controlled by quartz crystals. In the earlier models the readings were obtained in time units of milli-microseconds, now to be referred to as nanoseconds: a nanosecond is 1/1 000 000 000 second and was a convenient unit for British users because it is just about the time required for light to travel one foot! In some tellurometers, the outgoing modulation wave is caused to generate a circular trace on a cathode-ray tube, and the returning 'pattern' is converted to a pulse which causes a break in the circular trace: the position of the break round the circle indicates the phase difference. Some modern instruments give the readings directly in distance units; see Section 4.09.

FIG. 54. TELLUROMETER MRA3.
(photograph supplied by Tellurometer (U.K.) Limited)

Of the enormous number of geodetic distances measured in the past 15 years, a very large proportion has been measured by Tellurometers or instruments using the same technique, since this technique has been found to give adequate accuracy at distances up to 100 km or even more.

Of instruments using a light beam as carrier, the Geodimeter was the pioneer. This was invented by Prof. Bergstrand a few years before the Tellurometer. It was first used in the reverse role for finding the speed of light by measuring the time taken to travel along known geodetically measured distances. The early instruments used the mechanical delay path as described in Section 4.04, Fig. 52, but almost all instruments now use electronic delays.

One way to impress a modulation pattern on a beam of light is to polarise it and send it between the electrodes in a Kerr Cell containing nitrobenzene. When strong changes of voltage are applied to the electrodes the polarisation state of the light is correspondingly varied, and this constitutes the pattern on the beam.

Fig. 55 illustrates a modern Geodimeter instrument. It is to be

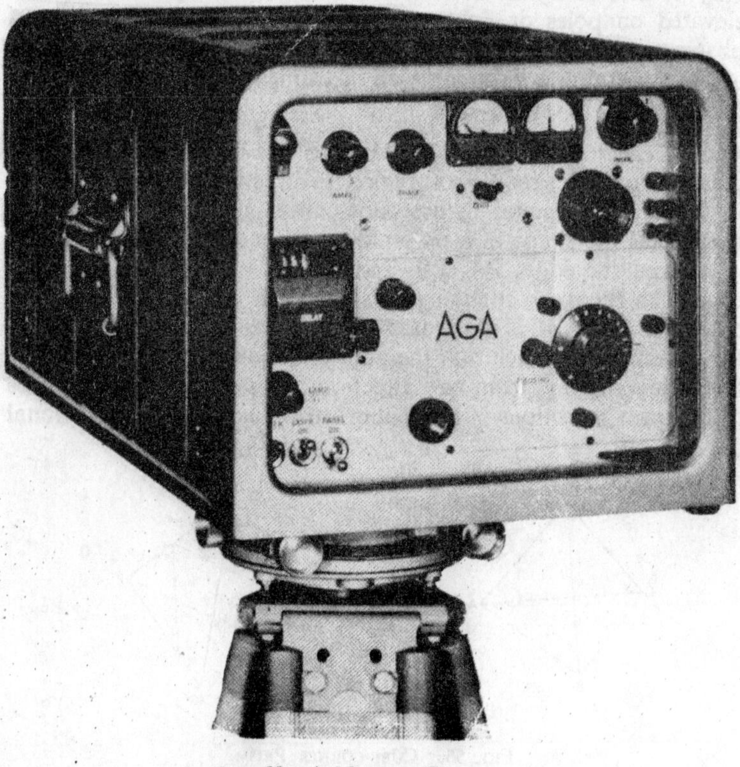

FIG. 55. A MODERN GEODIMETER.
(photograph supplied by Aga (U.K.) Limited)

expected that an instrument using a beam of ordinary light will have a rather restricted range in daytime, normally a few kilometres in fact, because of the confusion of the measuring signals by external light. Longer ranges can be measured at night, and the range can also be increased by using light from a mercury vapour lamp.

A source of light that is now widely used in instruments for measuring very short distances, up to 1 km perhaps, is the gallium arsenide diode. This is a piece of solid material of such composition and design that it emits a strong red light when electric voltage is applied to it. The wavelength is about 0·95 micrometre (a micrometre is a thousandth part of a millimetre). The advantage of this device is that the modulation voltage can be applied directly to it because its response to voltage changes is practically instantaneous; this simplifies the electronics.

A useful feature of some EDM equipments, especially for surveyors, is that the components (aerials or lenses) that actually transmit and receive the signals are separate from the main body of the electronics, only connected by flexible cables. So these components can be elevated on poles or light structures, to get the beams clear of obstructions. It is also possible to have these components attached above a theodolite, producing an instrument for measuring angles and distances at one set-up.

4.08 Reflectors. The early Geodimeter got its beam back from a plane mirror. This required very precise directional setting, and now it is universal practice to use cube-corner glass prisms. Each reflector has the shape of a pyramid on an equilateral triangle base ABC, and the edges PA, PB, and PC are perpendicular to each other like three rectangular axes. The path of a ray of light that enters the face ABC is illustrated in Fig. 56, and it is easy to see that the reflections are such that the path of the outgoing ray is parallel to the ingoing ray, from any direction: the same geometry applies to the three reflections in the real prism. Thus, accurate directional

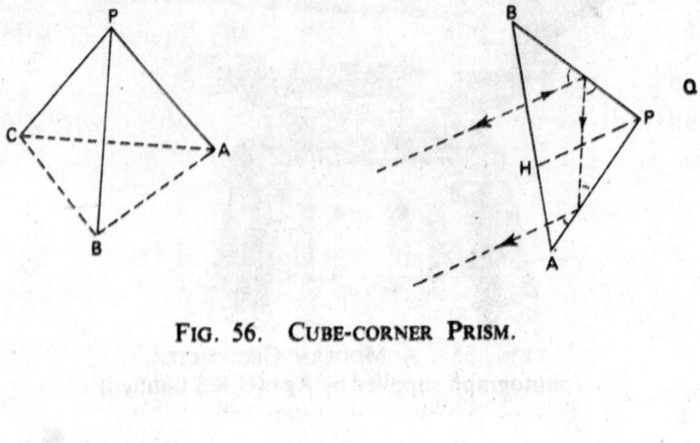

FIG. 56. CUBE-CORNER PRISM.

setting of the prism is not required. It is also easy to see that the length of the path of the beam within the glass is equal to twice the length PH. Now the speed of light in glass is about $\frac{2}{3}$ of its speed in air, so the time taken in the glass is the same as if the beam was in air and reflected at the point Q, where HP $= \frac{2}{3}$HQ. Reflectors may be mounted so that the point Q is centred over their supports, otherwise a correction can be applied. Some instruments are designed so that the reflected beam is required to come back in the direction exactly the reverse of the outgoing beam and is received by lenses or mirrors coaxial with the transmitting system: in others, the transmitting and receiving systems are alongside, and for measuring very short distances it is necessary to cover part of the reflector surface with a prism of small deflection so that the reflected beam comes back along a slightly different alignment.

Cube-corner reflectors have a diameter of about 75 mm, and for measuring longer lines it may be necessary to use a bank of as many as 19 reflectors, to get enough light back into the receiver.

4.09 Refractive Index. If V is the speed of the measuring signal through the air at the time of measurement and V_0 is the speed in vacuum, then the ratio V_0/V is the Refractive Index: in ordinary air it is about 1·0003. Thus the actual length of a modulation pattern is its vacuum length divided by the refractive index. The exact value of the index depends on the pressure, temperature and humidity of the atmosphere, and since the introduction of EDM there has been much research towards finding appropriate formulas for calculating the index from meteorological observations taken at the time of measurement. Meteorological field instruments must be included with the equipment for EDM work. It is usually impracticable to take meteorological readings anywhere but at the ends of the line. The hygrometer and barometer are seen in the picture, Fig. 54.

The meteorological corrections are significant in the measurement of medium and long lines, particularly for instruments using radio micro-wave carrier. Some instruments give readings directly in length units, such as metres: in that case, the instrument is adjusted to show readings based on an adopted value for the refractive index. These readings may be sufficiently accurate for the purposes of the survey, but it is always possible to take meteorological observations and apply corrections for the differences from the assumed standard atmosphere.

If we denote the refractive index by N then $(N-1)10^6$ is the number of parts per million by which the speed in air differs from that in vacuum.

The formula given by Froome and Essen for use with radio-wave instruments is:

$$(N-1)10^6 = \frac{103{\cdot}49}{T}(P-E)+\frac{86{\cdot}26}{T}\left(1+\frac{5748}{T}\right)E,$$

where P = total atmospheric pressure in millimetres of mercury barometer,
E = partial pressure of the water vapour,
T = absolute temperature, which may be taken as °C + 273.

For example, with $T = 300$ we have

$$(N-1)10^6 = 0{\cdot}345P+5{\cdot}46E$$

and with $P = 760$, $E = 10$ this gives $N = 1{\cdot}000\,317$. The value of E can be obtained from hygrometer readings of wet-bulb and dry-bulb thermometers, by reference to standard Tables. For best results, an aspiration hygrometer is used.

A little study of the refractive index formula shows that, if an accuracy of a few parts per million is sought, the meteorological conditions must be measured with reliable calibrated instruments: the effect of humidity is considerable at radio wavelengths, and this is the quantity which is least easy to measure accurately in the field. In any case, measurements taken at the ends of the line may not be representative of the whole ray path, but experience has shown that meteorological uncertainties may account for some 3 parts per million of errors in EDM on long lines. See Section 4.11.

For instruments using light-beam carrier, the refractive index formula is somewhat different and depends slightly on the wavelength of the light, but the effect of humidity is much less. The Barrell and Sears formula is:

$$(N-1)10^6 = \frac{273\,.\,P}{T\,.\,760}\left(287{\cdot}604+\frac{4{\cdot}886}{\lambda^2}+\frac{0{\cdot}068}{\lambda^4}\right),$$

where λ is the wavelength in micrometres. The term depending on humidity is $\left(-\frac{15{\cdot}02}{T}E\right)$ and is usually too small to be worth considering.

4.10 Ground Swing. The beams radiated by microwave instruments have a spread of some 5°–10° and this has the result that a small part of the radiation received by the instruments may have come from reflection from the intervening ground. Any route by reflection is longer than the direct line, and the receiver will be confused by two sets of patterns arriving with the same frequency but not coincident in time, i.e. out of phase.

This effect is called *ground swing* and it usually gives rise to erroneous readings. It has been found that by making small changes of the carrier wave frequency the instrument readings usually show a cyclic series of variations somewhat as shown in Fig. 57. The range of the variations may be a metre or even more.

The theory indicates that the average over a complete cycle, as shown by the broken line, is the correct reading. In exceptional

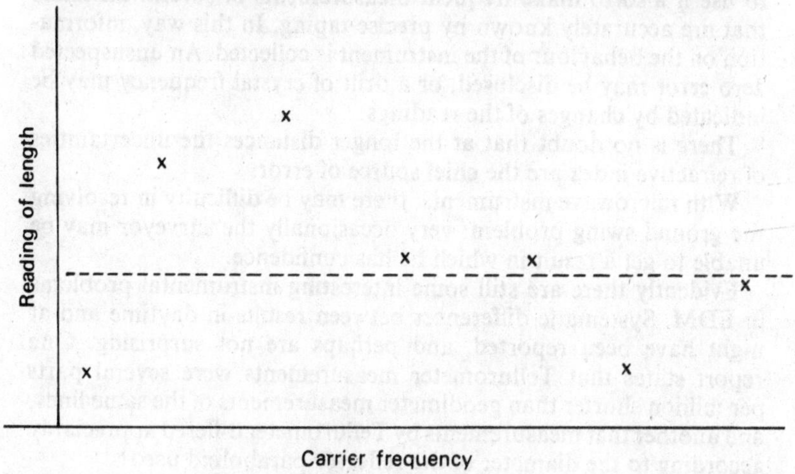

FIG. 57. GROUND SWING.

cases, the cyclic structure may not appear and a reliable result cannot be obtained: this difficulty can sometimes be overcome by raising or lowering the instrument, or perhaps by moving it back or forwards a few metres along the line and applying a correction length measured by tape.

There is no problem of reflection with instruments using light as carrier because the beams are very narrow and the reflecting power of intervening ground is practically zero.

4.11 Errors in EDM. In an EDM instrument, it is not obvious where is the point from which the measured distance is to be reckoned. The makers of the instrument should supply a zero correction if the measurement is not effectively from the centre of the tripod mount. There may also be a correction arising from the way the reflectors are fixed on their tripod. The surveyor should know these corrections for the instrument he is using.

The manufacturers usually quote accuracy by using two numbers, for instance '± 10 mm ± 2 parts per million': the 2 ppm is mainly to allow for meteorological uncertainties, and the other figure gives the supposed limits of a kind of sporadic unpredictable zero error which is a function of the electronics and independent of the distance being measured.

For most survey purposes, errors of the order mentioned above are tolerable.

Where an EDM instrument is in constant use, it is desirable to use it also to make frequent measurements of several distances that are accurately known by precise taping. In this way, information on the behaviour of the instrument is collected. An unsuspected zero error may be disclosed, or a drift of crystal frequency may be indicated by changes of the readings.

There is no doubt that at the longer distances the uncertainties of refractive index are the chief source of error.

With microwave instruments, there may be difficulty in resolving the ground swing problem: very occasionally the surveyor may be unable to get a result in which he has confidence.

Evidently there are still some interesting instrumental problems in EDM. Systematic differences between results in daytime and at night have been reported, and perhaps are not surprising. One report states that Tellurometer measurements were several parts per million shorter than geodimeter measurements of the same lines, and another that measurements by Tellurometer differed appreciably according to the diameter of the reflector paraboloid used!

The accuracy of EDM based on crystal-controlled modulation is directly dependent on accurate knowledge of the actual crystal frequency. Over short periods the constancy of a good crystal is quite reliable, but it is known that frequency may have a steady drift. Hence frequency should be tested at a suitably equipped laboratory at intervals of a year or so.

Many types of EDM instrument are now available, and new models appear frequently. The laser is being used as the carrier. Direct readout of the distance is a common feature, and complete automation of the programme of frequency changes has appeared. It has been found possible to measure the phase difference by means of a high-speed electronic counter that is triggered on and off by the outgoing and returning wave fronts of the modulation. The manufacturers of the instruments supply detailed instructions and by following these carefully the surveyor can expect to get satisfactory results so long as the instrument continues to work properly.

4.12 Power Supplies. EDM instruments are operated on electrical storage batteries of one kind or another. Some instruments are provided with special small rechargeable batteries which will provide a few hours of operation. When working in areas remote from electrical power supply, surveyors will have to be equipped with petrol-electric generator sets for battery charging.

4.13 Geometrical Corrections. In the Earth's atmosphere, the measuring beam will usually be curved upwards above the straight line, following the refracted path due to the decrease of density with increasing height. This path is shown dotted in Fig. 58, but the

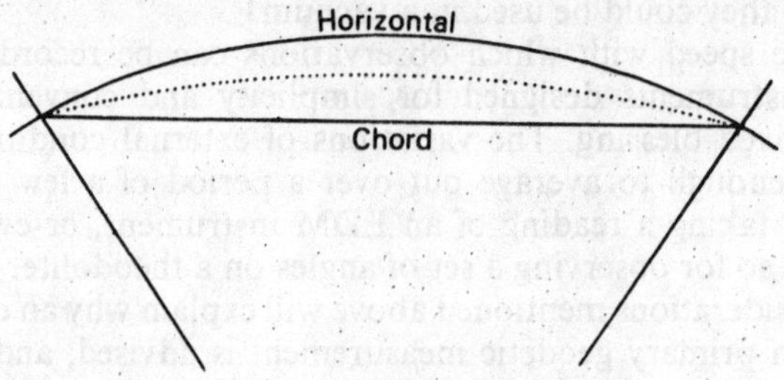

FIG. 58. RAY PATH.

length that is required in describing the geometry of a survey system and carrying out ordinary geodetic calculations of positions is the length of the curve having the Earth's radius.

If S is the length of an arc of radius R, then the length of the chord is

$$2R\sin\left(\frac{S}{2R}\right) = 2R\left(\frac{S}{2R}-\frac{S^3}{48R^3}\cdots\right) = S-\frac{S^3}{24R^2}\cdots$$

The radius of the ray path for radio waves is not easily determined in the field; if we take a typical value of $4R$, then to get the geodetic length the measured length must be reduced to chord by subtracting $S^3/24(4R)^2$ and then increased by adding $S^3/24R^2$, where R is the radius of the Earth. The combined correction is thus about $+S^3/26R^2$.

For instance, if $S = 30$ km, $R = 6370$ km, the correction is +26 millimetres, which is less than one part in a million. Thus, the curvature correction is usually negligible, but it should be applied to lines over 50 km long.

Fig. 58 shows the line as at sea level; in practice the points will generally be at different heights, and the usual corrections for slope and for height will be necessary. For measurement of slopes the surveyor will need a theodolite and should measure the slope of each line in both directions.

4.14 Spreading the Observations. It is a feature of modern geodetic surveying instruments that their internal accuracy has reached a standard at which the principal sources of errors in observations are external. Provided that proper observational procedures are carried out, the limits on accuracy are set partly by the observer's skill but mainly by influences such as temperature changes and atmospheric irregularities, and the impossibility of estimating the effects of these at the time of measurement.

This source of uncertainty is of course particularly apparent in the use of EDM instruments. These instruments may well have an internal accuracy of 1 in 10^6 or better, and they would achieve that accuracy if they could be used in a vacuum!

Even the speed with which observations can be recorded with modern instruments designed for simplicity and convenience in use is a mixed blessing. The variations of external conditions are not rapid enough to average out over a period of a few minutes needed for taking a reading of an EDM instrument, or even over an hour or so for observing a set of angles on a theodolite.

The considerations mentioned above will explain why an observer engaged on primary geodetic measurement is advised, and indeed should be instructed, to spread the observations over a period of two or three days, although he could occupy the stations more quickly. Such procedure would probably not be necessary in the observation of a survey work that is controlled by an existing framework.

The accuracy attainable in first-order measurement of horizontal angles is about half a second, about 1/400 000 radian or $2\frac{1}{2}$ parts per million. By EDM it seems that long lines can be measured to almost this accuracy. An obvious advantage of the increased use of length measurement in modern surveying is that uniform accuracy of scale can be maintained throughout the most extensive of survey systems.

4.15 Line Pairs. If the lengths of two lines AB and AC, Fig. 59, are measured by EDM instrument, with a period of only a few

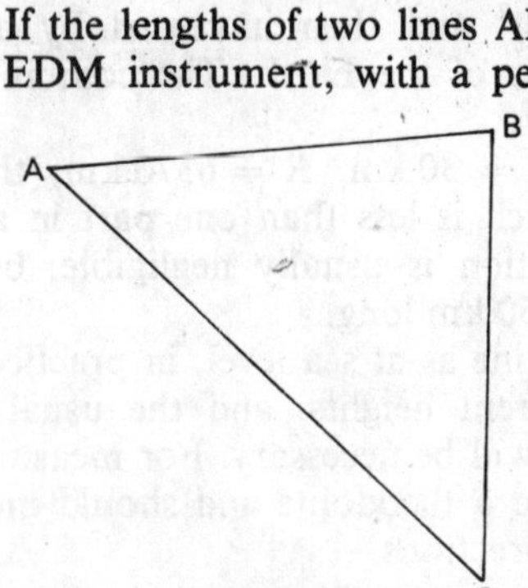

FIG. 59. TRILATERAL.

minutes between the two measurements, it is reasonable to suppose that errors due to uncertainties of refractive index will be similar for the two results, and the *ratio* of the two measurements will have a relatively higher accuracy than either measurement of itself.

Some recent experiments based on this idea have indicated that it has real value in this way.

To measure the triangle, the instrument would have to occupy the three points at considerable intervals of time, but if the three ratios $\frac{AB}{AC}$, $\frac{BC}{BA}$, and $\frac{CA}{CB}$ are derived in the way mentioned, it is to be expected that the effects of gross changes of refractive index will be largely avoided.

This mode of use of EDM makes it rather similar to conventional triangulation in which the geometrical shape of the figure is determined by measurement of angles and the fixing of the absolute scale is a separate operation, akin to base-measurement, with its peculiar problems of constant and systematic errors.

In further analogy with triangulation, the three measured ratios should satisfy one condition—that their product should be unity, or the sum of their logarithms should be zero. They are therefore subject to consistency requirements and adjustment processes

REFERENCES

ANGUS-LEPPAN P. V. 'A study of refraction in the lower atmosphere'. *Empire Survey Review* No. 120, April 1961.

BOMFORD A. G. 'Precise tellurometer traversing'. *Empire Survey Review* Nos. 117 and 118, July, October 1960.

BOSSLER J. D. and LAURILA S. H. 'Zero error of MRA3 Tellurometer'. *Bulletin Géodésique* No. 79, 1965.

CHRZANOWSKI A. and WILSON P. 'Underground surveys with the Tellurometer'. *Canadian Surveyor*, June 1966

DALE P. F. 'The establishment of control by tellurometer radiations'. *Survey Review* No. 147. January 1968.

DIXON J. L. 'Various aspects of the computation of tellurometer work'. *Survey Review* No. 133, July 1964.

FROOME K. D. and ESSEN L. *The Velocity of Light and Sound Waves*. Academic Press, 1969.

FROOME K. D. 'Precision distance measurement based on the velocity of light'. *Science Progress* Vol. 59 No. 234, 1971.

HODGES D. J. 'Errors in Model 6 Geodimeter measurements and a method for increasing accuracy'. *Mining Engineer* No. 99, December 1968.

HOLEJKO K. 'Microwave distance meter "Telemeter OG-1"'. *Canadian Surveyor*, June 1967.

HÖPCKE W. 'On the curvature of electromagnetic waves and its effect on measurement of distance'. *Survey Review* No. 141, July 1966.

HUMPHRIES G. J. 'First-order traversing with the Tellurometer'. *Empire Survey Review* No. 109, July 1958.

JAAKKOLA M. 'A survey with the Tellurometer MA 100'. *Survey Review* No. 159, January 1971.

JONES H. E. 'Systematic errors in tellurometer and geodimeter measurements'. *Canadian Surveyor*, September 1971.

KELSEY J. 'A method for reducing ground swing in tellurometer measurements'. *Empire Survey Review* No. 112, April 1959.

LESLEY G. B. 'Laser Geodimeter'. *Surveying and Mapping*, December 1967.

MACDONALD A. S. 'An inexpensive tower for tellurometer traverses'. *Empire Survey Review* No. 121, July 1961.

MACDONALD A. S. 'Field trials of the MRA 3 Tellurometer in Bechuanaland'. *Conference of Commonwealth Survey Officers, 1963*.

MACKENZIE I. C. C. 'The geodimeter measurement of the Ridgway and Caithness Bases 1953'. *Ordnance Survey Professional Paper*, New Series No. 19.

MCLEAN R. E. 'Accurate measurement by geodimeter: a practical application of the differential technique'. *Survey Review* No. 153, July 1969.

MCVILLY R. B. 'The use of the geodimeter on routine civil engineering projects'. *Survey Review* No. 138, October 1965.

MEADE B. K. 'High precision geodimeter traverse surveys in the United States'. *Bulletin Géodésique* No. 90, 1968.

MEADE B. K. 'Precision in electronic distance measuring'. *Surveying and Mapping*, March 1972.

MUNCK J. C. DE 'The theory of dispersion applied to electro-optical distance measurement and angle measurement'. *Netherlands Geodetic Commission*, New Series Vol. 3, No. 4; 1970.

NATIONAL PHYSICAL LABORATORY *Measurement of Humidity*. H.M. Stationery Office, 1958.

NATIONAL PHYSICAL LABORATORY *The Refractive Index of Air for Radio Waves and Micro-waves*. H.M. Stationery Office, 1960.

OWENS J. C. *The Use of Atmospheric Dispersion in Optical Distance Measurement*. ESSA Boulder, Colorado, 1967.

PARM T. *Zero Corrections for Tellurometers of the Finnish Geodetic Institute*. Finnish Geodetic Institute, Helsinki, Publication No. 70, 1970.

RICHARDS M. R. 'The use of thermistors for temperature measurement in EDM'. *Survey Review* No. 143, January 1967.

ROBERTSON K. D. 'The measurement of small movements in large structures'. *Survey Review* No. 160, April 1971.

ROBERTSON K. D. 'The use of line pairs in trilateration and traverse'. *Survey Review* No. 165, July 1972.

ROBINSON A. J. 'Field calibration of the Model MRA 101 Tellurometer'. *Australian Surveyor*, December 1969.

ROSS J. E. R. 'Canadian Shoran Effort 1949–53'. *Empire Survey Review* No. 92, April 1954.

SAASTAMOINEN J. 'Curvature correction in electronic distance measurement' *Bulletin Géodésique* No. 73, 1964.

SAASTAMOINEN J. 'On the path curvature of electromagnetic waves'. *Bulletin Géodésique* No. 78, 1965.

SAASTAMOINEN J. 'On the determination of the refractive index of electromagnetic waves in mountainous terrain'. *Survey Review* No. 135, January 1965.

SAASTAMOINEN J. (Ed.) *Surveyors' Guide to Electromagnetic Distance Measurement*. University of Toronto Press, 1967.

SCHOLDSTROM R. 'The AGA Model 8 laser geodimeter'. *Australian Surveyor*, June 1969.

SONNENBERG G. J. 'Radar and electronic navigation'. (4th Edn.) *Newnes-Butterworths*, 1971.

TAYLOR W. R. 'The remeasurement of the Lough Foyle base line, Northern Ireland'. *Survey Review* No. 126, October 1962.

TENGBORG E. 'Some experience with the Wild DI 50 Distomat in the establishment of a triangulation network'. *Stadsbyggnad* 2/69, Stockholm.

TUTTLE A. C. 'Aerodist geodetic surveying in Canada'. *Canadian Surveyor*, June 1967.

WADLEY T. L. 'The Tellurometer system of distance measurement'. *Empire Survey Review* Nos. 105, 106, 107, 109, 1957–58.

WRIGHT J. W. 'Electronic distance measurement in land surveying'. *Survey Review* No. 140, April 1966.

WRIGHT J. W. 'Tests on MRA 3 and MRA 101 Tellurometers in Southern Nigeria'. *Conference of Commonwealth Survey Officers*, 1967.

Electromagnetic Distance Measurement (Proceedings Symposium, Oxford, 1965). Hilger & Watts, London, 1967.

ORDNANCE SURVEY. *The History of the Retriangulation of Great Britain 1935–62*. H.M. Stationery Office, 1967.

CHAPTER V

SURVEY ADJUSTMENT

5.01 The Need for Adjustment. In almost all surveying operations, the number of quantities actually measured is greater than the geometrically necessary minimum: the simplest example of this is the measurement of all three angles of a plane triangle although the geometrical shape is determined by measuring only two.

There are very sound reasons for having a superabundance of measurement. One obvious general principle is that the more measurement contributed to a job the more accurate the results should be. Further, the excess observations provide checks against mistakes: thus if the three measured angles of a triangle total 175° a mistake is disclosed and the observations must be repeated, but if only two angles had been measured a mistake of 5° in either would not immediately show up, and the use of the erroneous angle would sooner or later have disturbing or alarming consequences.

We know that all observations are subject to small errors not to be described as mistakes, but however small these errors are they give rise to inconsistencies. For instance, if the total of the three observed angles of a plane triangle is 179° 59′ 56″ the three values cannot all be correct, indeed it is highly likely that they are all wrong, but the discrepancy in this case may not be considered large enough to warrant re-observation of the angles. Nevertheless, if the three observed values are used in trigonometrical calculations, confusion will arise because the three values are not the actual angles of the triangle on the ground, nor are they the angles of any possible triangle, and the processes of trigonometry do not apply to such sets of angles.

5.02 The Aim of Adjustment. An adjustment operation is a procedure for deriving a set of quantities that are consistent among themselves and differ by acceptably small amounts from observed values. In the case mentioned above, alterations totalling +4″ would produce angles having the correct total: trigonometrical formulas could then be applied to those angles without causing confusion. It must be emphasised that the essential requirement of an adjustment is to produce *consistency*, and we must not suppose that an adjustment will end in giving us *correct* values. Thus it is unsound to to use the word 'correction' often seen in the literature. However, implicit in any reasonable adjustment method is the requirement that the alterations applied should in some sense be 'as small as possible'; for instance, if the angles in the triangle mentioned above are all increased by $1\frac{1}{3}''$ consistency will be achieved, or it might be considered reasonable enough to make adjustments of $+1''$, $+1''$, and $+2''$ with perhaps an arbitrary rule that the largest adjustments should be applied to the largest angles. An adjustment of $+4''$ to one angle would not be reasonable; it would imply that the observed values of the other two angles were correct and the measurement of the third angle was a waste of time, or at best a precautionary check against large mistakes.

5.03 Conditions: Variations. There are two kinds of approach to an adjustment problem:

(i) by altering the observed values so as to achieve consistency;

(ii) by starting with a 'model' of the survey system that is necessarily self-consistent, and amending or varying the model so as to approach the observed quantities as closely as possible while remaining self-consistent. These methods are best explained by simple examples.

In the case of the triangle mentioned above, alterations, or 'adjustments', θ_1, θ_2, and θ_3 are applied to the angles. It is obvious that these must satisfy a condition $\theta_1+\theta_2+\theta_3 = +4''$, and any set that satisfies this condition will produce three angles that are angles of a plane triangle.

To illustrate the other method, let us suppose that the observed angles were 51° 18′ 12″, 59° 43′ 50″, and 68° 57′ 54″. We start with the model triangle having angles 51° 18′ 12″, 59° 43′ 50″, and 68° 57′ 58″; this is amended by making the first two angles (51° 18′ 12″ + a'') and (59° 43′ 50″ + b'') so that the third angle will be (68° 57′ 58″ − a'' − b''). The differences between these angles and the observed angles are a'', b'', and ($4'' - a'' - b''$). These are residuals, a residual being defined as 'adopted value *minus* observed value'. The residuals are also adjustments but they are expressed as functions of the variations. It is important to note that the variations a and b are arithmetically independent but the formulas for the residuals are such that they satisfy the condition mentioned in the previous Section, whatever may be the values of the variations. It is also to be noted that other sets of angles (totalling 180°) could be adopted as the initial 'model' triangle, but an adjustment procedure based on sound principles must produce the same final values of the angles, whatever initial values are chosen.

A rather less simple example will illustrate the adjustment methods in a more realistic way.

The traverse of three lines, AJKB, is surveyed between points A and B which have fixed coordinates in the plane rectangular

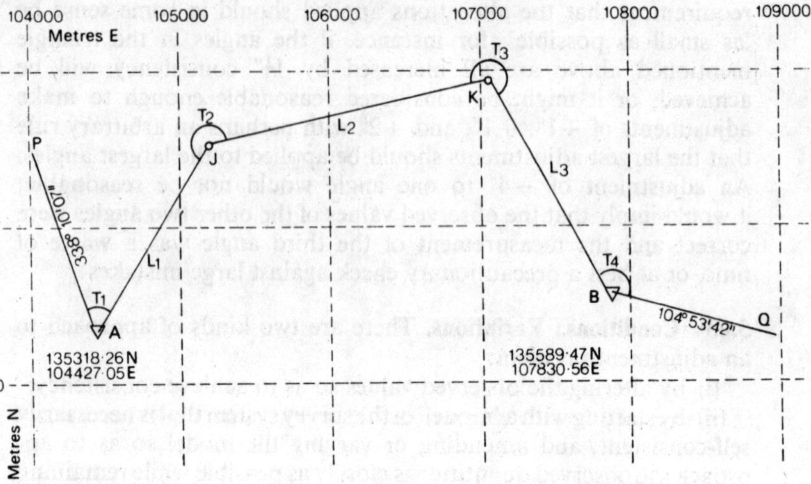

FIG. 60. SIMPLE TRAVERSE.

system shown in Fig. 60. The fixed reference bearings are also shown, and the survey involves measuring the four angles T_1, T_2, T_3, T_4 and the three lengths L_1, L_2, L_3. The observed values appear in columns (1) and (3) on the computation sheet, Fig. 61a. Starting from the reference bearing at A, bearings of the traverse lines are calculated as shown in column (2), and at the second reference

(1) Angles	(2) Bearings	(3) Lengths m	ΔN m	(4) ΔE m	(5) Coordinates m	m	Point
	AP 338° 10′ 07″						
A 53° 19′ 51″					135 318·26	104 427·05	A
	AJ 31 29 58	1439·58	+1227·45	+752·17			
J 226 19 26					136 545·71	105 179·22	J
	JK 77 49 24	1916·01	+404·14	+1872·90			
K 252 24 15					136 949·85	107 052·12	K
	KB 150 13 39	1567·91	−1360·95	+778·56			
B 134 40 24					135 588·90	107 830·68	B′
	104 54 03				135 589·47	107 830·56	B
	BQ 104 53 42						

FIG. 61a. Traverse Computation.

	AP 338° 10′ 07″						
A 53° 19′ 51″					135 318·26	104 427·05	A
	AJ 31 29 58	1439·58	+1227·45	+752·17			
J 226 19 26					136 545·71	105 179·22	J
	JK 77 49 24	1916·01	+404·14	+1872·90			
K *252 23 51*					136 949·85	107 052·12	K
	KB *150 13 15*	*1567·36*	*−1360·38*	*+778·44*			
B *134 40 27*					135 589·47	107 830·56	B
	BQ 104 53 42						

FIG. 61b. Model Traverse.

bearing a discrepancy of 21″ is disclosed. Ignoring this, the differences of coordinates are calculated and then the coordinates of the points in the usual way, with the results shown on the sheet. Discrepancies of 0·57 m and 0·12 m appear.

It would of course be possible to stop at this point and regard the three discrepancies as small, indicating only that the surveyor had made no gross error in his work. The coordinates of J and K as given by this computation are exactly consistent with the observed values of T_1, T_2, L_1, and L_2, but are inconsistent with the observed values of T_3, T_4, and L_3. This is of course unsatisfactory: it is desirable to take account of all seven measured quantities in a balanced sort of way, and the application of the two methods of adjustment to this problem is explained below.

5.04 Condition Method. In this system, adjustments $\theta_1\theta_2\theta_3\theta_4$ are applied to the observed angles, and adjustments $l_1l_2l_3$ to the observed lengths. Then clearly the adjusted bearings will be:

AJ 31° 29′ 58″ $+\theta_1$	KB 150° 13′ 39″ $+\theta_1+\theta_2+\theta_3$
JK 77 49 24 $+\theta_1+\theta_2$	BQ 104 54 03 $+\theta_1+\theta_2+\theta_3+\theta_4$.

One requirement is immediately evident—that $\theta_1+\theta_2+\theta_3+\theta_4 = -21''$. Changes of bearings and lengths will change the coordinates. Consider a line of length L, bearing B, as in Fig. 62(a). If the bearing is increased by θ and the length by l, the end of the line will move from P to P′ as shown: in the enlarged diagram, Fig. 62(b), it is obvious that (assuming both the changes to be small) the length of the perpendicular PM is $L \,.\, \theta_r$, where θ_r is the radian value of θ; and the length of MP′ is l. The bearing of PM is $(B + 90°)$ and

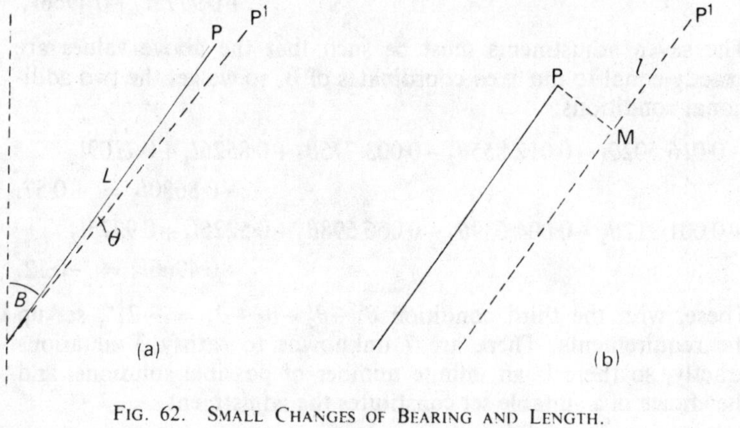

FIG. 62. SMALL CHANGES OF BEARING AND LENGTH.

clearly the changes of coordinates from P to P′ will be:

$$dN = -L\,.\,\theta_r\,.\,\sin B + l\,.\,\cos B = -\theta_r\,.\,\Delta E + l\,.\,\cos B$$
$$dE = +L\,.\,\theta_r\,.\,\cos B + l\,.\,\sin B = +\theta_r\,.\,\Delta N + l\,.\,\sin B.$$

Applying these formulas to the first traverse line AJ, for which $\sin B = 0{\cdot}5225$ and $\cos B = 0{\cdot}8526$, we get the changes:

$$dN = -0{\cdot}003\,647\,.\,\theta_1 + 0{\cdot}8526\,.\,l_1$$
$$dE = +0{\cdot}005\,951\,.\,\theta_1 + 0{\cdot}5225\,.\,l_1$$

in which the angle adjustment is now in seconds and its coefficients are therefore $-(L\,.\,\sin B)/206\,265$ and $+(L\,.\,\cos B)/206\,265$. Now the bearing of the line JK increases by $\theta_1+\theta_2$ and the additional changes of coordinates due to the changes in this line will be:

$$dN = -0{\cdot}009\,080(\theta_1+\theta_2) + 0{\cdot}2109\,.\,l_2$$
$$dE = +0{\cdot}001\,959(\theta_1+\theta_2) + 0{\cdot}9775\,.\,l_2.$$

The bearing of the third line is greater than 90°, and care must be taken about signs: the changes due to this line will be:

$$dN = -0{\cdot}003\,775(\theta_1+\theta_2+\theta_3) - 0{\cdot}8680\,.\,l_3,$$
$$dE = -0{\cdot}006\,598(\theta_1+\theta_2+\theta_3) + 0{\cdot}4966\,.\,l_3.$$

Including all the effects of the changes of angles and lengths, the computed coordinates of point B will be:

$$135\,588{\cdot}90 - 0{\cdot}016\,502\theta_1 - 0{\cdot}012\,855\theta_2 - 0{\cdot}003\,775\theta_3 + 0{\cdot}8526l_1 + 0{\cdot}2109l_2 - 0{\cdot}8680l_3$$
$$107\,830{\cdot}68 + 0{\cdot}001\,312\theta_1 - 0{\cdot}004\,639\theta_2 - 0{\cdot}006\,598\theta_3 + 0{\cdot}5225l_1 + 0{\cdot}9775l_2 + 0{\cdot}4966l_3.$$

The seven adjustments must be such that the above values are exactly equal to the fixed coordinates of B, so we get the two additional conditions:

$$-0{\cdot}016\,502\theta_1 - 0{\cdot}012\,855\theta_2 - 0{\cdot}003\,775\theta_3 + 0{\cdot}8526l_1 + 0{\cdot}2109l_2 - 0{\cdot}8680l_3 = +0{\cdot}57,$$
$$+0{\cdot}001\,312\theta_1 - 0{\cdot}004\,639\theta_2 - 0{\cdot}006\,598\theta_3 + 0{\cdot}5225l_1 + 0{\cdot}9775l_2 + 0{\cdot}4966l_3 = -0{\cdot}12.$$

These, with the third condition $\theta_1+\theta_2+\theta_3+\theta_4 = -21''$, set up the requirements. There are 7 unknowns to satisfy 3 equations exactly, so there is an infinite number of possible solutions, and the choice of a suitable set constitutes the adjustment.

5.05 Variation Method. One way to form a model traverse for this method is to obtain, somehow, coordinates of J and K that are known to be nearly correct, and then the lengths and bearings of the lines, and the traverse angles, are calculated exactly from these coordinates. In practice, a good way to find initial coordinates is to carry out a calculation as already done, as far as getting the coordinates of K: then the bearing and length of KB are calculated from these coordinates and the fixed coordinates of B. The angles T_3 and T_4 are then calculated. The working would be as shown on the second computation sheet, Fig. 61(b), where the quantities calculated simply to complete the 'model' traverse are printed in italic.

This computation represents a traverse with no internal discrepancies, but unsatisfactory because some of its geometrical elements (lengths and angles) agree exactly with the observed values and the rest do not, and obviously a different traverse could be constructed by choosing a different initial set of observed quantities, e.g. by starting at B and using the observed values of T_4, T_3, L_3, and L_2.

The model is now to be modified, or varied, and in this case the variations are small changes in the coordinates of J and K. Their effects on the lengths and angles are calculated, and the procedure is aimed at finding a set of variations that produce a traverse in which the angles and lengths are in close unbiased agreement with the observed values. The interpretation of 'close unbiased agreement' will be discussed below.

It is therefore necessary to have some formulas for the effect on the length and bearing of a line, due to making small changes of the coordinates of its ends.

In Fig. 63, the line OP has length L and bearing B. The point P

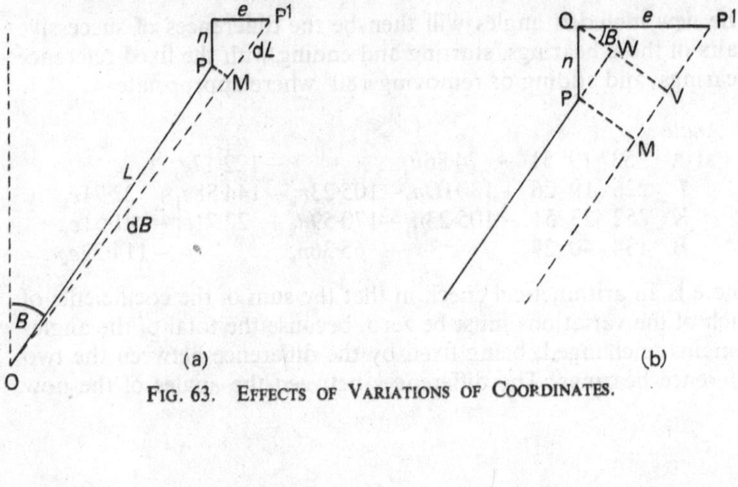

FIG. 63. EFFECTS OF VARIATIONS OF COORDINATES.

is moved to P′ by small changes of coordinates n and e as shown. From the enlarged detail at (b), it is obvious that the change in length is:

$$dL = MV+VP = n \,.\, \cos b + e \,.\, \sin B.$$

The change of bearing dB is (PM/L) radian, and

$$PM = (QV-QW) = e \,.\, \cos B - n \,.\, \sin B.$$

Converting to seconds by multiplying by 206 265, we get the formula:

$$dB'' = -n \,.\, \frac{206\,265 \,.\, \sin B}{L} + e \,.\, \frac{206\,265 \,.\, \cos B}{L}$$

If both ends of a line are shifted, for instance if point O has shifts $n_1 e_1$ and point P has shifts $n_2 e_2$, then obviously the effect on the length and bearing are the same as if O remained fixed and P had shifts (n_2-n_1) and (e_2-e_1).

Now, starting from the model as given on the computation sheet, let J have shifts (variations) n_j and e_j and let K have shifts (variations) n_k and e_k. After making these variations, the new lengths will be:

AJ	$1439{\cdot}58 + 0{\cdot}8526n_j$		$+0{\cdot}5225e_j$
JK	$1916{\cdot}01 + 0{\cdot}2109(n_k-n_j)$		$+0{\cdot}9775(e_k-e_j)$
KB	$1567{\cdot}36 - 0{\cdot}8679(-n_k)$		$+0{\cdot}4967(-e_k)$

The new bearings will be:

AJ	31° 29′ 58″	$-\ \ 74{\cdot}86n_j$	$+122{\cdot}17e_j$
JK	77 49 24	$-105{\cdot}23(n_k-n_j)$	$+\ \ 22{\cdot}71(e_k-e_j)$
KB	150 13 15	$-\ \ 65{\cdot}36(-n_k)$	$-114{\cdot}22(-e_k)$

The new included angles will then be the differences of successive pairs of these bearings, starting and ending with the fixed reference bearings, and adding or removing 180° where appropriate:

Angle					
at A	53° 19′ 51″	$-\ \ 74{\cdot}86n_j$		$+122{\cdot}17e_j$	
J	226 19 26	$+180{\cdot}09n_j$	$-105{\cdot}23n_k$	$-144{\cdot}88e_j$	$+\ \ 22{\cdot}71e_k$
K	252 23 51	$-105{\cdot}23n_j$	$+170{\cdot}59n_k$	$+\ \ 22{\cdot}71e_j$	$+\ \ 91{\cdot}51e_k$
B	134 40 27		$-\ \ 65{\cdot}36n_k$		$-114{\cdot}22e_k$

There is an arithmetical check in that the sum of the coefficients of each of the variations must be zero, because the total of the angles remains unchanged, being fixed by the difference between the two reference bearings. The differences between the angles of the now varied model, and the observed angles are:

at A		$-74{\cdot}86n_j$		$+122{\cdot}17e_j$	
J		$+180{\cdot}09n_j$	$-105{\cdot}23n_k$	$-144{\cdot}88e_j$	$+22{\cdot}71e_k$
K	$-24''$	$-105{\cdot}23n_j$	$+170{\cdot}59n_k$	$+22{\cdot}71e_j$	$+91{\cdot}51e_k$
B	$+3''$		$-65{\cdot}36n_k$		$-114{\cdot}22e_k$

The differences between the model lengths after variation, and the measured lengths are:

AJ		$+0{\cdot}8526n_j$		$+0{\cdot}5225e_j$	
JK		$-0{\cdot}2109n_j$	$+0{\cdot}2109n_k$	$-0{\cdot}9775e_j$	$+0{\cdot}9775e_k$
KB	$-0{\cdot}55$		$+0{\cdot}8679n_k$		$-0{\cdot}4967e_k$.

The above 7 expressions are the residuals or adjustments, and are linear functions of the 4 variations: the object of an adjustment procedure is to work out values of the variations so that the values of the residuals are in some sense 'as small as possible'. Again, it must be emphasised that the variations are independent; any one of them can be changed without necessitating a change in any other.

To put it in another way: whatever values are taken for the 4 variations, the four small angles and three small lengths calculated from the foregoing 7 formulas will, if applied to the observed quantities, automatically produce a self-consistent traverse. The 7 quantities expressed as functions of the independent parameters $n_j n_k e_j e_k$ automatically satisfy the 7 conditions discussed in the previous Section.

5.06 The Adjustment Problem. In preceding Sections, a simple traverse has been used to illustrate in two ways the setting up of an adjustment problem. It is not a problem that has a unique solution, therefore some guidance is required to finding a satisfactory solution, and in general this means that discrepancies and misclosures must be compensated in some way that spreads out the effects of errors, and yet causes minimum disturbance to observed values. To be more specific, an adjustment should lead to a set of final values none of which differs by an intolerable amount from the corresponding field measurement.

Another desirable feature of an adjustment procedure is that it should not depend on guesses or opinions of the computer, but should be described by a set of precise rules and formulas.

Many adjustment operations are based on the guidance derived from the Theory of Accidental Errors, about which there is of course an enormous literature, and which gives us the well-known Principle of Least Squares. This theory is described in Sections 5.10 to 5.18. However, not all adjustments are tied to a formal theory: some well-known simple procedures are designed to achieve the requirements already mentioned by what may be called, for want of a better word, arbitrary methods. These methods have even been called 'semi-rigorous' whatever that means, but such descriptions may be misleading, since these methods, however arbitrary, do remove inconsistencies by unbiased procedures, and surely this is the prime object of all adjustment.

5.07 Arbitrary Methods. Perhaps the most obvious example of arbitrary adjustment comes also from traversing. It is quite common practice to adjust a traverse by first distributing the bearing misclosure evenly among the angles, then using these adjusted bearings to calculate coordinates. Misclosures generally appear, and adjustments are applied to the coordinates, increasing along the traverse until the adjustments at the closing point exactly cancel the misclosures. In the Bowditch method, for instance, the adjustment at each point is proportional to the accumulated length of the traverse lines up to that point. In any case, the lengths and bearings of the adjusted traverse, if required, must then be calculated from the adjusted coordinates of the stations. Some more examples of arbitrary methods of adjustment are described later.

5.08 The Idea of Weights. Reference has been made to 'unbiased' application of adjustments. This does not necessarily mean that misclosures must be equally divided. It may happen that some of the measurements used in a survey are considered to be more reliable than others, and it is easy to take account of this fact. Consider the triangle mentioned in Section 5.01, and suppose that one of the angles is known to be more reliable than the other two: the surveyor might reasonably decide that this angle should receive a smaller share of the total adjustment.

Perhaps he thinks that adjustments in the ratios 1:2:2 would be appropriate. A little arithmetical thought indicates adjustments of 0·8″, 1·6″, and 1·6″. This process is called weighting. See Section 5.17.

Again, in a traverse it might reasonably be thought that angles measured between short lines will be less reliable than those measured where the lines are longer. Suppose that there are 12 measured angles and 5 of these are thought to be less reliable than the rest; the bearing misclosure could be divided by 17, and 2/17 of the misclosure applied to the 5 weaker angles, 1/17 to each of the others.

In levelling, it is obvious that the reliability of a difference of level will decrease as the length of the line increases: this can be taken into account by a system of weighting. See Section 5.30.

5.09 Types of Errors. It is usual to classify errors as: (1) mistakes, (2) systematic errors, (3) accidental or random errors.

(1) Mistakes may be quite large, in which case their existence is easily detected; very small mistakes, such as a mis-reading of 2 or 3 seconds on a micrometer theodolite, may remain undetected and will go in with the random errors of class 3. The surveyor should always arrange his observational methods so that serious mistakes will inevitably show up, and the erroneous observation can be rejected. Obvious precautions against mistakes are the repetition of observations, measuring lengths in two different units, changing zero on the arc of a theodolite so that numerical slips are not likely to be repeated, and many others.

(2) Systematic errors often produce effects that are numerically related to the value of the quantity measured. This is obvious, for instance, in the effect of an erroneous length of a measuring tape or of a levelling staff. However, a systematic error may not show up: a carefully surveyed traverse will close back on itself even if the steel tape is uniformly stretched by 1%! Evidently, the detection of systematic error may be difficult. The experienced surveyor will know what sources of systematic error may exist, and should assume that they do exist; in many cases the observational procedures can be arranged so that these errors automatically cancel out—changing face on a theodolite is a simple example. The calibration of scales against reliable standards is another obvious precaution.

(3) Accidental or random errors are due to those uncontrollable effects which cause the repeated measures of the same quantity to disagree among themselves, but the discrepancies are not large enough to warrant rejection of any of the values. Any properties discernible in random errors must be of a statistical nature, properties that are meaningful if ascribed to groups but not if applied to particular values. Thus we may assert that small errors are more frequent than large ones, or that negative and positive errors are equally likely to occur.

5.10 Accidental Errors. We are concerned in this chapter with errors of the kind mentioned in (3) above. These errors present the surveyor with several practical problems, three of which are:

(i) when several different observed values of a quantity are obtained, some particular value, not necessarily one of the observed values, must be adopted for use in the next stage of the work:

(ii) when the adopted values of several quantities are inconsistent, as for instance in the case of the triangle misclosure situation described in Section 5.01, some reasonable method of adjusting the values must be devised. The surveyor needs some general principle on which to base his adjustment procedure in situations far more complicated than those already discussed.

(iii) since all observations are subject to error, all results calculated from them will be more or less erroneous, and it is very desirable to have some rules for calculating quantities to indicate standards of accuracy either of observations or of results derived from them. For instance, in a piece of triangulation it may be required that all triangles shall close better than 10″, or in a network of levelling the heights of all benchmarks shall be accurate to 0·02 metre. Such criteria give guidance on the question of field methods to be adopted.

Consider first the simple case of a measurement being repeated a number of times, and giving a collection of different values. The observed values of an angle, omitting the degrees and minutes which are the same in each case, might be 49·6″, 46·2″, 48·2″, 50·4″, 47·8″, 49·0″, 48·2″, 46·6″, 47·4″, 48·0″, 48·8″, 47·6″. A collection like this can be illustrated by points on a line as in Fig. 64a. It is usually found that such a collection has the property, which one intuitively expects, of being more densely concentrated around the centre of the range of values. The set of values illustrated in Fig. 64b is centred about the same part of the scale but is more widely dispersed

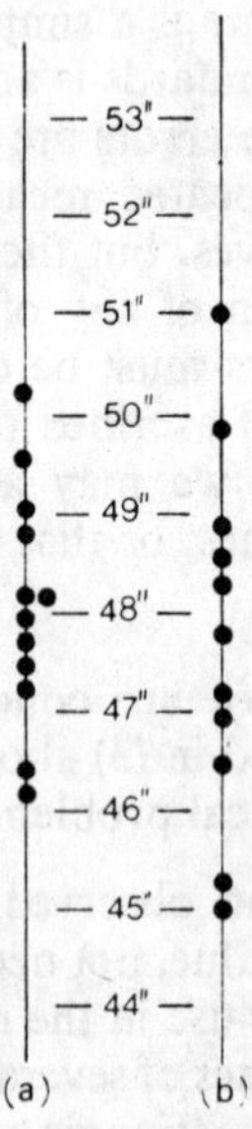

FIG. 64. OBSERVED VALUES.

and it is reasonable to assume that the final value derived from this set will be less trustworthy than that derived from the first set. Sets of repeated observations such as those described above give rise to two problems:

(i) it is necessary to have a rule for deciding on a single value of the observed quantity, the value to be used in the next stage of the work;

(ii) it is desirable to calculate an index of reliability to be associated with the adopted value.

If we had a much more numerous collection of values, some hundreds perhaps, it would be possible to group them in equal intervals and draw up a histogram as shown in Fig. 65. The number at the top of each column is the ratio that the number of values in the column bears to the whole number in the collection: obviously, all the fractions must total exactly 1. Going further, we see that if the number of items in the collection were enormous, we could group them in much narrower intervals, and then the top of the histogram would approach the form of a smooth curve. It is often found that the curve has the symmetrical bell-shape as indicated by the broken curve on Fig. 65. It is then natural to ask if there is a predictable mathematical formula for the curve.

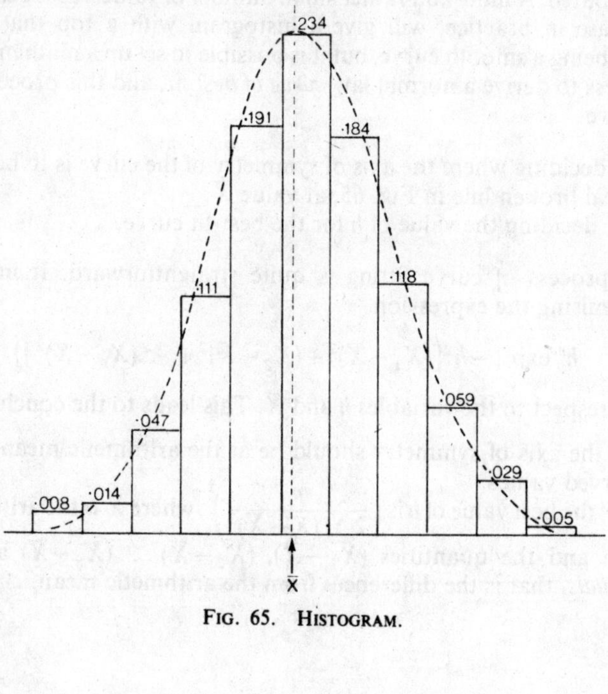

FIG. 65. HISTOGRAM.

There is indeed a mathematical theorem, the Central Limit Theorem, which states that if the measurements of a quantity are affected by a large number of sources of independent randomly varying errors, then the histogram of a large collection of observations will tend to the shape given by the formula:

$$dx \,.\, \frac{h}{\sqrt{\pi}} \,.\, \exp(-h^2x^2).$$

where x is the error of an observation, dx is the width of the grouping intervals, and h is a parameter that depends on the height of the histogram at its centre. The formula gives the fraction of the whole collection that will be found in the interval $(x-\frac{1}{2}\,dx)$ to $(x+\frac{1}{2}\,dx)$. Note that the parameter h must have the inverse dimension of x, because the expression inside the brackets must be a pure number. The above formula is referred to as the Normal Law of Error or the Gaussian Distribution Law.

5.11 Arithmetic Mean. Now let us suppose that we have a large collection of observed values $X_1, X_2, X_3, \ldots, X_n$ and we assume that they are 'normally' distributed. More precisely, we assume they are a good sample of a much larger collection that is normally distributed. A finite and rather small number of values, as is generally the case in practice, will give a histogram with a top that is far from being a smooth curve, but it is possible to set up a mathematical process to derive a normal law *curve of best fit*, and this process will involve:

(i) deciding where the axis of symmetry of the curve is to be—the vertical broken line in Fig. 65, at value $\bar{X}$,
(ii) deciding the value of h for the best-fit curve.

The process of curve fitting is quite straightforward. It involves maximising the expression

$$h^n \exp\{-h^2[(X_1-\bar{X})^2+(X_2-\bar{X})^2+\ldots(X_n-X)^2]\}$$

with respect to the variables h and $\bar{X}$. This leads to the conclusions:

(i) the axis of symmetry should be at the arithmetic mean of the observed values,
(ii) the best value of h is $\left(\frac{n}{2\sum(X-\bar{X})^2}\right)^{\frac{1}{2}}$, where $\bar{X}$ is the arithmetic mean and the quantities $(X_1-\bar{X}), (X_2-\bar{X}) \ldots (X_n-\bar{X})$ are the *residuals*, that is the differences from the arithmetic mean.

It is easily shown that the arithmetic mean of a number of quantities has the following properties:

(i) the algebraic sum of the residuals is zero,

(ii) the sum of the squares of the residuals from the arithmetic mean is less than the sum of the squares of the residuals from any other value.

This minimum property corresponds with the maximum of the exponential expression above, because of the minus sign before h^2.

5.12 Probable Error. The accuracy of a set of measurements will obviously be indicated by the shape of the histogram: high-quality observations will be packed more closely round the mean than observations of lower precision, and the histogram will be thinner and higher. An index of quality that is often quoted is the *probable error*: this is defined as the error p such that just 50% of the observations are inside the interval $(X-p)$ to $(\bar{X}+p)$. In the case of a normal distribution it can be shown that the formula for p is:

$$p = 0{\cdot}6745\left(\frac{\sum (X-\bar{X})^2}{n}\right)^{\frac{1}{2}} = \frac{0{\cdot}477}{h}.$$

5.13 Standard Deviation. Another index of precision that has long been in use is the *root mean square residual*, given by the formula:

$$\sigma = \left(\frac{\sum (X-\bar{X})^2}{n}\right)^{\frac{1}{2}}:$$

this is called the *standard deviation*. It is not based on any theory of the distribution of errors, but is simply an obvious and suitable formula to serve as an index of the spread of a group of observations. Clearly, it is functionally similar to the probable error, and in practice we can take the p.e. to be $\frac{2}{3}$ of the standard deviation. It seems unnecessary to have two such indices of precision, and the tendency nowadays is to use the standard deviation, or the *standard error* as described in the next Section.

5.14 Standard Error. The true value of a measured quantity is not known, and will in general differ from the arithmetic mean of a set of measurements. But errors are differences from the true value. Therefore, by a property of the arithmetic mean mentioned above, the sum of the squares of the errors will generally be greater than the sum of the squares of the residuals, and cannot be less. It can be shown by a simple statistical argument that the sum of the squares of the errors and the sum of the squares of the residuals will on average be in the ratio $n/(n-1)$. Thus, a more realistic index for error is:

$$\varepsilon = \left(\frac{\sum (X-\overline{X})^2}{n-1}\right)^{\frac{1}{2}},$$

and this is called the *standard error*. If n is large, ε will differ little from the standard deviation.

It is to be noted that the probable error and other indices of precision mentioned above are indicators of the precision of a particular measurement procedure: for instance, if we say that the probable error of a measured angle is 2 seconds, we mean that if many angle measurements are made by the method under consideration, then 50% of the observations will be within 2″ of the truth and the rest will be more than 2″ in error. In other words, the histogram of errors of a large (unspecified) number of measurements would have half of its area between the ordinates $\pm 2''$ from the centre of the curve. In practice we need another index of precision: if we measure a quantity a number of times we must adopt some particular value for use, and generally take the arithmetic mean. We require an index of precision of this mean, because it is the accuracy of the values that go into the calculations that determines the accuracy of the final results. Obviously we would expect this index to be in some inverse relation to the number of observations taken; the mean of 12 observations is presumably better than the mean of 2. A simple statistical argument shows that the standard error of the mean of a set of n observations is

$$\sigma_{\mathrm{m}} = \left(\frac{\sum (X-\overline{X})^2}{n(n-1)}\right)^{\frac{1}{2}}$$

Comparing this formula with that for the standard error of an observation, we have the additional factor of $\sqrt{n}$ in the denominator, and this shows that the standard error of the mean is inversely proportional to the square root of the number of observations. If we want to halve the standard error of the mean we must quadruple the number of measurements.

5.15 Combination of Observations. The value of some required quantity may be obtained by combining the measured values of two or more quantities: in such cases, we need to be able to estimate the precision of the derived quantity when we have knowledge of the precision of the measured values that have contributed to the operation. For instance, a difference of height may be found by measuring a distance and an angle, each subject to observational error: a much more complicated relationship is that between the measured angles and distances in a piece of triangulation, and the rectangular coordinates calculated from these observed values; yet the error of each measured quantity will contribute something to the error of each calculated coordinate.

Consider the simple case of a quantity found by adding together the measured values of a number of portions; the length of a traverse line obtained by laying a tape down successively and adding the results, or a difference of height found by adding the differences measured at each set-up of the Level. In general, suppose the actual errors existing in a set of the component measures are $\varepsilon_1, \varepsilon_2, \ldots, \varepsilon_k$; then the actual error E of the complete measurement will be $E = \varepsilon_1 + \varepsilon_2 + \ldots \varepsilon_k$. If the complete measurement is carried out n times, the square of the standard error of a complete measure will be

$$\left(\sum_n E^2\right)\Big/(n-1) \quad \text{or} \quad \left(\sum_n (\varepsilon_1+\varepsilon_2+\ldots\varepsilon_k)^2\right)\Big/(n-1).$$

Now

$$(\varepsilon_1+\varepsilon_2+\ldots\varepsilon_k)^2 = \varepsilon_1{}^2+\varepsilon_2{}^2+\ldots\varepsilon_k{}^2+2(\varepsilon_1\varepsilon_2+\varepsilon_1\varepsilon_3+\ldots\varepsilon_{k-1}\varepsilon_k).$$

In the expression on the right, the squares are necessarily positive, but the terms $\varepsilon_1\varepsilon_2$ etc. may be of either sign, depending on the signs of the two factors in each case. If the observational errors are truly random there will be no bias towards negative or positive products and statistically the average value of the expression in brackets, taken over a large number of sets, will be zero. Thus we can say, if the errors are random, that $(\sum \varepsilon)^2$ is on average equal to $\sum \varepsilon^2$ Hence, on average, $(\sum E^2)/(n-1)$ equals $(\sum \varepsilon_1{}^2)/(n-1)+(\sum \varepsilon_2{}^2)/(n-1)\ldots+(\sum \varepsilon_k{}^2)/(n-1)$. That is, the standard error of the total quantity measured is equal to the square root of the sum of the squares of the standard errors of the component parts.

A simple example of the combination of sources of error occurs in the effect of angle errors on triangle misclosures. Suppose that the standard error of a measured angle is 3″; then the misclosure of a triangle in which all three angles are measured is the sum of the errors of the three angles, so the standard error of closure will be $(3^2+3^2+3^2)^{\frac{1}{2}} = 5{\cdot}2''$.

A desired quantity may be found by calculation from two or more measured quantities: a height calculation from a measured horizontal length and a measured vertical angle is a simple example. If l is the measured horizontal length and A is the measured angle, the height is given by $h = l \,.\, \tan A$. Now if l has an error $\mathrm{d}l$, the result will be $(l+\mathrm{d}l)\tan A$, in error by $\mathrm{d}l \,.\, \tan A$: if A is in error by $\mathrm{d}A$ the result will be $l \,.\, (\tan A+\mathrm{d}A \,.\, \sec^2 A)$ in error by $l \,.\, \mathrm{d}A \,.\, \sec^2 A$. If the two sources of error are independent, and we now take $\mathrm{d}l$

to be the standard error of l, and dA to be the standard error of the measurement of A, then the standard error of the calculated height will be

$$(dl^2 \,.\, \tan^2 A + l^2 \,.\, dA^2 \sec^4 A)^{\frac{1}{2}}.$$

Take l = 100 metres with standard error 0·02 m = 20 mm, A = 10° with standard error 5″ (= 5/206 265 radian). Then tan A = 0·1763, $\sec^2 A$ = 1·031, $dl \,.\, \tan A$ = 3·5 mm, $l \,.\, dA \,.\, \sec^2 A$ = 2·5 mm, and the standard error of h is thus $((3{\cdot}5)^2+(2{\cdot}5)^2)^{\frac{1}{2}} = \pm 4{\cdot}3$ mm. The result is: h = 17·63 m with standard error 4·3 mm. We might reasonably conclude that the figure 3 in the quoted value of h has a fair chance of being correct, because to make it 4 or 2 the actual error would have to be somewhat greater than the estimated standard error.

5.16 Adjustment of Systems. The Theory of Errors is concerned with the pattern of the frequency distribution of errors of observation, and obviously such a pattern or formula can only be recognised to exist if the number of observations available for study is very great. In surveying practice, the number of observations of one quantity is usually quite small: an angle in a triangulation system may be measured 4, perhaps as many as 16, times; each leg of a traverse will probably be measured only twice. It is evidently absurd to think of drawing a histogram of only 6 values, and even more absurd to expect that they would comply with any mathematical formula of distribution. Nevertheless, an index of precision such as the standard error can be computed from the given formulas, even if only two observed values are available, and such indices can be accepted as giving some indication of relative reliability.

Though the measurements of a single quantity may never be numerous, measurements of a large number of similar quantities are familiar in surveying: in a triangulation system, for instance, the number of angles measured may be hundreds or even thousands, and it is not unreasonable to suppose, if all the measurements are made by the same observational procedure, that the errors of the observed angles will follow a formula of distribution such as the Normal Law. This idea that the errors of a set of similar quantities measured by a uniform procedure will tend to be 'normally distributed' is the basis of most adjustment procedures used in surveying and geodesy.

It is often necessary in practice to make use of measured values which are evidently not of uniform quality or reliability. For example, the angles in a triangulation system may have been measured at widely different times, by a variety of instruments, and in various favourable or unfavourable external conditions. Observers are usually instructed to record their opinions of the relative reliability of their observations. Again, the lines of a traverse will be of different lengths, and it is to be expected that the errors of the longer lines will on average be greater than the errors of the shorter lines.

5.17 Weighting. These variations of reliability can be expressed numerically by a system of *weighting*. The basic idea is illustrated by a very simple consideration that a measurement of weight 2 is as good as two measurements of weight 1. To put it numerically, if the value 18·9 has weight 2 and the value 17·7 has weight 1, then the value to be adopted is (18·9+18·9+17·7)/3 = 18·5. This calculated value is the *weighted arithmetic mean*. Obviously there is a general formula

$$\frac{w_1X_1+w_2X_2+\ldots w_nX_n}{w_1+w_2+\ldots w_n}.$$

Also obviously, weights are relative values; any set of weights in the same ratios will give the same weighted mean. In particular, the weighted mean could be expressed in the form:

$$f_1X_1+f_2X_2+\ldots f_nX_n,$$

in which the sum of the coefficients f is exactly 1. In the above numerical example, we have $\frac{2}{3}(18{\cdot}9)+\frac{1}{3}(17{\cdot}7) = 18{\cdot}5$.

The problem of assigning weights to quantities that are subject to random error has not always a simple solution. Particularly in the case of directly observed quantities, such as the angles of a piece of triangulation, information about reliability may come from several sources; (i) the observer's comments recorded at the time of observation, (ii) knowledge of the accuracy of similar work already done, (iii) a calculated index number, such as the standard error of the mean, derived from the actual observed values of each quantity that is measured a number of times.

It may be, however, that the only available information about the order of accuracy of some quantity is a quoted numerical index such as the standard error. Consider a simple case of an angle which has been measured on two occasions, one observer giving the value 31·4″ ±1·2″ (omitting degrees and minutes), and the other observer giving 33·3″ ±0·9″. If it is desired to make use of both of these results, some form of weighted mean must be devised to give the second measurement more weight than the first.

In general terms, suppose measurement X_1 has standard error s_1 and measurement X_2 has standard error s_2. Using the idea of fractional weights, let the weights assigned to the two measurements be w and $(1-w)$ respectively. The weighted mean is then $wX_1+(1-w)X_2$. Now the standard error of wX_1 will be ws_1 and the standard error of $(1-w)X_2$ will be $(1-w)s_2$ and, by the rule given above, the standard error of the sum $wX_1+(1-w)X_2$ will be the square root of $w^2s_1{}^2+(1-w)^2s_2{}^2$. This expression is a quadratic function of w and it is reasonable to choose that value of w which minimises the standard error of the weighted mean. By the usual procedure for finding a minimum, we have $2ws_1{}^2-2(1-w)s_2{}^2 = 0$, or $w:(1-w) = 1/s_1{}^2:1/s_2{}^2$. That is, the 'best' weighted mean is got by taking weights inversely proportional to the squares of the standard errors of the two measurements. The numerical example above gives:

$$\frac{\dfrac{31{\cdot}4}{1{\cdot}44}+\dfrac{33{\cdot}3}{0{\cdot}81}}{\dfrac{1}{1{\cdot}44}+\dfrac{1}{0{\cdot}81}} = 32{\cdot}26''$$

as the weighted mean; the denominator of the above fraction is the weight of the mean, so the standard error of the mean is the square root of the reciprocal of this, i.e. 0·72".

The foregoing result is a simple case of a general formula for taking account of weights. In Section 5.11, the best value to be adopted when we have a set of measurements $X_1, X_2, \ldots, X_n$ was found to be the arithmetic mean. But now suppose that the measurements are not of equal reliability. As shown in Section 5.12, the parameter 'h' is a measure of precision, and we must now suppose that the n measurements are characterised by having different parameters $h_1, h_2, \ldots, h_n$. It is then found that the expression to be minimised is

$$h_1{}^2(X_1-\bar{X})^2+h_2{}^2(X_2-\bar{X})^2 \ldots h_n{}^2(X_n-\bar{X})^2.$$

But the parameters h are inversely proportional to the probable errors p as explained in Section 5.12. Thus the expression to be minimised can be written

$$\frac{(X_1-\bar{X})^2}{p_1{}^2}+\frac{(X_2-\bar{X})^2}{p_2{}^2}\ldots+\frac{(X_n-\bar{X})^2}{p_n{}^2}.$$

This is the *weighted sum of squares*. In fact, the function to be minimised is the sum of squares of the ratios of the residuals to their individual probable errors: this function is dimensionless, so its numerical value is independent of the unit in which the measurements are made. As explained before, standard errors may just as well be used instead of probable errors.

5.18 Principle of Least Squares. The assumption that measurements are liable to random errors of the kind that produce the so-called normal distribution indicates that the 'best' value to adopt is that value which minimises the sum of squares of the residuals, or in the more general case, the sum of weighted squares. This Principle of Least Squares, as it is called, is generally accepted as the criterion which decides the set of adjustments to be applied. In any adjustment problem there are infinitely many solutions to the basic requirement that the adjusted quantities shall be internally consistent: the additional requirement of minimum sum of squares picks out a particular solution, the one which may be said to have a property of maximum probability in accordance with theory of random errors. In any case, the least-squares requirement has the advantage that it determines, quite objectively, a unique solution of a given adjustment problem, and thus also satisfies a requirement mentioned in Section 5·06.

5.19 Adjustment Algebra. The adjustment problem for a traverse has been set up for the two methods of treatment, with actual numerical values. The operation of each method is explained below in terms of general algebraic formulas: the applications to the numerical examples appear in Sections 5.22 and 5.23.

Condition Method. Let there be 7 measured quantities to receive the adjustments $\alpha_1, \alpha_2 \ldots \alpha_7$ in order to make them satisfy the 3 conditions:

$$a_1\alpha_1 + a_2\alpha_2 + \ldots a_7\alpha_7 = m_1$$
$$b_1\alpha_1 + b_2\alpha_2 + \ldots b_7\alpha_7 = m_2$$
$$c_1\alpha_1 + c_2\alpha_2 + \ldots c_7\alpha_7 = m_3$$

together with the requirement that the sum of weighted squares $w_1\alpha_1{}^2 + w_2\alpha_2{}^2 + \ldots w_7\alpha_7{}^2$ shall be as small as possible. In any particular problem, the coefficients $a_1 \ldots a_7$, $b_1 \ldots b_7$, $c_1 \ldots c_7$, the misclosures $m_1 m_2 m_3$, and the weights are known. This problem is invariably solved by use of a device invented by Gauss. Consider the function F defined as:

$$\begin{aligned} F = w_1\alpha_1{}^2 + w_2\alpha_2{}^2 \ldots + w_7\alpha_7{}^2 &- 2K_1(a_1\alpha_1 + a_2\alpha_2 \ldots + a_7\alpha_7 - m_1) \\ &- 2K_2(b_1\alpha_1 + b_2\alpha_2 \ldots + b_7\alpha_7 - m_2) \\ &- 2K_3(c_1\alpha_1 + c_2\alpha_2 \ldots + c_7\alpha_7 - m_3). \end{aligned}$$

The coefficients K are usuallv called correlatives.

As a function of the 10 variables $\alpha_1 \ldots \alpha_7$, K_1, K_2, and K_3, this expression will have a minimum value when the 10 partial differential coefficients are zero, i.e.:

$$\frac{\partial F}{\partial \alpha_1} = 2w_1\alpha_1 - 2K_1a_1 - 2K_2b_1 - 2K_3c_1 = 0$$

$$\cdot \quad \cdot \quad \cdot \quad \cdot$$

$$\cdot \quad \cdot \quad \cdot \quad \cdot$$

$$\cdot \quad \cdot \quad \cdot \quad \cdot$$

$$\frac{\partial F}{\partial K_3} = 2(c_1\alpha_1 + c_2\alpha_2 + c_3\alpha_3 + c_4\alpha_4 + c_5\alpha_5 + c_6\alpha_6 + c_7\alpha_7 - m_3) = 0$$

These 10 simultaneous linear equations in 10 unknowns have a unique solution. The last 3 equations ensure that the solution satisfies the conditions, and that the minimum value of F is just the sum of weighted squares of the adjustments. From the first 7 equations we have:

$$w_1\alpha_1 = a_1K_1 + b_1K_2 + c_1K_3$$

$$\cdot \quad \cdot \quad \cdot \quad \cdot$$

$$\cdot \quad \cdot \quad \cdot \quad \cdot$$

$$w_7\alpha_7 = a_7K_1 + b_7K_2 + c_7K_3.$$

These expressions for $\alpha_1, \alpha_2 \ldots \alpha_7$ in terms of K_1, K_2, and K_3 are substituted into the last 3 equations, which are in fact the three conditions, and the resulting equations are:

$$\left(\frac{a_1^2}{w_1} + \frac{a_2^2}{w_2} + \ldots \frac{a_7^2}{w_7}\right)K_1 + \left(\frac{a_1b_1}{w_1} + \frac{a_2b_2}{w_2} + \ldots \frac{a_7b_7}{w_7}\right)K_2 + \left(\frac{a_1c_1}{w_1} + \frac{a_2c_2}{w_2} + \ldots \frac{a_7c_7}{w_7}\right)K_3 = m_1$$

$$\left(\frac{b_1a_1}{w_1} + \frac{b_2a_2}{w_2} \ldots + \frac{b_7a_7}{w_7}\right)K_1 + \left(\frac{b_1^2}{w_1} + \frac{b_2^2}{w_2} + \ldots \frac{b_7^2}{w_7}\right)K_2 + \left(\frac{b_1c_1}{w_1} + \frac{b_2c_2}{w_2} + \ldots \frac{b_7c_7}{w_7}\right)K_3 = m_2$$

$$\left(\frac{c_1a_1}{w_1} + \frac{c_2a_2}{w_2} \ldots + \frac{c_7a_7}{w_1}\right)K_1 + \left(\frac{c_1b_1}{w_1} + \frac{c_2b_2}{w_2} + \ldots \frac{c_7b_7}{w_7}\right)K_2 + \left(\frac{c_1^2}{w_1} + \frac{c_2^2}{w_2} + \ldots \frac{c_7^2}{w_7}\right)K_3 = m_3$$

These are 3 simultaneous equations for the 3 correlatives, and they have a unique solution: when the values of K_1, K_2, and K_3

have been found, they are substituted into the set of 7 equations that give the desired values of the adjustments $\alpha_1, \alpha_2 \ldots \alpha_7$.

Variation Method. There are 4 independent variations $\lambda_1\lambda_2\lambda_3\lambda_4$ which modify an initially consistent model, and the final values of the 7 quantities that have been measured will differ from the measured values, the differences or residuals being expressed as linear functions of the variations:

$$\begin{array}{c} d_1\lambda_1+e_1\lambda_2+f_1\lambda_3+g_1\lambda_4+\delta_1 \\ d_2\lambda_1+e_2\lambda_2+f_2\lambda_3+g_2\lambda_4+\delta_2. \\ \vdots \\ d_7\lambda_1+e_7\lambda_2+f_7\lambda_3+g_7\lambda_4+\delta_7. \end{array}$$

In any practical problem, the 28 coefficients $d_1 \ldots g_7$, the 7 numerical terms $\delta_1 \ldots \delta_7$, and the weights, will be known. The numerical values of the above expressions will be the adjustments applied to the 7 measurements, and the principle of least squares requires that the sum of the weighted squares:

$$\begin{aligned} & w_1(d_1\lambda_1+e_1\lambda_2+f_1\lambda_3+g_1\lambda_4+\delta_1)^2 \\ & +w_2(d_2\lambda_1+e_2\lambda_2+f_2\lambda_3+g_2\lambda_4+\delta_2)^2 \ldots \\ & +w_7(d_7\lambda_1+e_7\lambda_2+f_7\lambda_3+g_7\lambda_4+\delta_7)^2 \end{aligned}$$

shall be as small as possible.

This is a quadratic algebraic function of the 4 independent variables $\lambda_1\lambda_2\lambda_3\lambda_4$, and by the standard procedure for finding a minimum the 4 partial differential coefficients should be set to zero and solved.

The partial differential coefficient with respect to λ_1 is

$$\begin{aligned} & 2w_1d_1(d_1\lambda_1+e_1\lambda_2+f_1\lambda_3+g_1\lambda_4+\delta_1) \\ & +2w_2d_2(d_2\lambda_1+e_2\lambda_2+f_2\lambda_3+g_2\lambda_4+\delta_2)+\ldots \\ & +2w_7d_7(d_7\lambda_1+e_7\lambda_2+f_7\lambda_3+g_7\lambda_4+\delta_7). \end{aligned}$$

Since this is to be equated to zero, the factor 2 can be ignored, and after rearrangement we get:

$$\begin{aligned} (w_1d_1{}^2+w_2d_2{}^2+\ldots w_7d_7{}^2)\lambda_1 &+(w_1d_1e_1+w_2d_2e_2+\ldots w_7d_7e_7)\lambda_2 \\ &+(w_1d_1f_1+w_2d_2f_2+\ldots w_7d_7f_7)\lambda_3 \\ &+(w_1d_1g_1+w_2d_2g_2+\ldots w_7d_7g_7)\lambda_4 \\ &+(w_1d_1\delta_1+w_2d_2\delta_2+\ldots w_7d_7\delta_7) = 0. \end{aligned}$$

It is easily seen that this equation can be constructed by the following 'recipe': *multiply each residual by its weight and by the coefficient of* λ_1*, add the results, and equate to zero.* The next equation, derived from partial differentiation with respect to λ_2, is obtained similarly except that the coefficients of λ_2 are multipliers; and so on.

The 4 equations are solved, giving the values of $\lambda_1\lambda_2\lambda_3\lambda_4$ which are then applied to the initial model and immediately provide the adjusted survey.

If the actual adjustments are required, they can be calculated by substituting the values of the variations into the original set of expressions, or by direct comparison of the measured quantities with their final counterparts.

In the literature, the residuals are often found written in such a form as $d_1\lambda_1+e_1\lambda_2+f_1\lambda_3+g_1\lambda_4+\delta_1 = r_1$ etc., and these are called *observation equations:* this could be misleading, because these are not *equations* from which the values of the unknowns are calculable; they are definitions of the new symbols $r_1 \ldots$, the introduction of which is unnecessary, though it may be convenient.

5.20 Normal Equations. Each of the methods of adjustment involves the solution of a set of simultaneous linear equations, which always have the property that the values of the coefficients of the unknowns are skew-symmetrical about the diagonal—for instance, the coefficient of λ_3 in the first equation is the same as the coefficient of λ_1 in the third equation. A little thought about the construction of the equations will show why this happens.

Equations of this type are usually called 'normal equations'.

In most adjustment problems, it is the solution of the normal equations that takes most of the computing effort and time. The adjustment, in a single process, of a survey such as the triangulation of Great Britain would theoretically require solution of a set of hundreds or perhaps thousands of equations. This was quite out of the question before the days of the electronic computer, and large adjustment jobs had to be divided into pieces small enough to be tractable in a reasonable time by use of logarithm tables or hand-operated calculating machines.

Methods for solving sets of equations will not be described in detail in this book: these are really matters of organised arithmetic and not peculiar to surveying operations.

Briefly, there are two possible ways of attack. From the first equation, the first unknown is expressed in terms of the others, and this expression is substituted for the first unknown in the rest of the equations: the same is done with the first equation of the new set, substituting an expression for the second unknown into the rest; and so on, until the last step gives a value for the last unknown, and this is substituted back step by step to derive the numerical values of all the unknowns. This method is successive elimination. In operating the method, advantage can be taken of the skew symmetry of the coefficients by ensuring that this feature is retained at each stage, thus considerably reducing the number of quantities that have to be calculated.

If the solution is being done by hand, it is essential to arrange automatic arithmetical checks all along the line, and this is easily done, so that the computer is quite sure that he will not waste a lot of tedious work because of a mistake made at an early stage in the process.

The other method of solution is successive approximation, or iteration. In most adjustment problems that arise in surveying, very many of the coefficients in the normal equations are zero, and the coefficients on the diagonal are usually much larger than the others because they are sums of squares. One can therefore imagine that rough approximations to the solution would be obtained by ignoring all the coefficients not on the diagonal, thus getting each unknown immediately by a simple division. In practice, the process is to go through the equations in succession, each time getting a value for the unknown on the diagonal and using the values last computed for the other unknowns in that equation. When all the equations have been dealt with, the cycle is repeated, starting again at the first equation. As a rule, the process will be quite rapidly convergent, that is the successive changes in the values of the unknowns will become smaller and a point will be reached eventually at which there will evidently be no more changes of significant amounts. An advantage of the iteration method for the human computer is that if an arithmetical mistake is made the process will still eventually reach the correct answer, though it will take more time: the operation is self-healing!

The number of normal equations that will appear in a given problem can be calculated from general considerations. Suppose there are N observed quantities, and m is the minimum number that would be needed if all the observations were perfect. The number of conditions to be satisfied by the adjustments will be $(N-m)$, and this is the number of correlative equations. On the other hand, if the variation method is contemplated, the number of independent variations is m, and this will be the number of normal equations.

For example, in a traverse of 10 lines between fixed points and reference bearings, there will be 21 measured quantities to fix 9 new points. These are controlled by 3 conditions, but by the variation method there will be 18 equations to be solved, two for each internal traverse station. The human computer will know which method to choose! But if a powerful electronic computer is accessible, the difference of computing time may be only a few seconds, and the choice of method may be made on other grounds, such as ease of programming, which generally seems to favour the variation method.

5.21 Arithmetical Considerations. When dealing with large numbers of expressions or equations with numerical coefficients, it is advisable to try to arrange for all the coefficients to be of about the same order of magnitude. It is true of course that if a set of linear equations has a solution, that solution can be obtained to any desired degree of precision by including a sufficient number of digits in the arithmetic. Nevertheless, for a specified order of accuracy, the arithmetic operations will be the more economically done the more limited the range covering the numerical values to be operated on.

5.22 Condition Method Computed. Consider the three conditions written out at the end of Section 5.04. The adjustments θ are in seconds and the linear unit in which the work has been calculated is the metre: the coefficients of the angle adjustments are, in effect, in the metre unit. If the linear unit were the centimetre, these three coefficients would be 100 times greater and of the same order of size as the other coefficients. The coordinate misclosures are 57 and 12 centimetres. So we rewrite the conditions:

$$-1{\cdot}6502\theta_1 - 1{\cdot}2855\theta_2 - 0{\cdot}3775\theta_3 + 0{\cdot}8526l_1 + 0{\cdot}2109l_2 - 0{\cdot}8680l_3 = +57$$

$$+0{\cdot}1312\theta_1 - 0{\cdot}4639\theta_2 - 0{\cdot}6598\theta_3 + 0{\cdot}5225l_1 + 0{\cdot}9775l_2 + 0{\cdot}4966l_3 = -12$$

$$\theta_1 + \theta_2 + \theta_3 + \theta_4 = -21$$

and the linear unit is now the centimetre. This adjustment will be completed, to illustrate the theory given in Section 5.19.

Since this computation includes quantities of two kinds, angles and lengths, it is necessary to make the arithmetic dimensionless by adopting estimates of the accuracy of the field measurements, and taking weights inversely as their squares. Let us take the standard error of an angle as 10″ and the standard error of a length as 15 cm. (These indicate poor quality observations, but they serve to illustrate the method; in practice, the three lengths might be given unequal weights.) The function F to be minimised is:

$$F = \frac{\theta_1^2}{100}+\frac{\theta_2^2}{100}+\frac{\theta_3^2}{100}+\frac{\theta_4^2}{100}+\frac{l_1^2}{225}+\frac{l_2^2}{225}+\frac{l_3^2}{225}$$
$$-2K_1(-1{\cdot}6502\theta_1-1{\cdot}2855\theta_2-0{\cdot}3775\theta_3 +0{\cdot}8526l_1+0{\cdot}2109l_2-0{\cdot}8680l_3-57)$$
$$-2K_2(+0{\cdot}13120\theta_1-0{\cdot}4639\theta_2-0{\cdot}6598\theta_3 +0{\cdot}5225l_1+0{\cdot}9775l_2+0{\cdot}4966l_3+12)$$
$$-2K_3(\theta_1+\theta_2+\theta_3+\theta_4+21).$$

By differentiating with respect to the θs and the ls, omitting factors of 2, and setting to zero, we get:

$$\begin{aligned}
\theta_1 &= 100(-1{\cdot}6502K_1+0{\cdot}1312K_2+K_3)\\
\theta_2 &= 100(-1{\cdot}2855K_1-0{\cdot}4639K_2+K_3)\\
\theta_3 &= 100(-0{\cdot}3775K_1-0{\cdot}6598K_2+K_3)\\
\theta_4 &= 100(K_3)\\
l_1 &= 225(+0{\cdot}8526K_1+0{\cdot}5225K_2)\\
l_2 &= 225(+0{\cdot}2109K_1+0{\cdot}9775K_2)\\
l_3 &= 225(-0{\cdot}8680K_1+0{\cdot}4966K_2)
\end{aligned}$$

and the differentials with respect to the Ks simply reproduce the conditions. Now these expressions are substituted into the 3 conditions to give the normal equations:

K_1	K_2	K_3	
+794·90	+112·51	−331·32	= +57
+112·51	+398·69	− 99·25	= −12
−331·32	− 99·25	+400·00	= −21

which have the solution;

$$K_1 = +0{\cdot}079\,14 \qquad K_2 = -0{\cdot}052\,42 \qquad K_3 = +0{\cdot}000\,03.$$

Substituting these into the 7 expressions for the adjustments gives the values, to the nearest second or centimetre:

$$\begin{aligned}
\theta_1 &= -14'' & \theta_3 &= +1'' & l_1 &= +9\text{ cm}\\
\theta_2 &= -8'' & \theta_4 &= 0 & l_2 &= -8\text{ cm}\\
& & & & l_3 &= -21\text{ cm}
\end{aligned}$$

The traverse computed with the adjusted angles and lengths closes satisfactorily, as seen on the computation form, Fig. 66.

Angles	*Bearings*	*Lengths* *m*	*N* *m*	*E* *m*	*Coordinates* *m*	*Point*
	AP 338° 10′ 07″					
A 53° 19′ 37″					135 318·26 104 427·05	A
	AJ 33 29 44	1439·67	+1227·58	+ 752·13		
J 226 19 18					136 545·84 105 179·18	J
	JK 77 49 02	1915·93	+ 404·32	+1872·78		
K 252 24 16					136 950·16 107 051·96	K
	KB 150 13 18	1567·70	−1360·69	+ 778·59		
B 134 40 24					135 589·47 107 830·55	B′
	BQ 104 53 42				135 589·47 107 830·56	B

FIG. 66. Adjusted Traverse.

5.23 Variation Method Computed. In Section 5.05 the 7 residuals are expressed as linear functions of the 4 variations. They are written below in tabular form, and again the coefficients have values appropriate to the centimetre as the unit of length for the working.

Angle	n_j	n_k	e_j	e_k		*Weight*
A	−0·7486		+1·2217			2·25
J	+1·8009	−1·0523	−1·4488	+0·2271		2·25
K	−1·0523	+1·7059	+0·2271	+0·9151	−24	2·25
B		−0·6536		−1·1422	+ 3	2·25
Line						
AJ	+0·8526		+0·5225			1·00
JK	−0·2109	+0·2109	−0·9775	0·9775		1·00
KB		+0·8679		−0·4967	−55	1·00

The weights adopted for the condition method were 1/100 and 1/225: as mentioned in Section 5.17, any numbers proportional to these may be used, so they have been multiplied by 225 for the purpose of this calculation.

The procedure for forming the normal equations has been explained in Section 5.19; the equations are:

$$\begin{array}{rrrrrl} n_j & n_k & e_j & e_k & & \\ +11{\cdot}8211 & -\ 8{\cdot}3472 & -7{\cdot}8144 & -1{\cdot}4528 & +\ 56{\cdot}284 & = 0 \\ -\ 8{\cdot}3472 & +10{\cdot}7972 & +4{\cdot}0958 & +4{\cdot}4296 & -144{\cdot}270 & = 0 \\ -\ 7{\cdot}8144 & +\ 4{\cdot}0958 & +9{\cdot}4256 & -1{\cdot}2282 & -\ 12{\cdot}263 & = 0 \\ -\ 1{\cdot}4528 & +\ 4{\cdot}4296 & -1{\cdot}2282 & +6{\cdot}1387 & -\ 29{\cdot}812 & = 0 \end{array}$$

The first coefficient in the above array is in fact

$$2{\cdot}25(-0{\cdot}7486)^2+2{\cdot}25(+1{\cdot}8009)^2+2{\cdot}25(-1{\cdot}0523)^2 +1{\cdot}00(+0{\cdot}8526)^2+1{\cdot}00(-0{\cdot}2109)^2,$$

and the second is

$$2{\cdot}25(+1{\cdot}8009)(-1{\cdot}0523)+2{\cdot}25(-1{\cdot}0523)(+1{\cdot}7059)$$
$$+1{\cdot}00(-0{\cdot}2109)(+0{\cdot}2109).$$

The solution of the normal equations, rounded to the nearest centimetre, is

$$n_j = +13 \qquad n_k = +31 \qquad e_j = -3 \qquad e_k = -15.$$

These variations, applied to the original model traverse, give the coordinates

J	136 545·84 m	105 179·19 m
K	136 960·16 m	107 051·97 m

in agreement with the results of the adjustment by condition method.

5.24 Review of Adjustment Methods. It may be helpful to review briefly the types of adjustment problem that may arise in surveying, and to distinguish the methods that can be used.

The examples already worked out show the distinction between the condition method and the variation method. The latter method is now much more commonly used: if the adjustment is guided by the principle of least squares, the two methods will produce the same answer. In earlier times, the amount of arithmetical work involved in an adjustment had to be very carefully considered. The condition method got off to a good start: it was used for almost all of the great national triangulation and levelling systems, and anyway for that sort of problem in those times it was in fact the better method to employ. The variation method is more suitable for many of the problems that arise nowadays, especially in fitting new surveys on to an existing system. However, the modern surveyor or computer may not have to choose between the method requiring solution of 10 equations and that leading to 14 equations; if he has access to an electronic computer the facility of programming will often be decisive, and probably on the side of variation. Anyway, he can calmly undertake a task of such magnitude that, by hand and paper, he would not have a hope of tackling it by either method.

A distinction can be made between 'free' and 'constrained' adjustment. An adjustment that is required to satisfy only internal geometrical conditions in a survey system is free. The adjustment of the initial surveys in any area, however large, is necessarily of this kind. A triangulation scheme can be adjusted to internal consistency without reference to its orientation or position on the Earth's surface: a network of levelling is usually adjusted so that there is exact closure round any complete circuit, before an actual 'height' is assigned to any benchmark. For an isolated survey, such as might be done for an engineering work, it may be quite sufficient to assign an arbitrary value to one of the benchmarks and calculate the others from this; similarly, the triangulation system might be placed on a quite arbitrary local coordinate grid to be used for that particular job only.

A great survey work recently completed is the primary levelling net of Australia. This was adjusted both as a free system of closed circuits, and as a constrained system with the additional requirements to fit exactly to the heights above sea level independently determined at about 30 tide gauge stations round the coast.

Nowadays, the surveyor will often be working in a 'constrained' situation. He may have to connect some new triangulation to the existing national system or some levels to the national benchmark datum. In such cases, the new survey must certainly be adjusted to internal consistency, but there will be further requirements of fitting to the previously fixed values. It will usually be found that in constrained adjustment problems the variation method is the more economical in arithmetic.

The two worked examples of traverse adjustment were done in a plane rectangular coordinate system. An isolated survey of small extent, up to a few hundred square kilometres, can be calculated directly in a plane system; that is, so far as the relative *horizontal* positions of the points are concerned, the adjustment requirements for internal consistency are taken to be those appropriate to plane geometry and trigonometry. For larger surveys, the curvature of the Earth must be taken into account. The survey points, marks on the Earth's physical surface, are represented by points directly below them on a reference surface of defined shape which is close to sea level and which in practice is a spheroid of revolution, a surface generated by rotating an ellipse about its minor axis. The geometry of this kind of surface is somewhat complex, but the surface actually used is so near to being spherical, and ordinary geodetic triangles are so tiny compared with the size of the Earth, that the adjustment conditions can be taken to be those applicable to a figure formed of spherical triangles. This means in practice that the only difference from the plane situation, so far as conditions on the angles are concerned, is that the required total of the angles of each triangle must include the spherical excess. (Section 6.02.)

An example of a constrained adjustment is illustrated in Fig. 67 showing a triangulation scheme for fixing the new points P and Q. The points A, B, and C belong to the existing survey, which means that the angle ABC is fixed and the lengths of AB and BC are known.

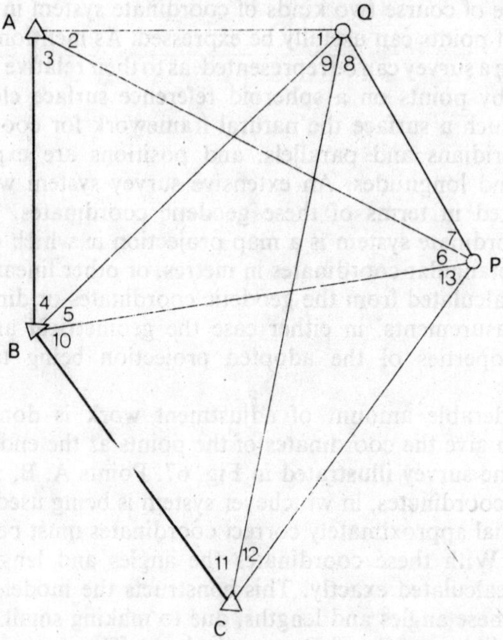

FIG. 67. TRIANGULATION EXTENSION.

The adjustment of the new work must therefore include not only the internal requirements of the geometry of the figure ABCPQ but in addition two conditions: one is that the angles 4, 5, and 10 must total to the sum already fixed, and the other is that if the known length of AB is taken, and through two adjusted triangles such as ABP and PBC the length of BC is calculated, the result must agree exactly with its already fixed value.

5.25 Adjustments in Coordinate Systems. It is to be noted that the adjustment described above is a process of making the geometry of the figure consistent: there is no reference to the positions of the points in a coordinate system. But of course it is usual practice to set up such a system as a framework for mapping and any other work involving land measurement. Therefore, after the work shown in Fig. 67 has been adjusted there will follow computations of positions of points P and Q in the same system as those of A, B, and C. We may imagine the lines CP, PQ, QA being treated as a traverse; in this case, however, the calculation would show no misclosure, or if it did there must be an error in the arithmetic. This is an illustration of the value of adjusting survey work.

There are of course two kinds of coordinate system in which the positions of points can usefully be expressed. As mentioned earlier, the points in a survey can be represented, as to their relative horizontal positions, by points on a spheroid reference surface close to sea level. On such a surface the natural framework for coordinates is one of meridians and parallels, and positions are expressed as latitudes and longitudes. An extensive survey system will usually be calculated in terms of these geodetic coordinates. The other kind of coordinate system is a map projection in which each point has two rectangular coordinates in metres, or other linear measure: these are calculated from the geodetic coordinates or directly from survey measurements, in either case the geometrical and mathematical properties of the adopted projection being taken into account.

A considerable amount of adjustment work is done in such a way as to give the coordinates of the points as the end products. Consider the survey illustrated in Fig. 67. Points A, B, and C will have fixed coordinates, in whichever system is being used: by some means, initial approximately correct coordinates must be found for P and Q. With these coordinates the angles and lengths of the figure are calculated exactly. This constructs the model. Then the effects on these angles and lengths, due to making small variations in the coordinates of P and Q, are calculated. The residuals between these angles and lengths, and the observed angles and lengths, are thus obtained as linear functions of the four variations. The adjustment then proceeds as already explained, Section 5.23, and the calculated values of the 4 variations lead to the adjusted positions of P and Q.

The immediately foregoing description applies to adjustment done in either system of coordinates: some distinctions must now be mentioned. If the geodetic system is in use, the calculation of the angles and lengths of the initial model must be based on the geometry of spheroid surfaces: this is explained in detail in the next chapter. If the adjustment is being done in a plane system, the angles and lengths calculated by plane trigonometry from the coordinates cannot be directly compared with the observed quantities. In no projection does the geometry of a plane triangle in the coordinate system agree exactly with that of the triangle formed by the points on the spheroid. The differences are properties of the projection and they are always very small in any projection used for a topographical mapping system. They can be calculated quite exactly when the approximate positions of the points are known, and they can be applied to the observed angles and lengths, thus giving what may be described as observed projection lengths and angles. These can then be compared with the model values, including the effects of the variations of coordinates, to give the true residuals. Projection properties are discussed in detail in Chapter IX.

In the remainder of this Chapter, a selection of types of problem will be used to illustrate practical methods of applying adjustment principles; and some numerical examples will be worked out. To save space, the purely arithmetical working will not, as a rule, be set out in full detail. Two problems of free adjustment and the use of the condition method will first be considered.

5.26 Isolated Triangulation Scheme. Fig. 68 represents a simple piece of triangulation in which there are 23 angles measured and the only other measure is a base line which provides the scale for the survey. This is the traditional triangulation method. For the present we assume the figure to be plane, which means in practice that the area it covers is so small that spherical excess can be ignored. This sort of work might be done for an engineering scheme, and the lines might typically be a few kilometres long. The necessary modifications of procedure if the scheme is so large that it should

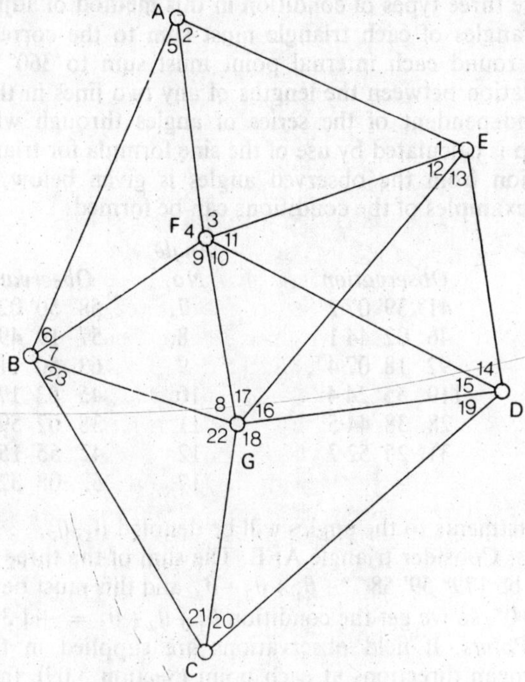

FIG. 68. TRIANGULATION SCHEME.

be treated as being on a spherical surface are really quite trivial so far as the adjustment is concerned, and will be explained later. There will be 23 adjustments to be applied to the angles so that the adjusted angles satisfy all the internal conditions for consistency.

The total number of conditions follows from the general rule, mentioned in Section 5.20, that there must be one condition for each point added two more angles must be observed. The reader necessary to determine the survey if all the observations were perfect. If one of the lines, say AF, is taken as a base on which to build up, point by point, the rest of the figure, it is obvious that for each point added two more angles must be observed. The reader will see that the whole figure could be built up, assuming errorless observations, by measuring just the 10 angles numbered 2 3 4 5 8 9 21 22 18 and 20. In each triangle, one of the angles would be calculated from the other two. The formula is obviously $2(p-2)$, where p is the total number of points in the scheme. Since 23 angles have been measured, there are 13 conditions to be satisfied by the adjusted angles.

There are three types of condition in this method of adjustment. The three angles of each triangle must sum to the correct total: the angles round each internal point must sum to 360° exactly: and the relation between the lengths of any two lines in the figure must be independent of the series of angles through which the relationship is calculated by use of the sine formula for triangles.

A selection from the observed angles is given below, so that numerical examples of the conditions can be formed:

Angle No.	*Observation*	*Angle No.*	*Observation*
1.	41° 39′ 07·2″	7.	58° 50′ 02·7″
2.	46 02 44·1	8.	57 19 49·8
3.	92 18 07·4	9.	63 50 11·0
4.	119 55 24·4	10.	45 53 17·5
5.	28 38 44·5	11.	38 02 59·7
6.	31 25 52·7	12	42 55 15·7
		17.	52 08 32·0

The adjustments to the angles will be denoted $\theta_1, \theta_2, \ldots, \theta_{23}$.

Triangles. Consider triangle AFE. The sum of the three *adjusted* angles will be $179° \ 59' \ 58{\cdot}7'' + \theta_1 + \theta_2 + \theta_3$, and this must be equal to $180° \ 00' \ 00{\cdot}0''$, so we get the condition $\theta_1 + \theta_2 + \theta_3 = +1{\cdot}3''$.

Centre Points. If field observations are supplied in the form of lists of mean directions at each point (Section 3.09), the angles calculated from them will necessarily sum to exactly 360°. Or perhaps a 'station adjustment' will have been done on the observed angles to make them satisfy this condition. Thus, it may well happen that the computer starts with a set of angles that total correctly round each internal point. However, this is not necessarily so, and in any case the condition must be taken care of in the adjustment. In the present example, the observed angles round point F do satisfy the condition, and the equation reads:

$$\theta_3+\theta_4+\theta_9+\theta_{10}+\theta_{11} = 0{\cdot}0''.$$

Side Equations. The necessity for the third type of condition is illustrated in Fig. 69. The 12 correct angles of this figure will satisfy

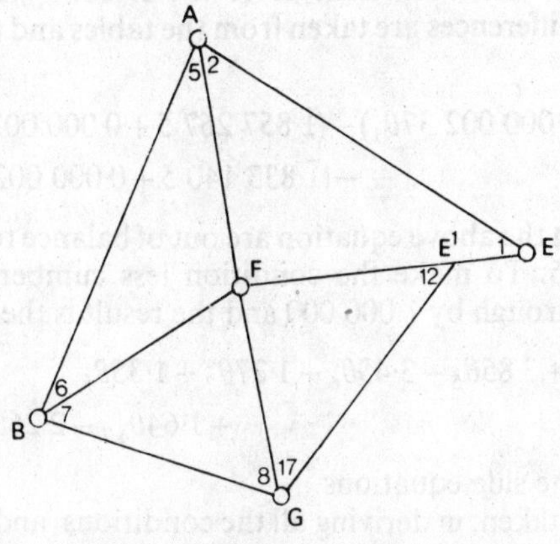

FIG. 69. SIDE CONDITION.

the closure conditions for the 4 triangles and the centre condition at point F, so these conditions are evidently not enough to ensure that E′ coincides with E. We introduce the notation:

observed angles	$1_0 \;\; 2_0 \;\; 3_0 \; \ldots\ldots \; 23_0.$
adjusted angles	$[1] \; [2] \; [3] \; \ldots\ldots \; [23].$

The Sine Formula applied to the four triangles gives

$$\frac{FA}{FE}=\frac{\sin[1]}{\sin[2]};\quad \frac{FB}{FA}=\frac{\sin[5]}{\sin[6]};\quad \frac{FG}{FB}=\frac{\sin[7]}{\sin[8]};\quad \frac{FE'}{FE}=\frac{\sin[17]}{\sin[12]}.$$

By multiplying all these together we get:

$$\frac{FE'}{FE}=\frac{\sin[1]\,.\,\sin[5]\,.\,\sin[7]\,.\,\sin[17]}{\sin[2]\,.\,\sin[6]\,.\,\sin[8]\,.\,\sin[12]}.$$

The condition to make FE′ = FE is therefore that the two products of sines shall be equal. In logarithmic form this is:

$$\log\sin[1]+\log\sin[5]+\log\sin[7]+\log\sin[17]$$
$$=\log\sin[2]+\log\sin[6]+\log\sin[8]+\log\sin[12].$$

That is, the logsines of the peripheral angles taken in order round the polygon with alternate signs must sum to zero.

Now $[1] = 1_0+\theta_1$, etc. For instance, $\log\sin[1] = \log\sin(41°\ 39'\ 07{\cdot}2''+\theta_1) = \bar{1}{\cdot}822\ 563\ 4+0{\cdot}000\ 002\ 37\theta_1$, the coefficient of θ_1 being the difference of log sin for 1″ at this point in the tables. Mathematically, the coefficient is $(0{\cdot}434\ 29\cot 1_0)/206\ 265$. The 8 logsines and differences are taken from the tables and the condition is found to be

$$(\bar{1}{\cdot}822\ 563\ 4+0{\cdot}000\ 002\ 37\theta_1)-(\bar{1}{\cdot}857\ 267\ 5+0{\cdot}000\ 002\ 03\theta_2)+$$
$$\ldots-(\bar{1}{\cdot}833\ 140\ 5+0{\cdot}000\ 002\ 26\theta_{12}) = 0.$$

The 8 logsines in the above equation are out of balance to the amount of +0·000 083 5. To make the condition less cumbersome, it can be multiplied through by 1 000 000 and the result is the neater form:

$$2{\cdot}37\theta_1-2{\cdot}03\theta_2+3{\cdot}85\theta_5-3{\cdot}45\theta_6+1{\cdot}27\theta_7-1{\cdot}35\theta_8$$
$$+1{\cdot}64\theta_{17}-2{\cdot}26\theta_{12} = -83{\cdot}5$$

This is one of the side equations.

Care must be taken, in deriving all the conditions, and particularly the side equations, to write the correct sign for the numerical term, the misclosure. Another point to remember is that when an angle is over 90° the difference of logsine is negative: for example, $\log\sin(98°\ 54'\ 43{\cdot}2''+\theta'') = \bar{1}{\cdot}994\ 725\ 0-0{\cdot}000\ 000\ 33\theta$.

The total of 13 conditions is made up by:

8 triangle equations: i.e. the 5 triangles that contain only three numbered angles, plus any 3 of the four triangles that make up the quadrilateral DEFG (if the adjustment gives correct totals for three of the triangles of the quadrilateral, the fourth triangle is automatically correct),

2 centre equations: at the points F and G,

3 side equations: as the figure is built up from a base, a side equation becomes necessary each time a new polygon is generated. The number of side equations can also be derived from a mechanical consideration: if the figure is thought of as made of rods (14) joined at the nodes (7), then the number of side equations is equal to the number of rods that could be removed without allowing the structure to collapse anywhere. In this case, the rods DE EG and DF, for example, could be removed, but no more.

The adjustment then proceeds on the same lines as the working in Section 5.22. There will be 13 correlatives to be found by solution of 13 normal equations. Weighting will probably be unnecessary, as in a survey of this kind the observations would usually be of uniform quality.

Spherical Figure. When the triangulation is so extensive that spherical excess must be taken into account, preliminary computations will be needed to find the areas of the triangles. This can be done by using any approximately known length in the figure, and computations with four-decimal-place functions using angles to the nearest minute will be of ample accuracy for this purpose. Enough lengths must be computed so that the area of each triangle can be found, and the spherical excess can be calculated at 1 second for each 200 square kilometres, for all practical purposes. Otherwise, the only difference from the plane situation is that the conditions must be expressed to ensure that the total of the adjusted angles of each triangle includes the spherical excess: for instance, the spherical excess of triangle AFE being 1·2″, the condition will read $\theta_1+\theta_2+\theta_3 = +2{\cdot}5''$. The derivation of the side equations in spherical trigonometry leads to exactly the same relationships as in a plane polygon, and of course the centre conditions will be the same on any kind of surface.

Missing Angles. In the best of triangulation schemes, it sometimes happens that one or two angles are not measured, perhaps due to poor visibility, or some mishap in arrangements for signal lights. As an example, suppose angle no. 1 is missing in the scheme just considered. There are now only 22 observed quantities, so there will be 12 conditions. Of course the missing one is the one for triangle AFE but angle 1 also appears in a side equation and that cannot be left out. The general principle is that the adjusted value of a missing angle must be expressed in terms of adjusted values of other angles. In this case, taking the triangle as spherical, the adjusted value of angle 1 is $180°\ 00'\ 01{\cdot}2''-(46°\ 02'\ 44{\cdot}1''+\theta_2)-(92°\ 18'\ 07{\cdot}4''+\theta_3) = 41°\ 39'\ 09{\cdot}7''-(\theta_2+\theta_3)$; its logsines are $\bar{1}{\cdot}822\,569\,3-0{\cdot}000\,002\,37(\theta_2+\theta_3)$ and in the side equation it will give rise to a term $-2{\cdot}37(\theta_2+\theta_3)$ instead of the term in θ_1.

5.27 Large Triangulation Systems. The adjustment of very extensive systems of survey framework is not discussed in detail in this book. The great national triangulations established in the nineteenth and early twentieth centuries were constructed almost entirely from angle observations. Laboriously measured base lines were few in number and widely separated—hundreds of kilometres. In some more recent systems there were also a few controls on bearing by Laplace Points (see Section 8.03). Limitations on computing power meant that large systems had to be divided into pieces which were successively adjusted to fit together as well as for internal consistency. Alternatively, as in India and the United States, the primary triangulation framework was built from chains of triangles and simple polygons, easily adjusted individually and then fitted together by a further adjustment in much the same way as a network of traverses or levelling lines.

In a chain of simple figures or any piece of triangulation that is adjusted in one operation it is quite easy to take account of extra base measurements. Suppose, for instance, that the lengths of the lines GB and ED have both been measured; a condition can be included to ensure that the length of either line, calculated from the other through the adjusted angles, will agree exactly with the measured length. The condition will look like a side equation and will be derived from the relationship:

$$\log \text{GB} - \log \sin [9] + \log \sin [7] - \log \sin [15] + \log \sin [10] - \log \sin [13] + \log \sin [16] = \log \text{ED}$$

Note that this is based on the assumption that measured lengths are not subject to adjustment. Maybe there was some justification for this in the days of invar tape base lines measured to 1/1 000 000 accuracy, but in practice the base lines generally had to be connected up to the main triangulation by a piece of local triangulation, the base extension figure, in which some accuracy was almost certainly lost on the way.

If measured lengths are to receive adjustments as well as angles, the relationship given above would be written as:

$$\log (\text{GB}_0 + a) - \ldots = \log (\text{ED}_0 + b),$$

where the measured lengths receive the small adjustments a and b. To form the side equation, the differential expansions $\log_{10} (\text{GB}_0 + a) = \log_{10} \text{GB}_0 + (0{\cdot}434\,29a)/\text{GB}_0$, etc. are used. Including the length adjustments will involve a weight system based on estimates of the accuracies of length and angle measurement.

When the adjustment of a large triangulation was done in pieces, the discrepancies between measured lengths (and azimuths, if any) were not discovered until the system was extended far enough to include them, so in theory a second adjustment would have to be made. What was sometimes done in practice was to take the angle adjustment as giving the shape of the geometry, and to adopt an overall scale that made the best general fit to all the measured lengths.

The electronic computer has banished some of the problems arising in the adjustment of large systems.

5.28 New Survey Schemes. In setting up a new survey framework for an area, the modern surveyor will probably choose either a triangulation structure with a number of lengths measured by EDM, or an initial framework of traverses, a type of surveying in which of course EDM comes into its own. It does not appear to be economical to set up a triangulation, or rather trilateration, system by measuring only lengths, as is theoretically possible. Although the measurement of an individual line by EDM is very quick, the measurement of 8 or 10 lines radiating from one station will raise obvious difficulties when there has to be a set of reflectors or an operator with another instrument at the other end of each line. Besides, EDM equipment is much more expensive than theodolites. The inclusion of a scattering of length measurements in a triangulation is desirable as a control on scale. Measurements made by EDM on lines in old-established triangulation systems have shown that systematic variations of scale of quite serious proportions have existed in some areas far away from measured bases.

With the addition of measured lengths to a triangulation system, the advantage of the variation method of adjustment becomes clear, since the number of variations depends only on the number of points to be fixed, not on the number of observations provided for fixing them.

5.29 Adjustment of Directions. For the measurement of horizontal angles, the readings taken on a theodolite are directional settings referred to the divided circle, and angles are derived as differences of settings. It is often argued that these settings ought to be adjusted, not the angles deduced from them: the results of the field work should be presented as lists of mean observed directions, as they often are (Section 3.09), and the adjustments should be applied to these.

This is quite easily done. Fig. 70 shows a portion of the triangulation scheme, with directions numbered. There are 28 directions, and we denote the adjustments to be applied to them as $v_1\ v_2\ v_3 \ldots v_{28}$. Since the geometrical conditions relate to angles, the misclosures have to be found by using angles as before, but the adjustment to each angle is the difference between two adjustments to directions. Obviously, the condition for closure of triangle AEF will now be written: $v_{12}-v_{11}+v_4-v_8+v_2-v_1 = +1{\cdot}3''$.

Centre conditions are not required: the angles round a point start off automatically with the correct total, and any alterations to any of the directions at the point cannot affect this state. The side equation derived in Section 5.26 will now read:

$$2{\cdot}37(v_{12}-v_{11})-2{\cdot}03(v_2-v_1)+\ldots-2{\cdot}26(v_{11}-v_{10}) = -83{\cdot}5$$

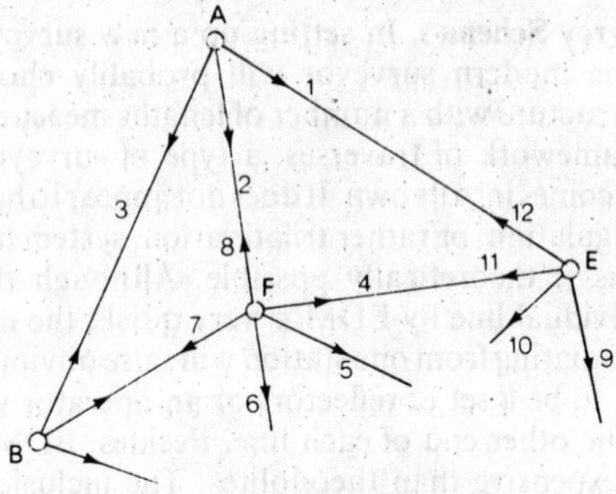

FIG. 70. DIRECTIONS.

and will have to be re-written because some of the adjustments appear twice and will have to be collected into one term, e.g. $-4{\cdot}63v_{11}$.

There will be 11 conditions. This agrees with the general rule, when we realise that directions are not absolute measurements; one direction at each point can be given an arbitrary value and only the others are then genuine measures. Thus the number of measured quantities in this case is $28-7 = 21$. If the figure is built up from a given base as explained before, two directions are needed for each additional point, so 10 directions are necessary and the other 11 introduce conditions.

Adjustment of directions reduces the number of conditions, but the total amount of work is probably much the same, because each condition now contains more of the adjustments.

5.30 Adjustment of Levelling. The Condition Method can conveniently be used for adjusting a network of levelling circuits such as is shown in Fig. 71. The 'observations' to be adjusted are the measured differences of level between junction points. Indeed, the adjustment of closures of circuits is a problem that arises in other kinds of work, such as gravity differences measured by pendulums or gravimeters, differences of height by vertical angles, and the differences of coordinates round circuits of traversing. As a rule, the original quantities to be adjusted in circuits are such as to require assignment of weights.

The observed differences of level along the 11 lines making up the 5 circuits shown on the figure are given in column (3) of the table below: it is assumed that these have received all necessary corrections for staff calibration, gravity effects (Section 7.07), etc., so that the circuit misclosures can be assumed to arise from effects of random observational errors. The arrows drawn on the lines in the Figure

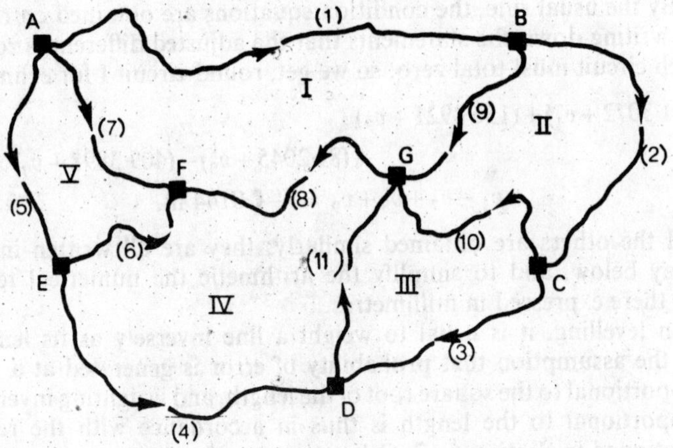

FIG. 71. LEVELLING NETWORK.

are in the direction of increasing height and the recorded observations are all positive; this is a useful rule to be adopted as it helps in getting signs right.

There are 7 junction points, but since levelling is a differential process, only 6 lines would be needed to determine the relative heights of the junction points, and the extra 5 observed differences will lead to 5 conditions: these are easily seen to be the requirement of closing each of the 5 circuits.

(1) Line	(2) Length km	(3) Observed Difference m	(4) Adjustment m	(5) Adjusted Difference m
AB	68	+171·3072	v_1 = −0·0043	171·3029
BC	41	+ 7·5043	v_2 = +0·0090	7·5133
CD	35	+ 38·6320	v_3 = +0·0056	38·6376
ED	47	+117·3439	v_4 = −0·0074	117·3365
AE	20	+100·1125	v_5 = +0·0048	100·1173
EF	22	+309·2602	v_6 = +0·0087	309·2689
AF	18	+409·3894	v_7 = −0·0032	409·3862
GF	31	+ 81·2045	v_8 = −0·0068	81·1977
BG	23	+156·8921	v_9 = −0·0065	156·8856
CG	29	+149·3707	v_{10} = +0·0016	149·3723
DG	26	+110·7346	v_{11} = +0·0001	110·7347

By the usual rule, the condition equations are obtained correctly by writing down the statements that the adjusted differences round each circuit must total zero: so we get, round circuit I for example:

$$(171{\cdot}3072+v_1)+(156{\cdot}8921+v_9)$$
$$+(81{\cdot}2045+v_8)-(409{\cdot}3898+v_7) = 0,$$

or;
$$v_1-v_7+v_8+v_9 = -0{\cdot}0144 \text{ m},$$

and the others are obtained similarly; they are all written in the array below, and to simplify the arithmetic the numerical terms are there expressed in millimetres.

In levelling, it is usual to weight a line inversely as its length, on the assumption that probability of error is generated at a rate proportional to the square root of the length, and weighting inversely proportional to the length is thus in accordance with the result mentioned in Section 5.17. Other systems of weights are, however, possible.

It is arithmetically convenient to take the lengths of the lines in tens of kilometres, as the reciprocal weights. The full set of 5 conditions is given below; in such arrays, a lot of space is saved, and the work looks much neater, if only the coefficients are written out.

v_1	v_2	v_3	v_4	v_5	v_6	v_7	v_8	v_9	v_{10}	v_{11}	
reciprocal weights											
6·8	4·1	3·5	4·7	2·0	2·2	1·8	3·1	2·3	2·9	2·6	
1	.	.	.	.	.	−1	+1	+1	.	.	+14·4 = 0
.	1	.	.	.	.	.	.	−1	+1	.	−17·1 = 0
.	.	1	.	.	.	.	.	.	−1	+1	− 4·1 = 0
.	.	.	1	.	−1	.	+1	.	.	+1	+22·8 = 0
.	.	.	.	1	+1	−1	.	.	.	.	−16·7 = 0

This adjustment will be carried out as an example of the method described in Section 5.22. The function F in this case is:

$$\frac{v_1^2}{6{\cdot}8}+\frac{v_2^2}{4{\cdot}1}+\ldots+\frac{v_{11}^2}{2{\cdot}6}-2K_1(v_1-v_7+v_8+v_9+14{\cdot}4)$$
$$-2K_2(v_2-v_9+v_{10}-17{\cdot}1)$$
$$-2K_3(v_3-v_{10}+v_{11}-4{\cdot}1)$$
$$-2K_4(v_4-v_6+v_8+v_{11}+22{\cdot}8)$$
$$-2K_5(v_5+v_6-v_7-16{\cdot}7).$$

Setting to zero the differential coefficient with respect to v_1 gives

$v_1/(6{\cdot}8) = K_1$ etc.; the first 11 relations are:

$$\begin{aligned} v_1 &= 6{\cdot}8K_1 & v_6 &= -2{\cdot}2K_4+2{\cdot}2K_5 & v_9 &= +2{\cdot}3K_1-2{\cdot}3K_2 \\ v_2 &= 4{\cdot}1K_2 & v_7 &= -1{\cdot}8K_1-1{\cdot}8K_5 & v_{10} &= +2{\cdot}9K_2-2{\cdot}9K_3 \\ v_3 &= 3{\cdot}5K_3 & v_8 &= +3{\cdot}1K_1+3{\cdot}1K_4 & v_{11} &= +2{\cdot}6K_3+2{\cdot}6K_4 \\ v_4 &= 4{\cdot}7K_4 \\ v_5 &= 2{\cdot}0K_5 \end{aligned}$$

The reader will easily see that these equations can be written down directly from the array of conditions: the coefficients in each v are its reciprocal weight multiplied by the numbers in the column below: it is quite unnecessary to write out the function F.

The other 5 differential coefficients simply reproduce the conditions: substituting the above expressions for the v's into these gives the normal equations:

K_1	K_2	K_3	K_4	K_5	
$+14{\cdot}0$	$-2{\cdot}3$	.	$+3{\cdot}1$	$+1{\cdot}8$	$+14{\cdot}4 = 0$
$-2{\cdot}3$	$+9{\cdot}3$	$-2{\cdot}9$	.	.	$-17{\cdot}1 = 0$
.	$-2{\cdot}9$	$+9{\cdot}0$	$+2{\cdot}6$	.	$-4{\cdot}1 = 0$
$+3{\cdot}1$	.	$+2{\cdot}6$	$+12{\cdot}6$	$-2{\cdot}2$	$+22{\cdot}8 = 0$
$+1{\cdot}8$	.	.	$-2{\cdot}2$	$+6{\cdot}0$	$-16{\cdot}7 = 0$

The solution of these is:

$$K_1 = -0{\cdot}630 \qquad K_2 = +2{\cdot}185 \qquad K_3 = +1{\cdot}613$$
$$K_4 = -1{\cdot}568 \qquad K_5 = +2{\cdot}397.$$

and the resulting values of the adjustments v are listed in column (4) of the table of observations: it is seen that the adjusted differences of level exactly close each circuit.

5.31 Fixing Additional Points. The foregoing descriptions of adjustments of triangulation and levelling are examples of 'free' adjustments done by the condition method. It is in the situation where the surveyor's job is to fix new points in an existing framework that the advantages of the variation method show up so markedly. There may also be points included in an observation scheme but not in the initially computed figures. For instance, if the points A B C D E of Fig. 68 were adjusted in the primary system, there would be 23 angles available for fixing the two points F and G: only 4 are geometrically needed, so there would be a set of 19 simultaneous equations to be solved in the condition method, as against 4 for calculating the variations to initial coordinates of F and G!

5.32 Variation of Coordinates: Triangulation. Adjustment procedures are best understood by use of worked numerical examples, and the variation method will now be applied to the adjustment of the triangulation shown in Fig. 72. Points A, B, C, D are already

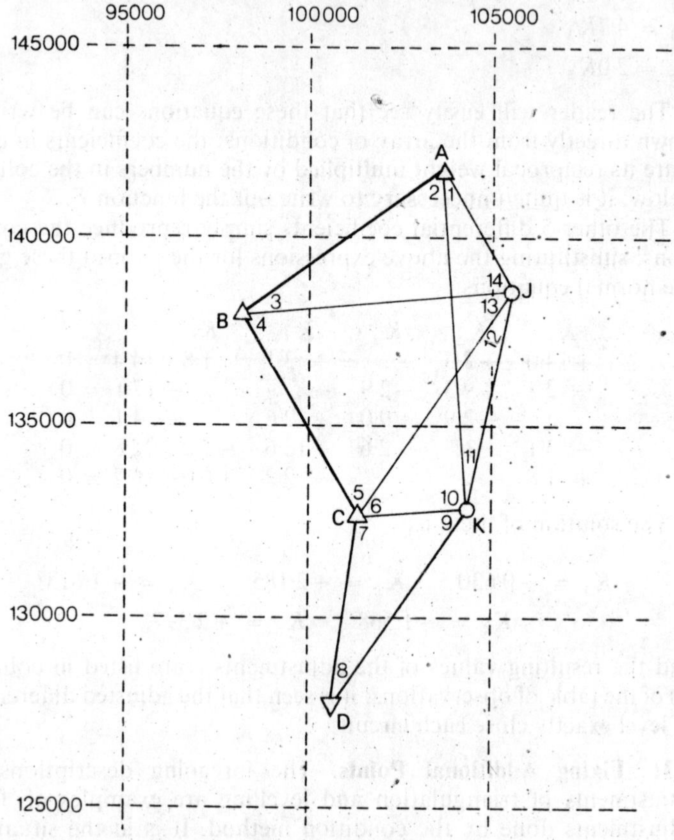

FIG. 72. ADDITIONAL POINTS FIXED.

fixed, and coordinates of the new points J and K are to be determined by use of the 14 observed angles. Only 4 new angles are needed to fix two new points, so the condition method would involve the solution of 10 simultaneous equations for the correlatives: there will be only 4 equations for the variations of coordinates of the two new points.

Basic data relating to the 4 fixed points are tabulated below: these values are of course self-consistent and have presumably resulted from adjustment and computation of previous work.

Point	*Coordinates* N (metres)	*Coordinates* E (metres)	*Line*	*Grid Bearing*	*Length* metres
A	141 826·17	103 630·92			
			AB	232° 40′ 25·6″	6714·11
B	137 755·05	98 291·88			
			BC	151 03 22·3	6220·02
C	132 311·94	101 302·07			
			CD	187 48 34·3	4836·90
D	127 519·90	100 644·83			

The new observations are:

Angle	*Observed Value*	*Angle*	*Observed Value*
1	20° 55′ 19·7″	8	30° 14′ 08·7″
2	58 15 25·5	9	48 37 09·7
3	31 15 59·1	10	87 45 03·9
4	67 07 01·7	11	12 41 38·0
5	61 45 24·6	12	25 42 08·3
6	53 51 07·2	13	51 07 39·2
7	101 08 38·6	14	69 33 17·6

By some means, preliminary coordinates are obtained for the new points: the method used for getting these is immaterial, but this would normally be done by using suitable items of the list of observations. In this case, the triangle ABJ will be used, taking the observed angles (1 + 2) and 3, solving the triangle, calculating the lengths and bearings of AJ and BJ, and thus obtaining coordinates for the preliminary position, which will be denoted J_0. Similarly, preliminary position K_0 will be got by using the triangle CDK and the observed angles 7 and 8. Details of these calculations need not be given here, only the results are tabulated below:

Triangle ABJ_0

Angles		*Sides*			*Coordinates*
1 + 2	79° 10′ 45·2″	AJ_0	bearing	153° 29′ 40·4″	J_0
3	31 15 59·1		length	3719·03 m	
14	69 33 15·7	BJ_0	bearing	83° 56′ 24·7″	138 498·04 N
			length	7038·11 m	105 290·66 E

Triangle CDK_0

Angles		*Sides*			*Coordinates*
7	101 08 38·6	CK_0	bearing	86° 39′ 55·7″	K_0
8	30 14 08·7		length	3246·07 m	
9	48 37 12·7	DK_0	bearing	38° 02′ 43·0″	132 500·75 N
			length	6324·70 m	104 542·64 E

The third angle of each triangle is deduced from the other two and there is no question of any adjustment at this stage.

The next step is to calculate the bearings and lengths of the three other lines, using the above coordinates, and the results are:

Line	*Bearing*	*Length*
AK_0	174° 24′ 58·0″	9369·88 m
CJ_0	32 48 45·2	7360·48
J_0K_0	187 06 34·5	6043·76

From the bearings of the lines, the angles of the figure $ABCDK_0J_0$, corresponding with the observed angles, are calculated, and are as follows:

Angles of figure $ABCDK_0J_0$:

1.	20° 55′ 17·6″	8.	30° 14′ 08·7″
2.	58 15 27·6	9.	48 37 12·7
3.	31 15 59·1	10.	87 45 02·3
4.	67 06 57·6	11.	12 41 36·5
5.	61 45 22·9	12.	25 42 10·7
6.	53 51 10·5	13.	51 07 39·5
7.	101 08 38·6	14.	69 33 15·7

The angles (1+2), 3, 7, and 8 are the same as the observed angles simply because these have been used in determining the geometry of the figure. The other angles differ from the observed angles, and it is important to note that the above listed 14 angles are calculated angles of a geometrical figure and are therefore automatically perfectly self-consistent. The initial model has been set up.

The differential formulas derived in Section 5.05 are now applied to the variations n_j e_j n_k e_k which will be made in the initial coordinates of J_0 and K_0, and as an example the changes to the bearing of AK_0 will be worked out.

$$\text{Bearing } V = 174° 24' 58{\cdot}0''; \quad \cos V = -0{\cdot}9953; \quad \sin V = +0{\cdot}0973$$

$$\text{Length} = 9370 \text{ m}; \quad 206\,265(\sin V)/L = 2{\cdot}14; \quad 206\,265(\cos V)/L = -21{\cdot}91.$$

Therefore the bearing of AK will be $174° 24' 58{\cdot}0'' - 2{\cdot}14'' n_k - 21{\cdot}91'' e_k$. The line J_0K_0 is the only one to be affected by all four variations, and the full list of the new bearings is given below:

AJ	153° 29′ 40·4″	$-24{\cdot}75''n_j - 49{\cdot}63''e_j$	
AK	174 24 58·0		$-\ 2{\cdot}14''n_k - 21{\cdot}91''e_k$
BJ	83 56 24·7	$-29{\cdot}14''n_j + 3{\cdot}09''e_j$	
CJ	32 48 45·2	$-15{\cdot}19''n_j + 23{\cdot}55''e_j$	
CK	86 39 55·7		$-63{\cdot}44''n_k + 36{\cdot}96''e_k$
DK	38 02 43·0		$-20{\cdot}01''n_k + 25{\cdot}68''e_k$
JK	187 06 34·5	$-\ 4{\cdot}22''n_j + 33{\cdot}87''e_j$	$+\ 4{\cdot}22''n_k - 33{\cdot}87''e_k$

If the fixed bearings of lines AB, BC, and CD are added to the above list, the angles of the final figure ABCDKJ are found by taking differences of appropriate pairs of bearings: for instance the new angle 12 is the difference (bearing JC – bearing JK), that is, the difference $(212^\circ\ 48'\ 45{\cdot}2'' - 15{\cdot}19''n_j + 23{\cdot}55''e_j) - (187^\circ\ 06'\ 34{\cdot}5'' - 4{\cdot}22''n_j + 33{\cdot}87''e_j + 4{\cdot}22''n_k - 33{\cdot}87''e_k)$, which comes to $25^\circ\ 42'\ 10{\cdot}7'' - 10{\cdot}97''n_j - 10{\cdot}32''e_j - 4{\cdot}22''n_k + 33{\cdot}87''e_k$. The difference between this last expression and the observed angle 12 is the adjustment which, in effect, will have been applied to the observed angle in order to get the angle of the final figure: this adjustment is thus: $(-10{\cdot}97''n_j - 10{\cdot}32''e_j - 4{\cdot}22''n_k + 33{\cdot}87''e_k + 2{\cdot}4'')$. The full list of the 14 adjustments arranged formally is:

Angle	n_j	e_j *Coefficients*	n_k	e_k	*Numerical*
1	+24·75	+49·63	− 2·14	−21·91	−2·1
2			+ 2·14	+21·91	+2·1
3	−29·14	+ 3·09			0
4	+29·14	− 3·09			−4·1
5	−15·19	+23·55			−1·7
6	+15·19	−23·55	−63·44	+36·96	+3·3
7			+63·44	−36·96	0
8			−20·01	+25·68	0
9			−43·43	+11·28	+3·0
10			+61·30	−58·87	−1·6
11	− 4·22	+33·87	+ 6·36	−11·96	−1·5
12	−10·97	−10·32	− 4·22	+33·87	+2·4
13	−13·95	−20·46			+0·3
14	+ 4·39	−52·72			−1·9

These are linear functions of the variations n_j, e_j, n_k, e_k: to apply the principle of least squares, the variations are to be chosen so that the sum of the squares of the adjustments is minimum. Since the variations are independent, the desired values are found by forming the sum of the squares of the 14 expressions above, differentiating with respect to each variation, setting each differential coefficient to zero and solving the 4 equations so obtained. As already explained (Section 5.19), the practical arithmetical procedure is to multiply each of the adjustments by the coefficient of n_j in it, add up the results and set the total to zero; then do similarly with the coefficients of the other 3 variations in turn: the normal equations so obtained are:

n_j	e_j	n_k	e_k		
+3124·3	+ 357·1	− 997·2	− 301·9	−128·0	= 0
+ 357·1	+8043·1	+ 1646·8	−2712·4	−190·8	= 0
− 997·2	+1646·8	+14 160·9	−9427·2	−448·4	= 0
− 301·9	−2712·4	− 9427·2	+9234·8	+441·2	= 0,

and the solution of these is:

$$n_j = +0{\cdot}04 \quad e_j = +0{\cdot}01 \quad n_k = +0{\cdot}01 \quad e_k = -0{\cdot}03$$

giving the final coordinates of the new points:

J	138 498·08 N	105 290·67 E
K	132 500·76 N	104 542·61 E

In connection with the foregoing adjustment, it is to be noted that:

(i) coefficients in the adjustment formulas could be brought to the same order of magnitude as the numerical terms by considering the 'unknowns' to be $10n_j$, $10e_j$, $10n_k$, and $10e_k$; this would have a similar effect on the normal equations. Or better still, the centimetre could be used as linear unit for the purposes of the adjustment, as was done in Sections 5.22 and 5.23.

(ii) bearings and angles of the preliminary model $ABCDK_0J_0$ must obviously be calculated to accuracy of 0·1″, the same precision as the observation of the angles. It will be seen, however, that for calculating the coefficients in the formulas for changes in the bearings and angles, the lengths of the lines are not needed precisely; the nearest metre is quite sufficient in this case.

(iii) as mentioned previously, one may sometimes see the residuals written as 'observation equations', even in the form $24{\cdot}75n_j + 49{\cdot}63e_j - 2{\cdot}14n_k - 21{\cdot}91e_k = 2{\cdot}1$. This is incorrect, because the values of n_j, e_j, n_k, e_k finally obtained do not satisfy this as an equation, and there may be confusion with the condition method in which the adjustments are required to satisfy conditions exactly.

5.33 Variation of Coordinates: Trilateration. The addition of electromagnetic distance-measuring apparatus to the surveyor's equipment has brought considerable changes to the field methods and to the computation work of surveying. While the principal change is perhaps the greatly increased use of traversing, it is also now a practicable possibility to determine the geometry of systems of points by means of linear measurements only, or by a combination of angles and distances.

Adjustment procedures now have to deal with large numbers of measured lengths as well as, or instead of, measured angles. The determination of the form of a geometrical figure from measured lengths will be referred to as 'trilateration', and a few problems involving the adjustment of measured lengths will now be considered.

For fixing the relative positions of a number of points by trilateration, a certain minimum number of measured lengths is necessary, and of course extra measurements should always be taken, so that an adjustment problem always arises. If there are three points, all three sides of the triangle must be measured, and there is no redundancy; after the triangle. fixation of each additional point into the system requires two more lengths. Therefore, the minimum number of sides to be measured to provide a mechanically rigid figure of N points is $3+2(N-3)$, or $2N-3$. For the quadrilateral with diagonals, Fig. 73, 5 measured lengths are necessary, as is

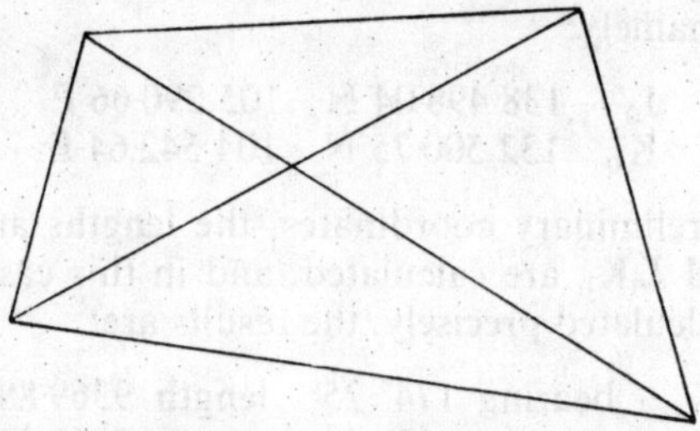

FIG. 73. QUADRILATERAL.

obvious enough, and the sixth line provides 1 extra. Evidently, the number of excess lengths in a trilateration system is equal to the number of side equation conditions that would have to be satisfied if the system were triangulation. To adjust the trilateration measurement of the quadrilateral, there would be only one condition on the six adjustments to the measured lengths. For this reason, trilateration is considered to be weaker than triangulation for the same figure.

To illustrate an adjustment of trilateration by variation of coordinates, the scheme shown in Fig. 72 will again be used, but this time the measured quantities are the lengths of the 7 new lines, as follows:

Line	*Measured Length*	*Standard Error*	*Weight*	*(Weight)*$^{\frac{1}{2}}$
AJ	3719·03 m	23 mm	18·9	4·3
AK	9369·79	34	8·7	2·9
BJ	7038·11	29	11·9	3·4
CJ	7360·44	30	11·1	3·3
CK	3246·07	22	20·7	4·5
DK	6324·70	26	14·9	3·8
JK	6043·88	24	17·4	4·2

The standard errors are calculated from a formula considered to be appropriate to the instrument. In accordance with the theory of errors, the weights are numbers inversely proportional to squares of the standard errors, and (weight)$^{\frac{1}{2}}$ is therefore inversely proportional to standard error, actually 100/(s.e.) in the table.

As before, the adjustment starts with calculation of preliminary positions J_0 and K_0: in this case, the measured lengths of AJ and BJ could be used to solve triangle ABJ and get coordinates for J_0 by straightforward geometry. Similarly, the measured values of CK and DK will give coordinates of K_0. Because the lengths calculated in the previous adjustment have been taken as the measured lengths for this example, the preliminary coordinates are the same, namely:

J_0	138 498·04 N	105 290·66 E
K_0	132 500·75 N	104 542·64 E.

Using these preliminary coordinates, the lengths and bearings of AK_0, CJ_0, and J_0K_0 are calculated, and in this case the bearings need not be calculated precisely, the results are:

AK_0	bearing	174° 25′	length	9369·88 m
CJ_0		32 49		7360·48
J_0K_0		187 07		6043·76

Bearings are required only for calculating coefficients in the adjustment formulas.

The effects of variations n_j e_j n_k e_k on the lengths of the lines are now calculated. The appropriate formula is given in Section 5.04: as an example, the length of AK_0 will change to $AK = 9369{\cdot}88 - 0{\cdot}9952n_k + 0{\cdot}0973e_k$, and the adjustment, that is the difference between final length and measured length of this line, will be $-0{\cdot}9952n_k + 0{\cdot}0973e_k + 0{\cdot}09$ metres. The full list of the 7 residuals arranged formally is:

	n_j	e_j	n_k	e_k		$(\text{wt.})^{\frac{1}{2}}$
AJ	−0·8949	+0·4462			0	4·3
AK			−0·9952	+0·0973	+0·09	2·9
BJ	+0·1055	+0·9944			0	3·4
CJ	+0·8404	+0·5419			+0·04	3·3
CK			+0·0581	+0·9983	0	4·5
DK			+0·7875	+0·6163	0	3·8
JK	+0·9923	+0·1237	−0·9923	−0·1237	−0·12	4·2

The sum of weighted squares of the above expressions is to be minimised by appropriate choice of the variations. As stated earlier, the differentiation of the weighted sum of squares shows that the desired values will be obtained by multiplying each adjustment by its weight and by the coefficient of n_j and adding the (four) expressions so obtained and setting the total to zero; then doing similarly with the coefficients of the other variations in turn. Arithmetically, this procedure is the same as first multiplying each adjustment by the square root of the weight and then forming the normal equations as if the weights were equal. The first step gives the following set of expressions:

	n_j	e_j	n_k	e_k	
AJ	−3·848	+1·919			+0
AK			−2·886	+0·282	+0·261
BJ	+0·359	+3·381			+0
CJ	+2·773	+1·788			+0·132
CK			+0·261	+4·492	+0
DK			+2·993	+2·342	+0
JK	+4·168	+0·520	−4·168	−0·520	−0·504

These give the normal equations:

n_j	e_j	n_k	e_k		
+39·998	+ 0·955	−17·372	− 2·167	−1·735	= 0
+ 0·955	+18·581	− 2·167	− 0·270	−0·026	= 0
−17·372	− 2·167	+34·727	+ 9·535	+1·348	= 0
− 2·167	− 0·270	+ 9·535	+25·913	+0·336	= 0

and the solutions: $n_j = +0{\cdot}034$, $e_j = -0{\cdot}003$, $n_k = -0{\cdot}021$, $e_k = -0{\cdot}002$, thus, the final coordinates are:

J 138 498·07 N, 105 290·66 E
K 132 500·73 N, 104 542·64 E.

5.34 Scale Constant. Check measurements by EDM have shown that a long-established triangulation system may have appreciable error of scale in parts remote from any precisely measured base line. When new points have to be fixed into such a system, practical considerations will generally preclude any thought of locally correcting the scale, and the object will be to fit the new work to the existing framework with the minimum of distortion. There will be no problem if the surveyor uses only a theodolite; adjustment of the new observations will produce the best possible shape, and the local scale error will remain. But if the modern surveyor uses EDM in the fixation of the new points, there will obviously be a conflict of scales.

One way to help in this situation is for the surveyor to measure with his EDM instrument some lines in the existing system, and so get an actual estimate of its scale error: his measured lengths for the new work would then be amended to agree with the actual prevailing scale. Use of the method of line-pairs, mentioned in Section 4.15 will have much the same effect.

Incidentally, the scale error may not be entirely in the old work. EDM results can have systematic error, either from instrumental causes such as change of crystal frequency, or from external causes such as incorrect allowance for atmospheric refractive index.

Another method for dealing with the scale problem is to include a 'scale constant' in the adjustment process, as one of the unknowns, the actual value of which will emerge as one of the results of the computation. Refer again to Fig. 72, and the adjustment of trilateration described in Section 5.33. Assume a scale constant of α parts per million; then each measured length will be multiplied by $(1+\alpha/1\,000\,000)$ before being used in the adjustment. For example, the length of AJ will be taken as $(371\,903+0{\cdot}372\alpha)$ cm, and the length of CJ as $(736\,044+0{\cdot}736\alpha)$ cm. In centimetre units, the final length of CJ will be $736\,048+0{\cdot}8404n_j+0{\cdot}5419e_j$, and its 'measured' length is as above, so the residual for CJ is now $(0{\cdot}8404n_j+0{\cdot}5419e_j-0{\cdot}736\alpha+4)$. There will be a term in α in each residual: α is an additional unknown, and there will be 5 normal equations, the solution of which will produce the 'best' value for α as well as the 4 variations of coordinates.

5.35 Adjustment in Spheroidal Coordinates. An extensive system of primary survey control is usually adjusted and computed as a figure of points on a spheroidal surface, the shape generated by rotating an ellipse around its shorter axis. The surface is close to sea level. The elliptical shape must be taken into account in precise geodetic work, though it is so near to being circular that if it were drawn to scale the size of this page the difference from a circle could not be seen by eye.

On such a surface, the natural coordinate system is one of latitude and longitude; each meridian is an ellipse and each parallel of latitude is a circle. The spheroidal coordinates of points in a geodetic survey are usually quoted to 0·001″. The size of the Earth is such that one second of latitude represents a distance of about 31 metres (~ 100 feet) along the meridian.

In adjustment problems by variation, it is necessary to calculate the changes of length and bearing of a line, due to making small changes of coordinates, $d\phi$ and $d\lambda$, at one or both of its end points. These changes are converted to linear quantities by use of the two radii of curvature ρ and ν, both approximately equal to a general 'radius of the Earth', about 6370 km. Accurate values of ρ and ν, which vary slightly with the latitude, are to be found in geodetic tables. The geometry of these radii is explained in more detail in Chapter VI.

A small change of latitude $d\phi$ means a linear shift of $\rho\,.\,d\phi$ along the meridian; and a change of longitude $d\lambda$ means a linear shift of $\nu\,.\cos\phi\,.\,d\lambda$ along the parallel, the radius of the parallel

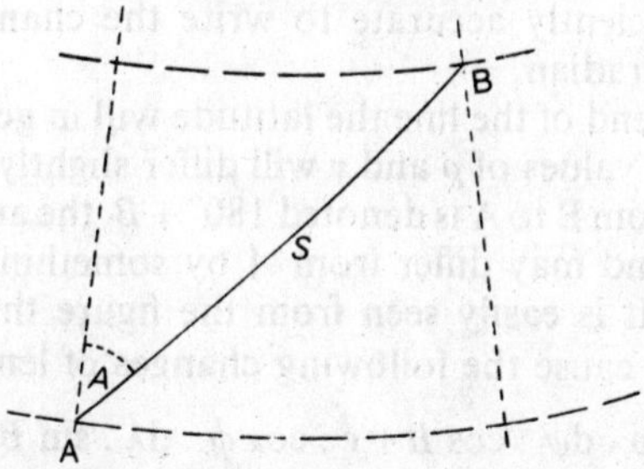

FIG. 74. LINE ON SPHEROID.

being $\nu \cdot \cos\phi$. These formulas hold if $d\phi$ and $d\lambda$ are in radian, but in practice they are always in angle seconds, so the appropriate formulas are $\rho \cdot d\phi''/206\,265$ and $\nu \cdot \cos\phi \cdot d\lambda''/206\,265$.

The geometry of a line on the spheroid is illustrated in Fig. 74. The broken lines are the meridians and parallels through A and B. A feature of spheroid geometry is that the forward and reverse bearings of a line do not differ by exactly 180°. Setting up a 'model' geometrical figure on a spheroid requires precise calculations of latitudes and longitudes from bearings and distances, and *vice versa;* the formulas for this are described in the next chapter. For the present we are concerned with the effects of small variations of coordinates.

The line AB starts off with bearing A and has length s. The detail Fig. 75(a) shows the shift AE equal to $\rho \cdot d\phi$ and the shift EA′ equal to $\nu \cdot \cos\phi \cdot d\lambda$. The geometry is similar to that of Fig. 63. The change of length is

$$ds = -\rho \cdot d\phi \cdot \cos A - \nu \cdot \cos\phi \cdot d\lambda \cdot \sin A.$$

The length of A′K is $\rho \cdot d\phi \cdot \sin A - \nu \cdot \cos\phi \cdot d\lambda \cdot \cos A$: and at ordinary geodetic distances, although the line is on a curved surface,

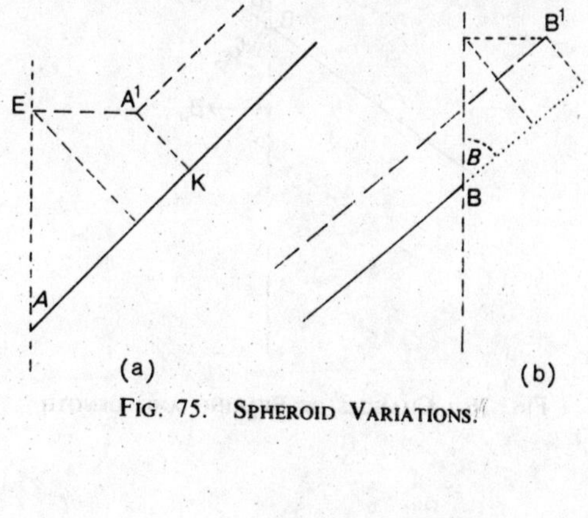

FIG. 75. SPHEROID VARIATIONS.

it is quite sufficiently accurate to write the change of bearing as $dA = +A'K/s$ radian.

At the other end of the line the latitude will in general be different from ϕ and the values of ρ and ν will differ slightly from those at A. If the bearing from B to A is denoted $180° + B$, the angle B is as shown on Fig. 75(b) and may differ from A by something of the order of half a degree. It is easily seen from the figure that small changes $d\psi$ and $d\lambda$ will cause the following changes of length and bearing:

$$ds = +\rho \,.\, d\psi \,.\, \cos B + \nu \,.\, \cos \phi \,.\, d\lambda \,.\, \sin B$$

$$dA = (-\rho \,.\, d\psi \,.\, \sin B + \nu \,.\, \cos \phi \,.\, d\lambda \,.\, \cos B)/s \text{ radian.}$$

In practice, it is quite sufficiently accurate to use average values of ρ and ν for the general latitude of the survey, in all the formulas. Indeed, within 1% accuracy, the factors $\rho/206\,265$ and $\nu/206\,265$ can be taken as 30·9 metres.

The inverse of the foregoing problem is to calculate the effects on coordinates, due to a small change of length ds and of bearing dA. The change ds, BB_1 in Fig. 76, has linear components $+ds \,.\, \cos B$ along meridian and $+ds \,.\, \sin B$ along the parallel: these are converted to differences of spheroid coordinates by dividing by the appropriate radius of curvature, ρ for latitude and $\nu \,.\, \cos \phi$ for longitude, and then multiplying by 206 265 to convert from radian to seconds. The change dA shifts the end of the line a distance $s \,.\, dA$, or $s \,.\, dA''/206\,265$, in the direction of bearing $B + 90°$, shown as BB_2 on the figure. The required formulas for the changes

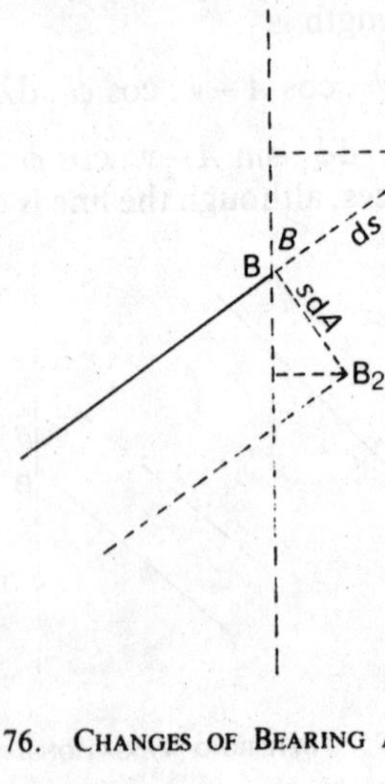

FIG. 76. CHANGES OF BEARING AND LENGTH.

of latitude and longitude in seconds are therefore:

$$d\psi'' = +\frac{ds \,.\, \cos B}{\rho} \,.\, 206\,265 - \frac{s \,.\, dA'' \,.\, \sin B}{206\,265 \,.\, \rho} \,.\, 206\,265$$

$$= +\frac{ds \,.\, \cos B}{\rho} \,.\, 206\,265 - \frac{s \,.\, dA'' \,.\, \sin B}{\rho}$$

$$d\lambda'' = +\frac{ds \,.\, \sin B}{\nu \,.\, \cos \psi} \,.\, 206\,265 + \frac{s \,.\, dA'' \,.\, \cos B}{206\,265 \,.\, \nu \,.\, \cos \psi} \,.\, 206\,265$$

$$= +\frac{ds \,.\, \sin B}{\nu \,.\, \cos \psi} \,.\, 206\,265 + \frac{s \,.\, dA'' \,.\, \cos B}{\nu \,.\, \cos \psi}$$

In most literature on geodetic computation the quantities $\rho/206\,265$ and $\nu/206\,265$ are written as $\rho \,.\, \sin 1''$ and $\nu \,.\, \sin 1''$, so the first term in the formula for $d\psi''$ would be written $(ds.\cos B)/(\rho.\sin 1'')$.

5.36 Adjustment of Traverse Systems. The advent of EDM has brought a revival of traversing as a method of providing primary framework for the surveys of extensive areas. The most notable manifestation of this is the geodetic survey of Australia established recently by the survey of about 54 000 km of tellurometer traverses: this is an exceptionally large system.

Some problems arising in the adjustment of a network of traverses in terms of spheroid coordinates will be discussed with reference to the modest system shown in Fig. 77, in which the traverses joining the lettered points may be thought of as having lengths of the order of 100–200 km.

The observational data available in such a system would be (1) lengths of traverse lines measured by EDM and reduced to sea level, (2) horizontal angles between the lines at all traverse stations, (3) control bearings, mostly at the junction points, and perhaps additional ones along the traverses as at A and E. Item (3) needs further explanation. A bearing to control computation on the spheroid must be a bearing referred to the spheroid meridian—such as the bearing A in Fig. 74. Astronomical observations for azimuth do not directly yield spheroid bearings, but by using special methods as described in Section 8.03 the azimuths can be converted to spheroid bearings, and points at which this is done are called Laplace Points. From the computer's point of view, a Laplace Point is a station where a reference bearing suitable for controlling surveys in the spheroid coordinate system is available.

Because of the large number of points in a traverse network—there could be 200 in the simple scheme of Fig. 77—adjustment of the whole system by variation method in one computation would

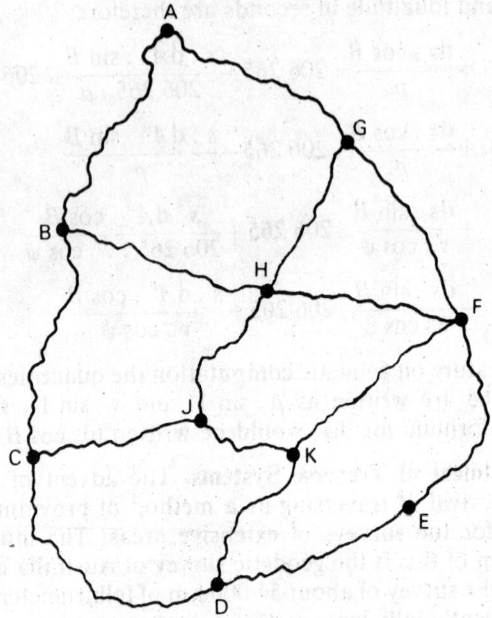

FIG. 77. TRAVERSE NETWORK.

require considerable computer power. The process, in outline, would be somewhat as follows:

(a) to one point in the system, say point B, origin values of latitude and longitude would be assigned. These would probably be derived from astronomical observations.

(b) initial spheroid coordinates for all the other traverse stations would be computed, using geodetic formulas described in Chapter VI, with observed angles and distances. Any angles and distances not used in these calculations would be computed from the coordinates. This would set up the initial model.

(c) the effects on the lengths and angles, due to making small variations of the coordinates, would be calculated, using methods described in Section 5.35; there would be two variations per point, except point B.

(d) the formulas for the residuals between the measured values and the model values including their variations would be written down as functions of the variations, for all the measured angles, lengths and bearings in the system. Each residual for a bearing or a length would be a function of 4 variations, and each residual for an angle would be a function of 6 variations (with the exception that there are no variations for point B).

(e) the usual procedure for finding variations such that the sum of the weighted squares of all the residuals is minimum would then be worked out. The number of normal equations would of course be equal to the number of variations—twice the number of points in the system other than the fixed origin. Weights would be assigned in accordance with estimated standard errors of the angles, lengths, and bearings contributed to the survey.

The computer would not only print out the required coordinates, but also could be programmed to work out interesting statistical quantities related to the final residuals. The programme could also be designed so that different systems of weighting could be input, thus making it possible to use estimates of precision derived from the adjustment itself.

5.37 Shorter Methods. There are simpler methods of adjustment of traverse networks, that can be used if available computing facilities are very modest.

Consider one of the traverses, say BH. At each end of it there is a control bearing. Starting with assumed values of latitude and longitude for B, the spheroid coordinates can be calculated through to H using observed angles and lengths. There will be a bearing misclosure at H: this can be distributed back through the traverse angles in a reasonable way, and the coordinates recomputed with the adjusted angles. The results of this computation will be a difference of latitude and a difference of longitude, say $\Delta\phi_{BH}$ and $\Delta\lambda_{BH}$. Differences of latitude and longitude can be separately obtained for each traverse in the network.

The adjustment problem now becomes similar to that of a levelling system. The differences $\Delta\phi$ for all the lines must now be treated as observed quantities and adjusted so as to close exactly round each of the closed circuits in the system—6 circuits in Fig. 77. This adjustment can be done by the condition method with 13 adjustments to the differences of latitude being required to satisfy exactly the 6 conditions of closure Suitable weights will have to be assigned, based on estimates of the standard errors of the various values of the $\Delta\phi$.

A similar adjustment of the 'observed' values of the $\Delta\lambda$ must also be made.

With these adjusted differences of latitude and longitude, and the assigned spheroid coordinates of the origin, the final spheroid coordinates of all the junction points are calculated. Each individual traverse must then be adjusted again to fit to the definitive values of its end points.

The method described above is very similar to the Bowie method used for adjustment of large grid systems of chains of simple triangulation.

5.38 Two-step Adjustment. There are other ways of simplifying methods of adjustment if it is essential to reduce labour of computation while maintaining the general principle of unbiased application of adjustment. Consider again the triangulation scheme, reproduced in Fig. 78. In adjusting this by the condition method, there are 2 centre conditions and 8 triangle conditions, all of which are simple equations with unity coefficients. Suppose the adjustments are found that satisfy exactly these 10 conditions under the principle of least squares.

Now suppose that each of the lines in the system receives a small rotation as shown. The effect on any angle will be the difference of two rotations; the rotation of an internal line will affect 4 angles

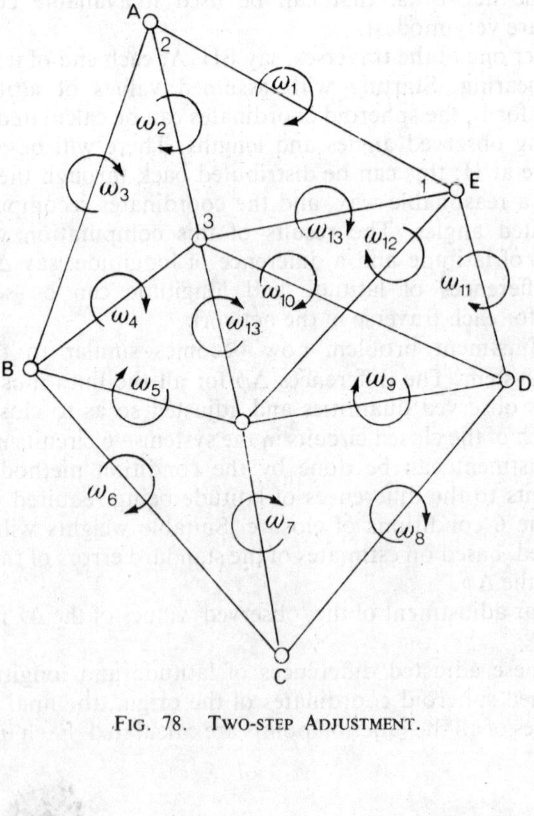

FIG. 78. TWO-STEP ADJUSTMENT.

and of a peripheral line such as AE will affect 2 angles. The effects on angles 1, 2, and 3 will be $(w_1 - w_{13})$, $(w_2 - w_1)$, and $(w_{13} - w_2)$ which total zero automatically, and it is easily seen that, whatever the values of the rotations may be, the centre and triangle conditions, which have already been satisfied, will remain satisfied. Now the rotations can be calculated so as to satisfy the 3 side equations. For instance, referring to Section 5.26, the side equation will start off:

$$2{\cdot}37(w_1 - w_{13}) - 2{\cdot}03(w_2 - w_1) \ldots$$

(The numerical term will be different because the angles have already received some adjustment.) For this part of the adjustment there will be only 3 normal equations to be solved, and the final angles will satisfy all 13 conditions.

5.39 Simpler Methods. Several references have been made to the so-called arbitrary methods of adjustment. This description could be used for any adjustment process that is simplified so as to facilitate computation, thereby giving results that do not have the property that the sum of the weighted squares of the residuals is minimum. The arbitrary methods are designed with the intention to produce unbiased adjustments, or to spread the errors evenly through the work, as far as possible. With modern computation facilities to hand, it is perhaps less necessary nowadays to devise these simple procedures.

Certainly in the case of traverse adjustment, some recent publications have drawn attention to the possible effects of conventional methods, in producing uneven incidence of adjustment. The reason for this can be seen by considering the geometry of a few cases.

In Fig. 79(a) the traverse is more or less straight, and it is supposed that the bearing misclosure is first distributed among the angles, then the traverse is computed, arriving at the point B′ near to the given position B. The Bowditch method, or indeed any other reasonable method for adjusting the coordinate misclosures, will produce the traverse shown in broken lines. In effect, the result of the method is what would be obtained by (i) systematic increase of scale in the ratio AB/AB′, plus (ii) rotation of the traverse bodily round A by the angle B′AB. Obviously this procedure has the effect of making additional alterations to the terminal angles only.

In Fig. 79(b) the traverse has a marked change of direction at point T. Suppose the first computation arrives at point Q′ and a Bowditch type of adjustment of coordinates is applied. The effect will be to 'stretch' the portion TQ of the traverse without much

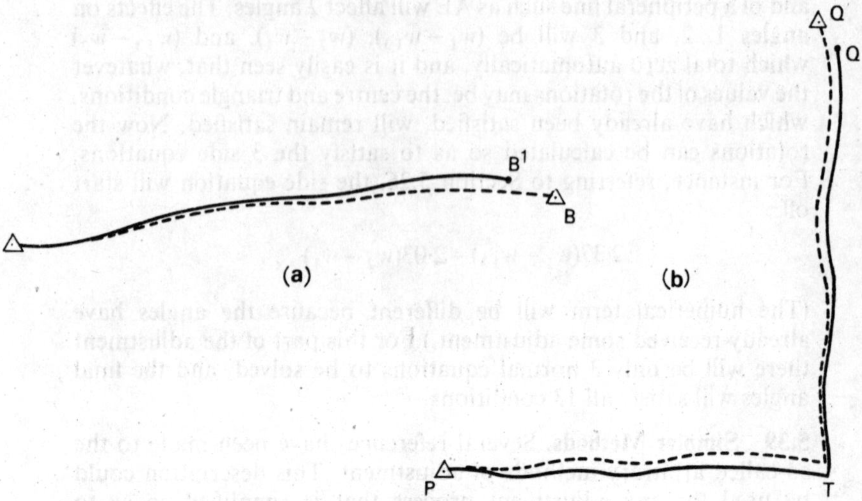

FIG. 79. TRAVERSE ADJUSTMENT EFFECTS.

change of bearings, to rotate the portion PT without change of scale, and to introduce large changes of angles at P and T.

With quite modest computing capacity, and using the condition method, it is possible, and appears desirable, to adjust traverses on the basis of the principle of least squares.

5.40 Plane Quadrilateral. An example of the arbitrary method of adjustment of the 8 angles of a plane quadrilateral is shown on the computation form Fig. 80. In this process, four equations controlling the adjustments are satisfied in succession by making numerically equal adjustments in such a way that each step does not disturb the conditions already satisfied.

The observed angles are given in column (2) on the form. The misclosure of 8·3″ is distributed as evenly as possible. In a plane quadrilateral, the sum of angles 1 and 2 must equal that of angles 5 and 6, so the difference between 27·6 and 25·2 is distributed as shown in column (4); similarly the difference between 32·7 and 34·5 is distributed among angles 3 4 7 and 8. These last two steps do not alter the correct total of 360°. Now, the angles in column (5) satisfy the three angle conditions exactly, and there remains the side equation which arises as explained in Section 5.26. The logsines show a discrepancy of 0·000 034 3 and the numbers in column (7) are the differences for 1″ in the seventh decimal place. Evidently, if the odd-numbered angles were each increased by 1″ and the even-numbered angles decreased by 1″, the discrepancy would be reduced by 200 in the seventh place. Therefore, the adjustments required are 343/200 = 1·7″: these are applied with alternate signs, and the final angles are given in column (9). The angle conditions remain satisfied.

The logsines of the final angles will be needed for computation of lengths of sides, assuming that the length of one of the lines is available. It will generally be found that the logsines of the final

(1) Angle No.	(2) Observed Angles	(3) 1st Adj.	(4) 2nd & 3rd Adj.	(5) Angle	(6) Logsines		(7) $10^7 d''$	(8) 4th Adj.	(9) Final Angles	(10) Total Adj. v.
1	46° 12′ 31·2″	−1·0″	−0·6″	29·6″	$\bar{1}$·858 4527		20·2	+1·7″	46° 12′ 31·3″	+0·1″
2	47 01 58·4	−1·0	−0·6	56·8		$\bar{1}$·864 3567	19·6	−1·7	47 01 55·1	−3·3
3	65 38 09·7	−1·1	+0·5	09·1	$\bar{1}$·959 4908		9·5	+1·7	65 38 10·8	+1·1
4	21 07 25·1	−1·0	+0·4	24·5		$\bar{1}$·556 7594	54·5	−1·7	21 07 22·8	−2·3
5	39 53 17·2	−1·0	+0·6	16·8	$\bar{1}$·807 0538		25·2	+1·7	39 53 18·5	+1·3
6	53 21 10·1	−1·1	+0·6	09·6		$\bar{1}$·904 3502	15·7	−1·7	53 21 07·9	−2·2
7	25 58 11·5	−1·0	−0·4	10·1	$\bar{1}$·641 3672		43·2	+1·7	25 58 11·8	+0·3
8	60 47 25·1	−1·1	−0·5	23·5		$\bar{1}$·940 9325	11·8	−1·7	60 47 21·8	−3·3
	360 00 08·3			00·0	$\bar{1}$·266 3645	$\bar{1}$·266 3988	199·7		360 00 00·0	$\sum v^2$ = 34·9
						3645				
						343				

Notes: (i) In practice, the angles would probably be taken to 0·01″ during the computation, but there is no point in going to higher precision. In any case, some rounding-off discrepancies are bound to occur, and may be removed by small, insignificant alterations to maintain numerical consistency.

(ii) for an angle over 90°, the difference of logsine for 1″ is negative.

(iii) the total adjustments are given in column (10) and the sum of their squares is 34·91.

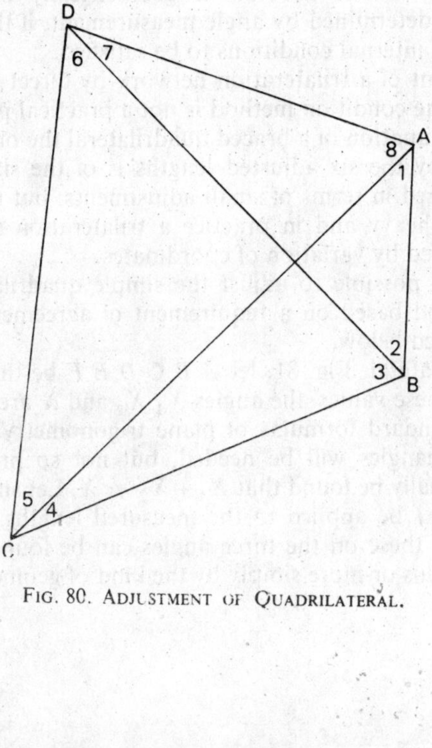

FIG. 80. ADJUSTMENT OF QUADRILATERAL.

angles in column (9) will still show a discrepancy of a few units in the 7th place. In that case, suitable alterations to the logsines should be made so that no discrepancy is left. It will be seen that a change of one unit in the 7th place means less than 0·1″ in all but one of the angles, and if the necessary changes are made in the logsines of the smaller angles the effects on the angles themselves will not be significant. The logsine of the final value of angle 4 could be anything from $\bar{1}$·556 750 0 to $\bar{1}$·556 750 4 inclusive, so far as can be deduced from 7-figure tables, so any value in this range is acceptable.

In an adjustment operation, it is most desirable to arrive eventually at exact internal consistency. Adjustment is usually worked out on values expressed to one or even two more decimal places than the real accuracy of the work warrants, so any final (really arbitrary) small alterations made to achieve consistency are arithmetical formalities with no real significance.

5.41 Simple Quadrilateration Figure. Though the formation of a framework of survey points by measurement of lengths only—pure trilateration—is theoretically possible, it seems that for covering a large area with many points the preferred system would be either a traverse network or triangulation with an adequate amount of scale control by use of EDM. And apart from any question of economics or convenience of surveying, there is the consideration that a trilateration structure is geometrically weaker than the same figure determined by angle measurement, if the criterion is the number of internal conditions to be satisfied.

The adjustment of a trilateration network by direct application of the ideas of the condition method is not a practical proposition. In the simplest situation of a braced quadrilateral the one equation to be satisfied by the six adjusted lengths is of the sixth degree; it can be linearised in terms of small adjustments, but the ensuing algebra is very heavy and in practice a trilateration system will always be adjusted by variation of coōrdinates.

However it is possible to adjust the simple quadrilateral by a condition method based on a requirement of agreement between angles, as outlined below.

In the quadrilateral, Fig. 81, let A B C D E F be the *measured* lengths. Using these values, the angles X_1 X_2 and X are calculated exactly from standard formulas of plane trigonometry: the other angles of the triangles will be needed, but not so precisely. Of course it will usually be found that $X_1 + X_2 \neq X$. Let small adjustments a b c d e f be applied to the measured lengths. Formulas for the effects of these on the three angles can be found by using differential calculus or more simply by the kind of geometry shown

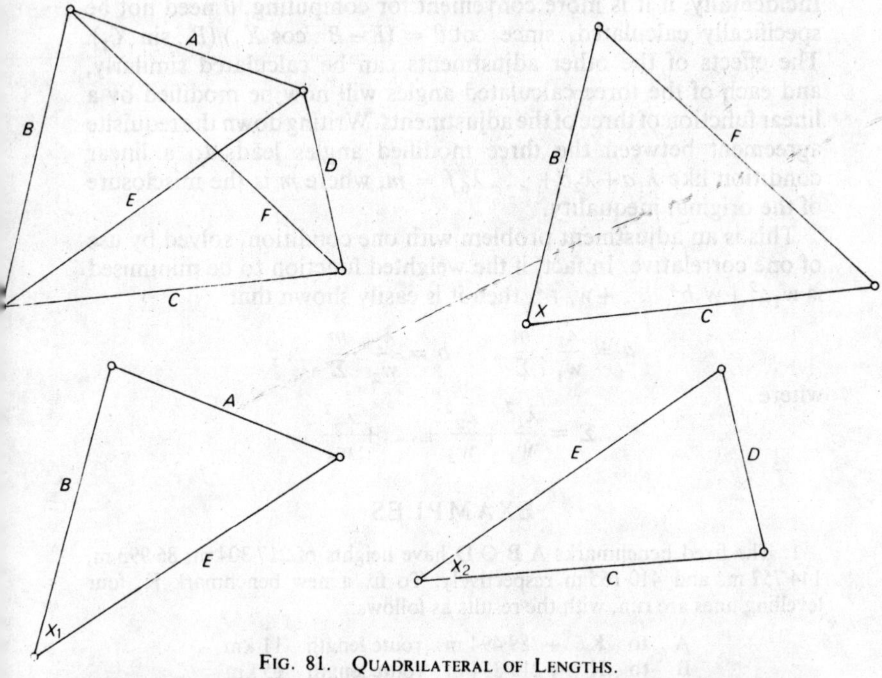

FIG. 81. QUADRILATERAL OF LENGTHS.

in Fig. 82. The length of MP′ is e, so MP is $e\ .\ \cot\theta$; hence δX_1 is $-\mathrm{MP}/E$ radian or $(-206\ 265\ .\ e\ .\cot\theta)/E$ seconds, the negative sign because obviously in this case an increase of E causes a decrease of X_1. For example, if $A = 771{\cdot}622$ m, $B = 965{\cdot}091$ m, $E = 1144{\cdot}850$ m then $X_1 = 41°\ 49'\ 30''$, $\theta = 56°\ 31'$, and the contribution of e to the change in X_1 is found to be $-119{\cdot}2e$ seconds.

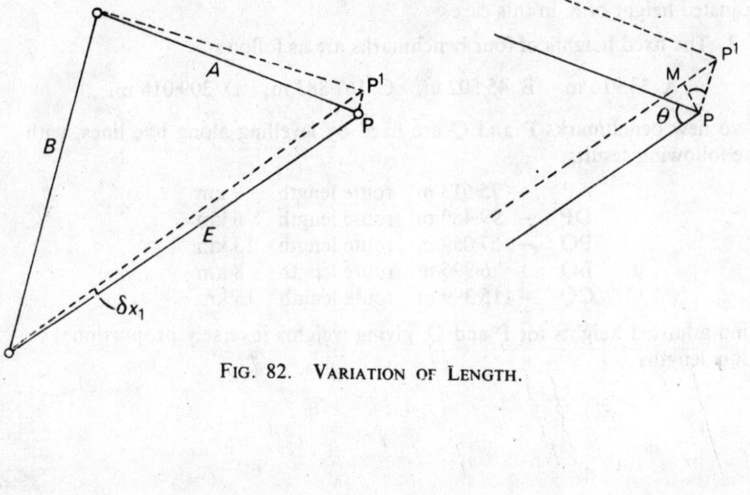

FIG. 82. VARIATION OF LENGTH.

Incidentally, if it is more convenient for computing, θ need not be specifically calculated, since $\cot\theta = (E-B\,.\cos X_1)/(B\,.\sin X_1)$. The effects of the other adjustments can be calculated similarly, and each of the three calculated angles will now be modified by a linear function of three of the adjustments. Writing down the requisite agreement between the three modified angles leads to a linear condition like $\lambda_1 a+\lambda_2 b+\ldots\lambda_6 f = m$, where m is the misclosure of the original inequality.

This is an adjustment problem with one condition, solved by use of one correlative. In fact, if the weighted function to be minimised is $w_1 a^2+w_2 b^2 \ldots +w_6 f^2$, then it is easily shown that

$$a = \frac{\lambda_1}{w_1}\cdot\frac{m}{\Sigma}, \qquad b = \frac{\lambda_2}{w_2}\cdot\frac{m}{\Sigma}\ldots,$$

where

$$\Sigma = \frac{\lambda_1{}^2}{w_1}+\frac{\lambda_2{}^2}{w_2}+\ldots+\frac{\lambda_6{}^2}{w_6}.$$

EXAMPLES

1. The fixed benchmarks A B C D have heights of 217·304 m, 86·993 m, 144·757 m, and 410·165 m respectively. To fix a new benchmark K, four levelling lines are run, with the results as follows:

A	to	K	+ 89·494 m	route length	11 km
B	to	K	+219·814 m	route length	15 km
C	to	K	+162·038 m	route length	28 km
D	to	K	−103·378 m	route length	20 km.

Calculate an adjusted height for K, giving weights inversely proportional to route lengths.

2. In the levelling described in question 1, the observations on the line BK were considered to be of inferior quality, so it was decided to allow for this by treating this observation as if the line were of length 30 km. Find the adjusted height of K in this case.

3. The fixed heights of four benchmarks are as follows;

A 33·916 m, B 45·502 m, C 167·883 m, D 209·018 m.

Two new benchmarks P and Q are fixed by levelling along five lines, with the following results:

AP	+ 75·613 m	route length	9 km
DP	− 99·489 m	route length	6 km
PQ	− 57·059 m	route length	13 km
BQ	+ 6·995 m	route length	8 km
CQ	−115·399 m	route length	16 km

Find adjusted heights for P and Q, giving weights inversely proportional to route lengths.

4. A small triangulation scheme has the form of a pentagon ABCDE with internal point F. The 15 observed angles are:

BAF	54° 04′ 39·9″	AEF	64° 10′ 19·3″
ABF	49 43 02·5	EAF	75 31 40·7
CBF	31 07 28·8	AFB	76 12 15·3
BCF	98 48 42·9	BFC	50 03 53·8
DCF	50 25 23·0	CFD	101 37 33·8
CDF	27 57 07·0	DFE	91 48 18·2
EDF	48 37 39·6	EFA	40 17 58·9
DEF	39 34 03·0		

Treating this as a plane figure, adjust the angles by the condition method, on the principle of least squares.

5. Suppose that four trigonometrical stations are situated in the form of an exact square, so that, with the two diagonals, all 8 angles are exactly 45°. Suppose then that the angles are 'observed', one being obtained as 45° 00′ 08″ and the rest exactly 45°. Adjust the 8 angles to satisfy the conditions, on the principle of least squares.

6. Show that if the 8 'observed' angles described in the previous question are adjusted by the simple method (Section 5.40) the same result is obtained as by the method of least squares.

7. The rectangular coordinates of two points are:

A	9110·78 m N	6509·90 m E,
B	8807·13 m N	10 008·47 m E.

To fix a new point P, the distances AP = 2645·17 m and BP = 1985·53 m, and the angles PAB = 34° 05′ 07″, PBA = 48° 17′ 40″, APB = 97° 37′ 16″ are measured. If the two measured lengths only are used, the coordinates of P are found to be 10 398·17 N and 8820·65 E. Calculate the angles at A, B, and P, and the effects on the angles and lengths due to small variations n and e to the above coordinates. Find the values of the variations that produce residuals satisfying the principle of least squares, taking the probable error of each angle as 2″, the probable error of length AP as $1\frac{1}{2}$ cm and the probable error of BP as 1 cm.

REFERENCES

ALLMAN J. S. 'Angles and directions'. *Survey Review* No. 139, January 1966.

ASHKENAZI V., WUDDAH-MARTEY E. E. L., and DODSON A. H. 'Rigorous adjustment of an EDM traverse on a desk calculator'. *Survey Review* No. 165, July 1972.

BELLAMY C. AND WATTERSON G. 'Least squares adjustment of networks'. *Survey Review* No. 156, April 1970.

BERTHON JONES P. 'The notion of a permissible misclose in traversing'. *Australian Surveyor*. September 1970.

BERTHON JONES P. 'A comparison of the precision of traverses adjusted by Bowditch Rule and by Least Squares'. *Survey Review* No. 164, April 1972.

BIESHEUVEL H. 'The adjustment and weighting of dissimilar quantities'. *Empire Survey Review* No. 126, October 1962.

BIRD R. G. 'Least squares adjustment of a traverse'. *Survey Review* No. 155, January 1970.

BIRD R. G. 'Least squares adjustment of EDM traverses'. *Survey Review* No. 165, July 1972.

BROWN H. C. 'Adjustment of traverse networks'. *Empire Survey Review* No. 120, April 1961.

CHUNG-CHI YING. 'On phased adjustment'. *Survey Review* No. 156, April 1970.

CIMIOTTI J. C. 'Adjustment of a quadrilateral'. *Survey Review* No. 162, October 1971.

COLE J. H. 'The adjustment of a first-order levelling network'. *Empire Survey Review* No. 100, April 1956.

DALE P. F. 'A note on scale errors in triangulation'. *Survey Review* No. 158, October 1970.

FUBARA D. M. J. 'Three-dimensional geodesy for terrestrial network adjustment'. *Journal of Geophysical Research* Vol. 77, No. 5, 1972.

HIRVONEN R. A. *Adjustment by Least Squares in Geodesy and Photogrammetry*. Ungar, New York, 1971.

HOMEWOOD E. T. 'The least squares adjustment of geodetic figures in which both angles and sides have been measured, and of geodetic traverses on the Transverse Mercator projection'. *Empire Survey Review* No. 124, April 1962.

JERIE H. G. 'Analogue computer for calculating and adjusting trilateration nets'. *Empire Survey Review* No. 126, October 1962.

KING C. W. B. 'Traverse network adjustments'. *Survey Review* Nos. 144 and 145, April and July 1967.

KUI-WAI LEUNG. 'Adjustment of a trilateration figure by frame structure analogue'. *Empire Survey Review* No. 123, January 1962.

LILLY J. E. 'Least squares adjustment of dissimilar quantities'. *Empire Survey Review* No. 121, July 1961.

MATHER R. S. 'The residual error distribution'. *Survey Review* No. 143, January 1967.

MURPHY B. T. 'The least squares adjustment of geodetic figures with observed angles and measured lengths'. *Empire Survey Review* No. 108, April 1958.

MURPHY B. T. 'The least squares adjustment of geodetic traverses'. *Empire Survey Review* No. 112, April 1959.

RAINSFORD H. F. 'The weighting of tellurometer lengths in relation to observed angles'. *Empire Survey Review* No. 126, October 1962.

RAINSFORD H. F. 'Triangulation adjustment—angles or directions?' *Survey Review* No. 141, July 1966.

RAINSFORD H. F. 'Combined adjustments of angles and distances'. *Survey Review* No. 150, October 1968.

SABBAH A. M. 'Adjustment of the braced geodetic quadrilateral by least squares'. *Survey Review* No. 140, April 1966.

SABBAH A. M. 'The use of an electrical analogue in adjusting triangulation problems'. *Survey Review* No. 146, October 1967.

HSUAN-LOH SU. 'Electric analogue for triangulation adjustment, Part I, Problems involving angles only'. *Empire Survey Review* No. 111, January 1959.

THORNTON-SMITH G. J. 'Least squares adjustment of dissimilar quantities'. *Survey Review* No. 136, April 1965.

TOPPING J. *Errors of Observation and Their Treatment*. Chapman & Hall London, 1962.

VINCENTY T. 'Least squares adjustments of triangulation: directions versus angles'. *Survey Review* No. 129, July 1963.

VINCENTY T. 'The scale problem in adjustments of linear measurements'. *Survey Review* No. 154, October 1969.

WARREN H. A. 'Adjustment of a traverse by relaxation'. *Empire Survey Review* No. 34, October 1939.

WINDSOR L. M. AND ASHKENAZI V. 'Current thinking on the solutions of large sets of simultaneous linear equations'. *Survey Review* No. 159, January 1971.

CHAPTER VI

GEODETIC COMPUTATION

6.01 Reference Surface. Surveying measurements are made at, and between, points marked on the Earth's physical surface—a set of points in three dimensions, illustrated by A_0 B_0 . . . on Fig. 83. To describe the positions of these points for mapping purposes it has been the practice to represent them by points A, B, C, . . . on a reference surface and thus effectively to reduce the geometry to two dimensions. The reference surface has a specified geometrical shape and is situated close to sea level.

Distances measured on the Earth's surface are converted to their horizontal components and reduced to sea-level lengths; horizontal angles are measured round the verticals at the points. These measurements are the raw data from which positions of points A, B, C, ... are to be calculated. Whatever kind of reference surface is used, an obvious requirement is a system for locating points on it.

In most survey work, there are redundant observations, and this leads to adjustment problems. Adjustment is treated in Chapter V, and for the present we assume that the adjustment has provided a set of mutually consistent angles and lengths which describe a geometrical figure. One of the functions of geodetic computation is to place this figure on a suitable reference surface, and specify the positions of the points in the recognised coordinate system.

Now the overall shape of the Earth is almost spherical, but the differences from a sphere are enough to make it impossible to use a sphere as reference surface for a reasonable fit to sea level over an area such as that covered by a large national mapping system. In any case, one of the aims of geodetic measurement is to find the form of a simple geometrical surface that would be a good fit over the whole Earth. A good fit can be achieved with a spheroid, the shape generated by rotating an ellipse about its shorter axis, as shown in Fig. 84.

It is therefore necessary to study the relevant geometrical and mathematical properties of such a surface. On illustrative diagrams, the elliptical form must be highly exaggerated; the appropriate shape is so nearly spherical that an ellipse of true shape, drawn to the width of this page, would be visually indistinguishable from a circle.

If the Earth were in fact an exact spheroid of rotation, with the

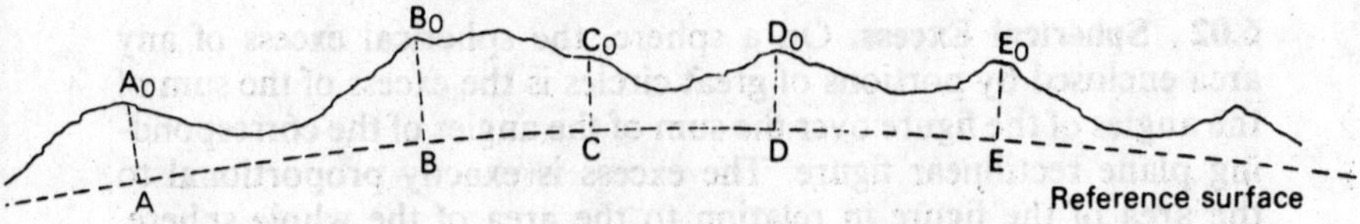

FIG. 83. GEODETIC REFERENCE SURFACE.

direction of the plumb line precisely perpendicular to its surface, instruments like theodolites would be set up with their rotation axes along normals to the surface.

In Fig. 84, the ellipse is shown as the shape obtained by reducing the scale of a circle in one direction: the circle of radius a is reduced to the fraction b/a in the direction perpendicular to the diameter QQ′. For a reasonable fit to the Earth, this ratio is about 0·9966, and the spheroid generated from the ellipse differs so little from a sphere that many of the geometrical properties of figures on a sphere can be assumed to hold for corresponding figures on the spheroid, to a quite adequate degree of accuracy. Any triangle used in a geodetic survey system covers such a small portion of the total surface area of the Earth that the sum of its three horizontal angles only exceeds 180° by a few seconds, and the spherical excess can be calculated from its area, using the simple formula for a spherical triangle and any reasonable mean radius.

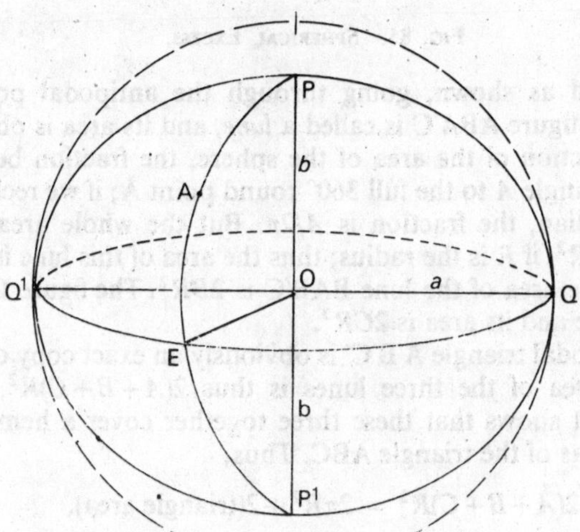

FIG. 84. SPHEROID.

6.02 Spherical Excess. On a sphere, the spherical excess of any area enclosed by portions of great circles is the excess of the sum of the angles of the figure over the sum of the angles of the corresponding plane rectilinear figure. The excess is exactly proportional to the area of the figure in relation to the area of the whole sphere. Consider the spherical triangle ABC in Fig. 85. The great circle AC

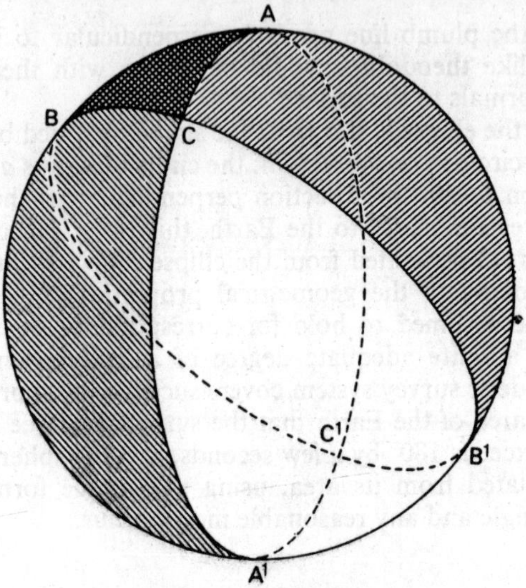

FIG. 85. SPHERICAL EXCESS.

is completed as shown, going through the antipodal points A′ and C′. The figure ABA′C is called a *lune*, and its area is obviously a simple fraction of the area of the sphere, the fraction being the ratio of the angle A to the full 360° round point A; if we reckon the angle in radian, the fraction is $A/2\pi$. But the whole area of the sphere is $4\pi R^2$, if R is the radius; thus the area of this lune is $2AR^2$. Similarly, the area of the lune BAB′C is $2BR^2$. The figure CA′C′B′ is also a lune and its area is $2CR^2$.

The antipodal triangle A′B′C′ is obviously an exact copy of ABC. The total area of the three lunes is thus $2(A+B+C)R^2$, and a little thought shows that these three together cover a hemisphere plus two areas of the triangle ABC. Thus,

$$2(A+B+C)R^2 = 2\pi R^2 + 2(\text{triangle area}),$$

or

$$(\text{triangle area}) = (A+B+C-\pi)R^2.$$

But $A+B+C-\pi = E$ is the spherical excess of the triangle: hence, $E = (\text{triangle area})/R^2$.

For example, a triangle of area 500 km^2 on a sphere of radius 6370 km will have an excess of $500/(6370)^2$ radian = 206 265 × $500/(6370)^2$ seconds = 2·57″. This is a moderate sized triangle in geodetic survey practice. On an Earth-size sphere the excess is practically 1″ for each 200 km^2.

6.03 Legendre's Theorem. One of the operations that will often have to be carried out in survey computation is the solution of triangles; this involves the relationships between the sides and angles of triangles.

Surveyors are concerned with the linear lengths of sides of triangles, rather than with angles they may represent at the centre of some sphere. In the early days of geodetic computation, triangles were treated as spherical, the linear lengths of the sides were converted to angle units using a reasonable radius for the sphere, and sides computed in angle measure were converted back to linear values: obviously a tedious process! It was Legendre who provided another solution, published as an exercise in his book on trigonometry in 1813.

Legendre's Theorem states, in effect, that if a spherical triangle that is small in relation to the size of the sphere is regarded as made of three pieces of string, and is removed from the sphere and stretched out flat, the resulting plane triangle will have angles that differ by almost exactly equal amounts from the spherical angles: the differences must therefore each be $\frac{1}{3}E$. This means that if the linear length of one side of a spherical triangle is known, and the angles are each reduced by $\frac{1}{3}E$, then the relationships

$$\frac{a}{\sin(A-\frac{1}{3}E)} \doteqdot \frac{b}{\sin(B-\frac{1}{3}E)} \doteqdot \frac{c}{\sin(C-\frac{1}{3}E)}$$

are so nearly true that they can be used for calculating the lengths of the other two sides in triangles of geodetic dimensions.

A proof of the theorem is outlined below; a different proof, based on the idea of stretching out the triangle, will be found in previous editions of this book, and elsewhere. The spherical triangle in Fig. 86(i) has sides expressed as usual in angle measure, α, β, γ: their lengths will be αR, βR, γR if R is the radius of the sphere. Consider a plane triangle, Fig. 86(ii), having angles as shown, which of course add up to the correct 180°. We have

$$\frac{x}{y} = \frac{\sin(A-\frac{1}{3}E)}{\sin(B-\frac{1}{3}E)} = \frac{\sin A-\frac{1}{3}E\,.\cos A\ldots}{\sin B-\frac{1}{3}E\,.\cos B\ldots}$$

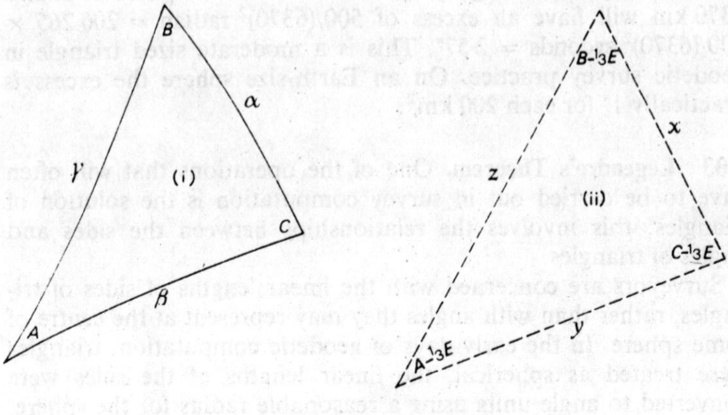

FIG. 86. LEGENDRE'S THEOREM.

Now, for calculating the spherical excess, which is very small, the area of the spherical triangle can be taken as

$$\tfrac{1}{2}\beta R \,.\, \gamma R \,.\, \sin A, \quad \text{or} \quad \tfrac{1}{2}\gamma R \,.\, \alpha R \,.\, \sin B \quad \text{or} \quad \tfrac{1}{2}\alpha R \,.\, \beta R \,.\, \sin C,$$

so, using the formula proved in Section 6.02, we have

$$E = \tfrac{1}{2}\beta\gamma \sin A = \tfrac{1}{2}\gamma\alpha \sin B \; (= \tfrac{1}{2}\alpha\beta \sin C, \text{ also}).$$

By the same token, the cosines which appear in the fraction x/y can be replaced by the formulas

$$\cos A = \frac{\beta^2 + \gamma^2 - \alpha^2}{2\beta\gamma} \quad \text{and} \quad \cos B = \frac{\gamma^2 + \alpha^2 - \beta^2}{2\gamma\alpha}$$

derived from plane trigonometry. Making the substitutions, we find

$$\frac{x}{y} = \frac{\sin A(1 - \frac{1}{12}(\beta^2 + \gamma^2 - \alpha^2))}{\sin B(1 - \frac{1}{12}(\gamma^2 + \alpha^2 - \beta^2))}.$$

$$\text{But} \quad \frac{\sin A}{\sin B} = \frac{\sin \alpha}{\sin \beta} \text{(exactly)} = \frac{\alpha - \frac{1}{6}\alpha^3 \ldots}{\beta - \frac{1}{6}\beta^3 \ldots} = \frac{\alpha(1 - \frac{1}{6}\alpha^2 \ldots)}{\beta(1 - \frac{1}{6}\beta^2 \ldots)}.$$

Substituting for $\dfrac{\sin A}{\sin B}$ and retaining only 2nd order terms in the series, we get

$$\frac{x}{y} = \frac{\alpha}{\beta} \cdot \frac{1 - \frac{1}{12}(\beta^2 + \gamma^2 + \alpha^2) \ldots}{1 - \frac{1}{12}(\gamma^2 + \alpha^2 + \beta^2) \ldots}.$$

To second order, the series cancel out and we are left with $\frac{x}{y} \doteqdot \frac{\alpha}{\beta}$, showing that the sides of the plane triangle are proportional to the lengths of the sides of the spherical triangle, to the order of accuracy mentioned.

If further study is made of the formulas, it can be shown that the discrepancies involved in using Legendre's Theorem are quite amazingly small: if the triangle is big enough to cover the whole of Great Britain, the ratios x/α, y/β, and z/γ are equal to within 2 parts per million!

It must be emphasised that the use of Legendre's Theorem is only a computing aid: the Legendre triangle associated with a spherical triangle cannot be shown geometrically related to it by any simple construction.

6.04 Ellipse Geometry. Some details of the geometry and mathematics of an ellipse must now be set out. In Fig. 87 the line EOE′ is a diameter of a circle of radius a and the ellipse is generated by reducing the scale to the fraction b/a in the direction perpendicular to this diameter: then POP′ of length $2b$ is the minor axis of the ellipse. Point A′ is on the circle where OA′ makes angle θ with OE, and A′T is the tangent to the circle. The scale reduction brings A′ to A but does not affect the property of tangency, so AT is tangent to the ellipse.

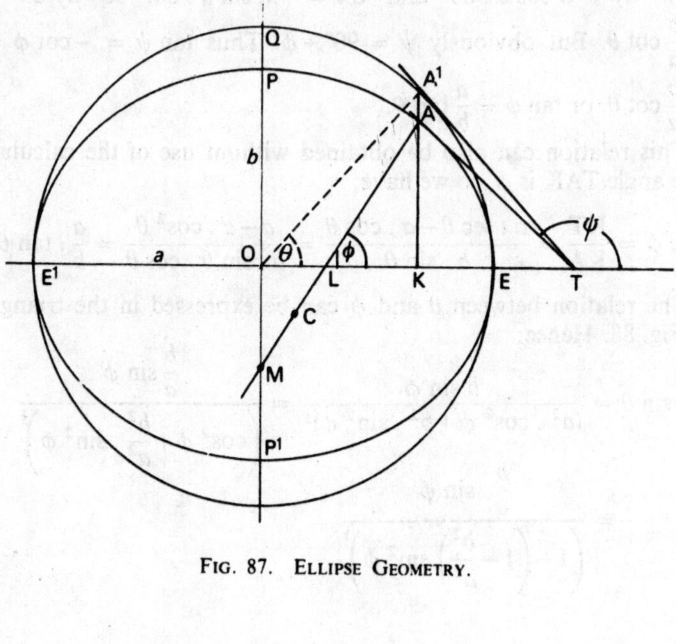

FIG. 87. ELLIPSE GEOMETRY.

Now we consider a rectangular coordinate system with axes along OE and OQ. With the angles denoted as shown, we have OK = $x = a \,.\, \cos\theta$, and

$$\text{KA} = y = \frac{b}{a}\,.\,\text{KA}' = \frac{b}{a}\,.\,a\,.\,\sin\theta = b\,.\,\sin\theta.$$

Hence

$$\frac{x^2}{a^2}+\frac{y^2}{b^2} = \cos^2\theta+\sin^2\theta = 1,$$

the familiar algebraic equation of an ellipse of semi-axes a and b. Perpendicular to the tangent AT is the normal ALM. On the ideal spheroidal Earth with POP′ as the polar axis, and the direction of the plumb line exactly perpendicular to the surface, the angle ϕ would be the astronomical latitude: however, as a feature of spheroid geometry this is to be distinguished by being called 'geodetic latitude' or 'spheroid latitude', and must not be confused with actual astronomical latitude observable on the real Earth. The angle ϕ locates the position of a point on the ellipse, and most geodetic computations related to spheroid geometry are expressed in terms of this angle, rather than by use of the angle θ.

In the usual notation of the differential calculus, $\mathrm{d}y/\mathrm{d}x = \tan\psi$, and ψ is the angle measured anticlockwise from the x axis round to the direction of the tangent TA, as indicated on the figure. Now $\mathrm{d}y = b\cos\theta\,.\,\mathrm{d}\theta$ and $\mathrm{d}x = -a\sin\theta\,.\,\mathrm{d}\theta$, so $\mathrm{d}y/\mathrm{d}x$ is $-\dfrac{b}{a}\cot\theta$. But obviously $\psi = 90°+\phi$. Thus $\tan\psi = -\cot\phi = -\dfrac{b}{a}\cot\theta$, or $\tan\phi = \dfrac{a}{b}\tan\theta$.

This relation can also be obtained without use of the calculus. The angle TAK is ϕ, so we have;

$$\tan\phi = \frac{\text{KT}}{\text{KA}} = \frac{a\,.\,\sec\theta-a\,.\,\cos\theta}{b\,.\,\sin\theta} = \frac{a-a\,.\,\cos^2\theta}{b\,.\,\sin\theta\,.\,\cos\theta} = \frac{a}{b}\,.\,\tan\theta.$$

The relation between θ and ϕ can be expressed in the triangle in Fig. 88. Hence:

$$\sin\theta = \frac{b\sin\phi}{(a^2\,.\,\cos^2\phi+b^2\,.\,\sin^2\phi)^{\frac{1}{2}}} = \frac{\dfrac{b}{a}\sin\phi}{\left(\cos^2\phi+\dfrac{b^2}{a^2}\,.\,\sin^2\phi\right)^{\frac{1}{2}}}$$

$$= \frac{\dfrac{b}{a}\,.\,\sin\phi}{\left(1-\left(1-\dfrac{b^2}{a^2}\right)\sin^2\phi\right)^{\frac{1}{2}}}.$$

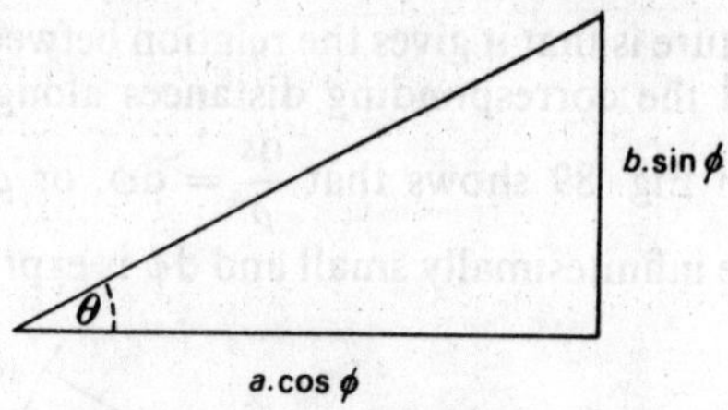

FIG. 88. SPHEROID LATITUDE.

The quantity $\left(1-\dfrac{b^2}{a^2}\right)$ is usually denoted by e^2, and e is called the excentricity of the ellipse.

By similar reasoning we get: $\cos\theta = \dfrac{\cos\phi}{(1-e^2\,.\,\sin^2\phi)^{\frac{1}{2}}}$

The coordinates x and y can now be expressed in terms of ϕ.

$$x = a\,.\cos\theta = \frac{a\,.\cos\phi}{(1-e^2\,.\,\sin^2\phi)^{\frac{1}{2}}}$$

$$y = b\,.\sin\theta = \frac{\dfrac{b^2}{a}\,.\sin\phi}{(1-e^2\,.\,\sin^2\phi)^{\frac{1}{2}}} = \frac{a\,.\dfrac{b^2}{a^2}\sin\phi}{(1-e^2\,.\,\sin^2\phi)^{\frac{1}{2}}}$$

$$= \frac{a(1-e^2)\sin\phi}{(1-e^2\,.\,\sin^2\phi)^{\frac{1}{2}}}.$$

Now it is obvious from the diagram that x = OK = AM $\cos\phi$, hence AM $= a/(1-e^2\sin^2\phi)^{\frac{1}{2}}$. This length AM is of fundamental importance in spheroid geometry: it is always denoted by the Greek letter ν, so we have the simple formulas:

$$x = \nu\,.\cos\phi: \qquad y = \nu(1-e^2)\sin\phi.$$

for the coordinates of points on the ellipse.

It then follows that AL $= \nu(1-e^2)$, LM $= \nu\,.\,e^2$, and thus all normals of the ellipse are divided in the ratio $e^2:(1-e^2)$ at the points of intersection with the major axis.

Radius of Curvature. An important length expressing the shape of any curve is the radius of curvature: this may be visualised as the radius of a circle that best represents the curve over a short length in the vicinity of a point. From the formula for an ellipse, the radius of curvature, always denoted by ρ, can be calculated by using standard calculus methods. The significant property of the radius of curvature is that it gives the relation between small changes of latitude and the corresponding distances along the curve. The thin triangle in Fig. 89 shows that $\frac{ds}{\rho} = d\phi$, or $\rho = ds/d\phi$ if the differentials are infinitesimally small and $d\phi$ is expressed in radian.

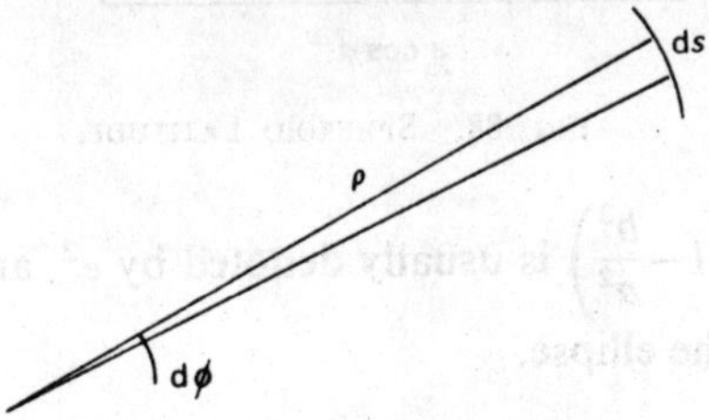

FIG. 89. RADIUS OF CURVATURE.

Another way to get the formula for ρ is as follows: We have $x = v \,.\, \cos\phi$ and $v = a(1-e^2 \,.\, \sin^2\phi)^{-\frac{1}{2}}$. By standard differentiation procedures we find that $dv = a.e^2.\sin\phi.\cos\phi.d\phi/(1-e^2\sin^2\phi)^{\frac{3}{2}}$. Then $dx = dv \,.\, \cos\phi - v\sin\phi \,.\, d\phi$, and after some algebra and trigonometry we find that $dx = a(1-e^2)\sin\phi \,.\, d\phi/(1-e^2 \,.\, \sin^2\phi)^{\frac{3}{2}}$. By similar work we get $dy = a(1-e^2)\cos\phi \,.\, d\phi/(1-e^2\sin^2\phi)^{\frac{3}{2}}$. But $ds^2 = dx^2+dy^2$ and thence

$$ds = \frac{a(1-e^2) \,.\, d\phi}{(1-e^2 \,.\, \sin^2\phi)^{\frac{3}{2}}}.$$

Referring to the expression for ρ we see that $\rho = \frac{a(1-e^2)}{(1-e^2 \,.\, \sin^2\phi)^{\frac{3}{2}}}$. The centre of the circle of radius ρ and touching the ellipse at point A is at C, which is in fact always situated on the portion of the normal between L and M.

It is sometimes mathematically helpful to use the relationships:

$$\frac{v}{\rho} = \frac{1-e^2\sin^2\phi}{1-e^2} = 1+\frac{e^2}{1-e^2} \,.\, \cos^2\phi.$$

Arc Length. Another linear quantity that appears mostly in formulas for map projections is the length along the curve from E to A. There is no finite formula for this in terms of simple mathematical functions. From the relation $ds = \rho \,.\, d\phi$ the length S can theoretically be obtained by an integration process; in practice, this integral must be expressed as an infinite series, and we take advantage of the fact that for any spheroid resembling the shape of the Earth the value of e^2 is only about 1/150. To get the formula for

S the expression for ρ is expanded to:

$$a(1-e^2)(1+\tfrac{3}{2}e^2 \,.\, \sin^2\phi+\tfrac{15}{8}\,.\,e^4\,.\,\sin^4\phi+\tfrac{35}{16}\,.\,e^6\,.\,\sin^6\phi\ldots)$$

and the powers of $\sin\phi$ are replaced by the trigonometrical identities: $\sin^2\phi = \tfrac{1}{2}-\tfrac{1}{2}\cos 2\phi$

$$\sin^4\phi = \tfrac{3}{8}-\tfrac{1}{2}\cos 2\phi+\tfrac{1}{8}\cos 4\phi$$

$$\sin^6\phi = \tfrac{5}{16}-\tfrac{15}{32}\cos 2\phi+\tfrac{3}{16}\cos 4\phi-\tfrac{1}{32}\cos 6\phi \text{ etc.}$$

then we find that

$$\begin{aligned}\rho = {} & a(1-\tfrac{1}{4}e^2-\tfrac{3}{64}e^4-\tfrac{5}{256}e^6\ldots)\\ & -a(\tfrac{3}{4}e^2+\tfrac{3}{16}e^4+\tfrac{45}{512}e^6\ldots)\cos 2\phi\\ & +a(\tfrac{15}{64}e^4+\tfrac{45}{256}e^6\ldots)\cos 4\phi\\ & -a(\tfrac{35}{512}e^6\ldots)\cos 6\phi\ldots\end{aligned}$$

and the integration gives:

$$\begin{aligned}S = {} & a(1-\tfrac{1}{4}e^2-\tfrac{3}{64}e^4-\tfrac{5}{256}e^6\ldots)\phi\\ & -a(\tfrac{3}{8}e^2+\tfrac{3}{32}e^4+\tfrac{45}{1024}e^6\ldots)\sin 2\phi\\ & +a(\tfrac{15}{256}e^4+\tfrac{45}{1024}e^6\ldots)\sin 4\phi\\ & -a(\tfrac{35}{3072}e^6\ldots)\sin 6\phi\ldots\end{aligned}$$

in the first term of which the ϕ must be in radian measure.

To get a numerical idea of the small terms in this series, it will be found that the coefficient of $\sin 6\phi$, using reasonable values for a and e^2, is about 22 millimetres: evidently, the term in e^8 will be negligible.

If we put $\phi = \tfrac{1}{2}\pi$, all the trigonometrical terms are zero and we are left with the term in ϕ itself. This gives the length of the quarter-ellipse from E to P, and it is interesting to recall that the original definition of the metre was a ten-millionth of this length: a recent estimate of the dimensions of an Earth-spheroid makes the quadrant 10 002 001 metres.

Flattening. Although the quantity e^2 is a measure of the departure of an ellipse from the circular shape, a perhaps more obvious indicator is the fraction by which b is shorter than a: this is $(a-b)/a$ or $\left(1-\dfrac{b}{a}\right)$, usually denoted by f and called the flattening or ellipticity. For the Earth it is about 1/300. The mathematical relation $e^2 = f(2-f)$ is easily proved.

Geodetic Tables. For carrying out extensive geodetic computations, by use of logarithms or on a small computer, it is helpful to have tables of S, ν, and ρ: for some of the widely used spheroids, such tables have been published. The quantities mentioned are functions of the latitude, and it will be found convenient to have them tabulated at intervals of 1 minute. Various other functions of latitude may also be found useful, especially in connection with calculations of map projections.

A short length s along the meridian corresponds with a small change of latitude of approximately s/ρ radian. This is usually required in seconds, and then the formula is $(s/\rho)206\,264{\cdot}8\ldots$, that is, s divided by $\rho/206\,264{\cdot}8$. This last ratio is usually written $\rho\,.\,\sin 1''$ (which is not strictly correct) and it is the length of a circular arc of radius ρ subtending an angle of 1″. Some geodetic tables give the values of $\rho\,.\,\sin 1''$ and $\nu\,.\,\sin 1''$, or the logarithms of these. A typical portion of a geodetic table is given below.

Portion of GEODETIC TABLE for a = 637 816 0 m, $1/f$ = 298·255

ϕ Latitude	S_ϕ Meridian Length	ρ Meridian Radius	ν Cross Radius
50° 00′	554 086 6·814 m	637 297 8·864 m	639 072 5·182 m
49 59	553 901 2·993	296 0·458	071 9·030
58	715 9·177	294 2·051	071 2·878
57	530 5·366	292 3·642	070 6·724
56	345 1·560	290 5·231	070 0·570
55	553 159 7·760	288 6·819	069 4·416
54	552 974 3·965	286 8·405	068 8·261
53	789 0·176	284 9·988	068 2·105
52	603 6·392	283 1·571	067 5·949
51	418 2·613	281 3·151	066 9·792
49 50	232 8·839	279 4·730	066 3·634
49	552 047 5·071	277 6·307	065 7·476
48	551 862 1·308	275 7·883	065 1·317
47	676 7·550	273 9·457	064 5·158
46	491 3·798	272 1·029	063 8·998
49° 45	551 306 0·051	637 270 2·599	639 063 2·837
49° 00′	542 964 7·006		

When a large electronic computer is available, functions can of course be computed as required from their basic formulas.

6.05 Spheroid Geometry. Rotation of the ellipse about POP′ produces the spheroid surface. Any position of the semi-ellipse PAEP′ (Fig. 84) is a geodetic meridian, and the angle between any two meridians is a geodetic difference of longitude.

With an adopted meridian as zero of longitude, the longitude L and latitude ϕ constitute a coordinate system defining positions of points on the spheroid.

The rotation of the semi-ellipse causes a point A to trace out a circle, which is a parallel of geodetic latitude and has radius $\nu \cdot \cos\phi$.

Geodetic computation is concerned with angles and lines on a spheroid surface, but what angles and what lines? Some definitions are needed here, because the answers to the questions are not obvious, as they are in relation to a sphere.

It is helpful to consider an ideal surveyor on an ideal spheroidal Earth as mentioned in Section 6.01. A theodolite set up at point A will have its axis of rotation along the normal AM, Fig. 90. If the

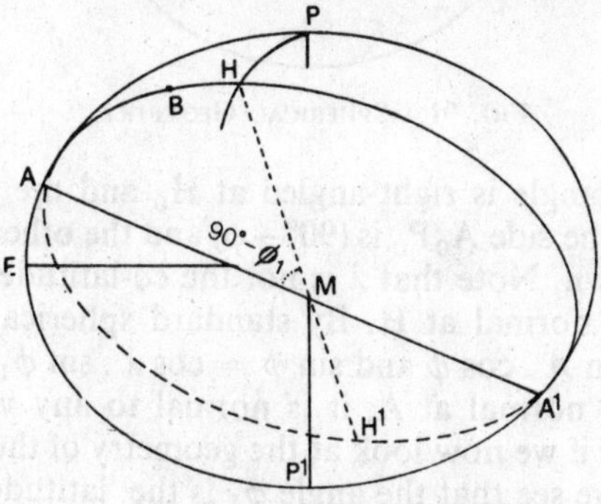

FIG. 90. VERTICAL PLANE SECTION.

telescope is directed on point B and the horizontal clamps are tight, the telescope transit plane is the plane defined by the point B and the line AM. This plane is a vertical plane at A. Any plane section of a spheroid is an ellipse; the elliptic curve shown joining A and B is the line which would be set out by the surveyor if he sent an assistant out from A to peg out the line sighted in the telescope. The geodetic bearing of the line is the angle A between the plane BAM and the meridian plane PAEP′. If two or more points are sighted from A, the measured horizontal angles are the angles between the vertical planes through the observed points.

Given the latitude ϕ and the bearing A, it is possible to compute the dimensions of the vertical section ellipse ABA′ as follows. Somewhere on this ellipse, perhaps just beyond B in Fig. 90, the section must reach a point of highest latitude, at H, where it obviously cuts the meridian at right angles, and the planes AMH and PMH must be at right angles to each other. In fact the geometry is symmetrical about the meridian PH, so the line HM must be the minor axis of the plane section.

Now consider a sphere centred at M, and the spherical triangle generated by the directions MA MP MH indicated by $A_0P_0H_0$ in

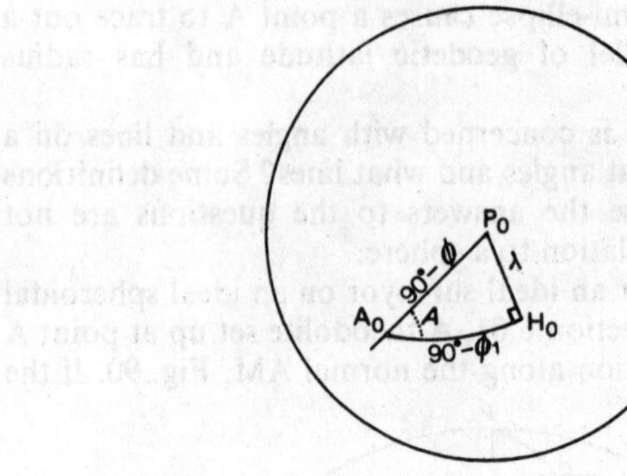

FIG. 91. SPHERICAL GEOMETRY.

Fig. 91. The triangle is right-angled at H_0 and the angle at A_0 is the bearing A: the side A_0P_0 is $(90° - \phi)$ and the other two sides are denoted as shown. Note that λ is not the co-latitude of H because HM is not the normal at H. By standard spherical formulas we have $\sin \lambda = \sin A \,.\, \cos \phi$ and $\sin \phi = \cos \lambda \,.\, \sin \phi_1$. Because AM is the spheroid normal at A, it is normal to any vertical section through A, and if we now look at the geometry of the plane section itself, Fig. 92, we see that the angle ϕ_1 is the 'latitude' of A on this ellipse. As noted above, HM is the minor axis of the section ellipse, so the section ellipse and the spheroid ellipse have two lengths in common—the chord AA′ and the 'normal' AM. From these two relationships, the two parameters of the section ellipse can be derived.

First, it is necessary to get a formula for the chord AA′. Referring to Fig. 87, consider a point on the normal AM at distance c from A: its rectangular coordinates will be $x = \nu \,.\, \cos \phi - c \,.\, \cos \phi$, and

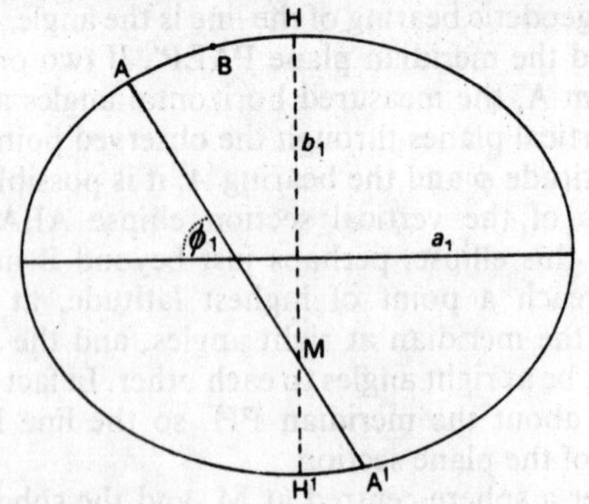

FIG. 92. PLANE SECTION ELLIPSE.

$y = \nu(1-e^2)\sin\phi - c \,.\, \sin\phi$. We now express the condition that this point shall be on the ellipse. For this purpose, the ellipse formula $(x^2/a^2)+(y^2/b^2) = 1$ is transformed to $(1-e^2)x^2+y^2 = b^2 = a^2(1-e^2) = \nu^2 \,.\, (1-e^2)(1-e^2 \,.\, \sin^2\phi)$. Substituting the coordinate formulas into this will give the required value of c: the resulting equation, obtained after some rearrangements and cancellations, is strictly quadratic in c but one root must be zero because the point A itself is one of the solutions. The other root is

$$c = \text{AA}' = 2\nu(1-e^2)/(1-e^2\cos^2\phi).$$

Now let the parameters of the plane section ellipse be a_1, b_1 and e_1; we have to express the two relationships mentioned above in terms of the latitude ϕ_1 on this ellipse and latitude ϕ on the spheroid.

From the identity of normal lengths:

$$a_1{}^2/(1-e_1{}^2\sin^2\phi_1) = a^2/(1-e^2\sin^2\phi),$$

and from the identity of chord lengths, omitting the factor ν which is the same on both ellipses;

$$(1-e_1{}^2)/(1-e_1{}^2\cos^2\phi_1) = (1-e^2)/(1-e^2\cos^2\phi).$$

From this last equation, and making use of the spherical triangle formulas given above, we get:

$$\begin{aligned} e_1{}^2 &= e^2\cos^2\lambda/(1-e^2\sin^2\lambda) \\ &= e^2(1-\sin^2 A \,.\, \cos^2\phi)/(1-e^2\sin^2 A \,.\, \cos^2\phi), \end{aligned}$$

and

$$e_1{}^2\sin^2\phi_1 = e^2\sin^2\phi/(1-e^2\sin^2\lambda).$$

From the other relation we get:

$$\begin{aligned} a_1{}^2 &= \frac{a^2(1-e^2\sin^2\lambda - e^2\sin^2\phi)}{(1-e^2\sin^2\lambda)(1-e^2\sin^2\phi)} \\ &= a^2 - \frac{a^2e^4\sin^2\lambda \,.\, \sin^2\phi}{(1-e^2\sin^2\lambda)(1-e^2\sin^2\phi)}. \end{aligned}$$

This shows that the difference between a_1 and a is of 4th order in e, and in numerical terms it is not greater than 38 metres for any values of ϕ and A on any reasonable Earth-spheroid.

Two Plane Sections. Now suppose the surveyor goes to B and sets up the instrument; its rotation axis will be along BN the normal at B, Fig. 93. The vertical plane containing A is ABN. If the surveyor sets out the line from B to A it will not coincide with that set out

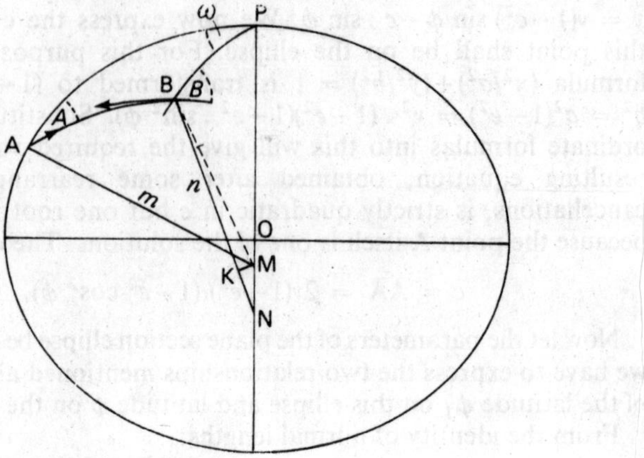

FIG. 93. SKEW NORMALS.

from A to B. The two planes have the chord AB as their common line of intersection. The geometry is shown, exaggerated, in Fig. 94.

To resolve the ambiguity of having two lines between A and B, one can carry the idea of the surveyor a step further. Suppose he takes his instrument to some place between A and B, and finds a point G where A and B are in the same vertical plane: this is the same as saying that the bearings to A and B differ by exactly 180°, and it is the same as saying that the normal at G intersects the chord

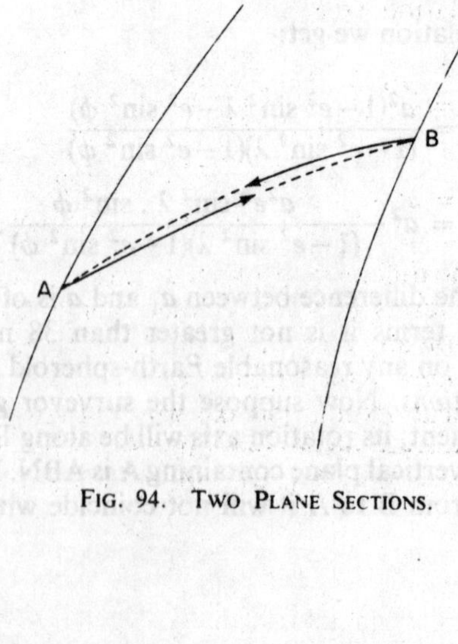

FIG. 94. TWO PLANE SECTIONS.

AB. The locus of points G is the *curve of alignment*, Fig. 95; it obviously starts from A and B in the same directions as the respective setting-out curves.

So we can now say that horizontal angles observed at points on the spheroid are angles between the curves of alignment to the various points sighted. Furthermore, it is reasonable to define a spheroidal triangle as the figure formed by the three curves of alignment joining the three points.

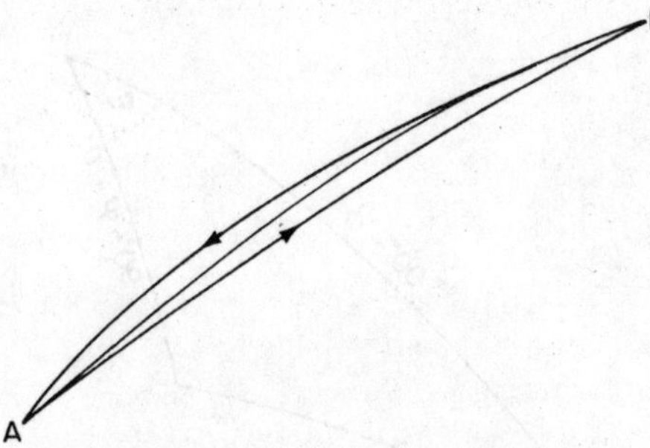

FIG. 95. CURVE OF ALIGNMENT.

Geodesic. Another curve joining two points is the geodesic, defined as the shortest line between the points and lying on the surface. On a spheroid, the geodesic has the property that the value of $v \cdot \cos\phi \cdot \sin A$ is the same at all points on it. The geodesic joining two points does not leave the points in the same directions as the curve of alignment, and its direction cannot readily be set out by use of instruments. The distance between two points could be defined as the length of the geodesic; however, for even the longest distances used in geodetic surveying the lengths of the four curves that have been mentioned are so nearly equal that their differences are far smaller than any errors likely to arise in practical surveying of the highest precision, so it is not really necessary to specify the curve when referring to the distance.

Spheroid Bearings. Computation of plane section bearings between given points on a spheroid is quite straightforward. Points A and B have latitudes ϕ and ψ respectively, and their difference of longitude is ω. Referring to Fig. 93, let AM $= m$ and BN $= n$. Join BM and draw MK perpendicular to BN. Consider the plane triangle MBN. From the geometry of the ellipse, Section 6·04, we have OM $= m \cdot e^2 \cdot \sin\phi$, ON $= n \cdot e^2 \cdot \sin\psi$, so MN $= e^2(n \cdot \sin\psi - m \cdot \sin\phi)$. The angle BNM is the co-latitude of B so angle NMK is ψ. Hence, MK = MN . cos ψ and NK = MN . sin ψ. Denoting the angle NBM by v, we have

$$\tan v = \frac{\text{MK}}{\text{BN}-\text{NK}} = \frac{e^2(n \,.\, \sin\psi - m \,.\, \sin\phi)\cos\psi}{n - e^2(n \,.\, \sin\psi - m \,.\, \sin\phi)\sin\psi}.$$

The angle BMO is then $(90° - \psi + v)$.

Now consider a sphere centred at M: the lines MA, MB, MP intersect it at points that form the spherical triangle shown in Fig. 96.

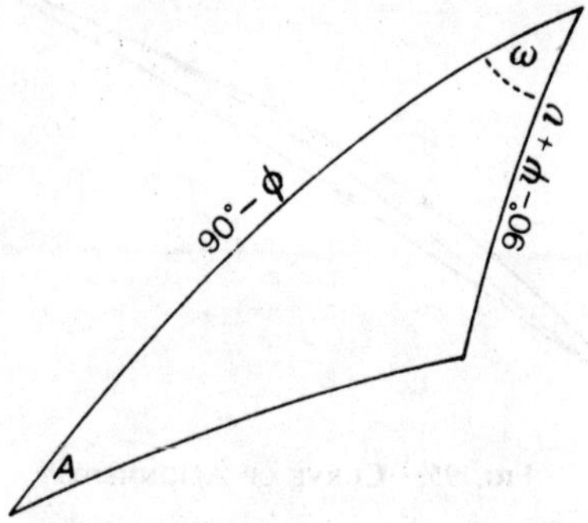

FIG. 96. SPHEROID BEARING.

A standard spherical formula gives

$$\sin\phi \,.\, \cos\omega = \cos\phi \,.\, \tan(\psi - v) - \sin\omega \,.\, \cot A.$$

But, using the expression above for tan v, we find that

$$\tan(\psi - v) = \frac{n \,.\, \sin\psi - e^2(n \,.\, \sin\psi - m \,.\, \sin\phi)}{n \,.\, \cos\psi},$$

and the spherical formula transforms to

$$\sin\omega \,.\, \cot A = \tan\psi \,.\, \cos\phi - \sin\phi \,.\, \cos\omega - \frac{e^2(n \,.\, \sin\psi - m \,.\, \sin\phi) \,.\, \cos\phi}{n \,.\, \cos\psi},$$

whereby cot A is expressed as a finite function of ϕ, ψ, and ω.

It is to be noted that the derivation of this formula is in no way dependent on the excentricity of the spheroid being small or on the positions of the two points being close together.

A similar piece of working, by joining AN and using a sphere centred at N, produces the formula

$$\sin w \,.\, \cot B = \sin \psi \,.\, \cos w - \tan \phi \,.\, \cos \psi - \frac{e^2(n \,.\, \sin \psi - m \,.\, \sin \phi) \,.\, \cos \psi}{m \,.\, \cos \phi}$$

in which the angle B is as indicated on Fig. 93, that is the clockwise bearing from B to A is (B + 180°). This is a convenient notation

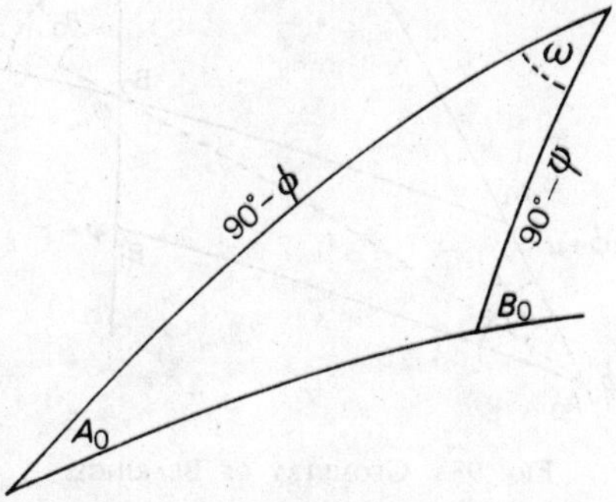

FIG. 97. SPHERICAL BEARINGS.

because then the difference $(B-A)$ can be regarded as the change of bearing (of the curve of alignment) between points A and B. This difference is sometimes referred to as 'convergence'.

It is interesting to compare angles on the spheroid with corresponding angles of the figure formed by points of the same latitudes and longitudes on a sphere. This spherical triangle is shown in Fig. 97 and the bearings are A_0 and B_0. Formulas for A_0 and B_0 in terms of ϕ, ψ, and ω are obtained from standard spherical formulas, or of course by putting $e^2 = 0$ in the formulas already given. For instance,

$$\sin \omega \,.\, \cot A_0 = \tan \psi \,.\, \cos \phi - \sin \phi \,.\, \cos \omega.$$

Therefore

$$\sin \omega \,.\, (\cot A_0 - \cot A) = \frac{e^2(n \,.\, \sin \psi - m \,.\, \sin \phi) \,.\, \cos \phi}{n \,.\, \cos \psi}.$$

The bearings on sphere and spheroid can all be shown on one spherical diagram, Fig. 98. Corresponding to angle v the angle NAM is u. The points on the two meridians have latitudes as shown.

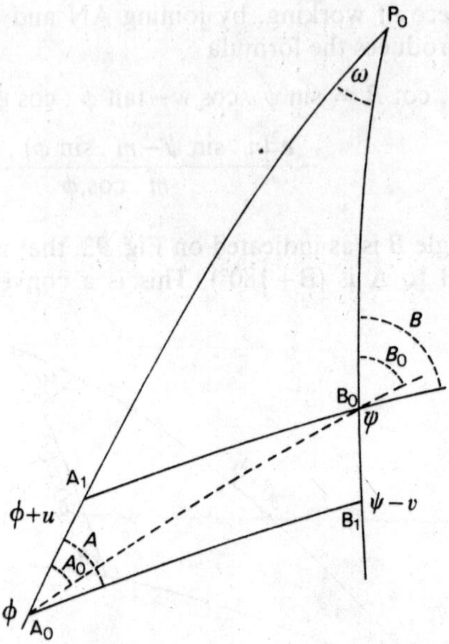

FIG. 98. GEOMETRY OF BEARINGS.

Then angle $P_0A_0B_1$ is the spheroid plane section bearing at A to B: the external angle $P_0B_0A_1$ is the spheroid plane section bearing at B to A. The triangle $P_0A_0B_0$ gives the spherical bearings shown on Fig. 97.

For a numerical example, take $\phi = 49\frac{1}{2}°$, $\psi = 50°$, $\omega = \frac{1}{2}°$. The bearings between these points on a sphere are 32° 40′ 36·9″ and 213° 03′ 30·7″. Using the spheroid with parameters a = 6 378 160 m, f = 1/298·255, we find that u = 5·076″, v = 5·024″, and the spheroid plane section bearings are 32° 45′ 01·2″ and 213° 07′ 55·0″. The differences spheroid-sphere are each 4′ 24·3″. See Section 6.06.

The formula for $(\cot A_0 - \cot A)$ can be derived directly from the geometry of Fig. 99 in which the bearings A_0 and A both appear. From B the perpendicular BK is dropped on to the meridian plane of A. BH and KH are perpendiculars to the spheroid axis, so the triangle BKH is right-angled at K and the angle BHK is ω. BH is a radius of the parallel of latitude ψ so its length is $n \,.\, \cos\psi$: therefore BK $= n \,.\, \cos\psi \,.\, \sin\omega$.

Through N the line NF is drawn parallel to MA in the meridian plane of A; the distance between these lines is MN . $\cos\phi$ that is $e^2(n \,.\, \sin\psi - m \,.\, \sin\phi)\cos\phi$. From K in the meridian plane of A, line KEF is perpendicular to AM and FN. BE and BF are joined.

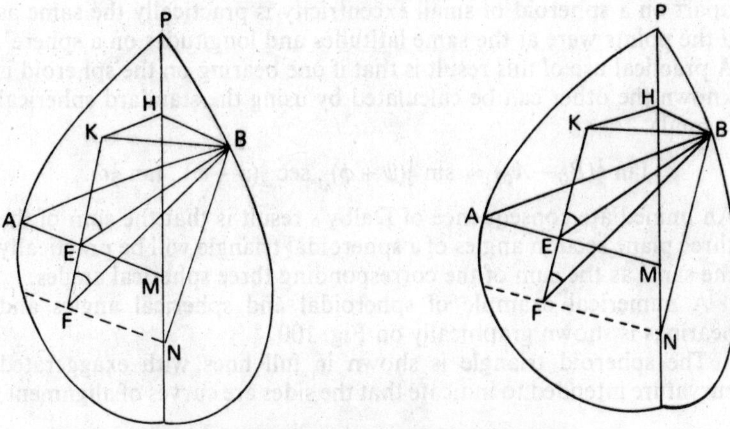

FIG. 99. SPHEROID AND SPHERICAL BEARING.
(this diagram is a stereo-pair)

Angle KEB is simply the angle between the meridian plane of A and the plane section ABM, so it is the bearing A, the spheroid bearing at A to B. The angle FNP is $90° - \phi$ and the angle BNP is $90° - \psi$, and a sphere centred at N will be cut by the lines NP NB NF in points forming the spherical triangle $A_0B_0P_0$ of Fig. 97, hence angle KFB is A_0, the spherical bearing.

Now $\cot A_0 = \text{KF/KB}$ and $\cot A = \text{KE/KB}$ so

$$\cot A_0 - \cot A = \frac{\text{KF}-\text{KE}}{\text{KB}} = \frac{e^2(n\,.\sin\psi - m\,.\sin\phi)\cos\phi}{n\,.\cos\psi\,.\sin\omega}$$

as calculated previously. If we put $A = (A_0 + \alpha)$, then by expansion $\cot A = \cot A_0 - \alpha\,.\operatorname{cosec}^2 A_0 \ldots$ and we get

$$\alpha\,.\sin\omega = \frac{e^2(n\,.\sin\psi - m\,.\sin\phi)\cos\phi\,.\sin^2 A_0}{n\,.\cos\psi}.$$

But $\dfrac{\cos\phi}{\cos\psi} = \dfrac{\sin B_0}{\sin A_0}$ by the sine formula, hence

$$\alpha = e^2(n\,.\sin\psi - m\sin\phi)\,.\sin A_0\,.\sin B_0/n\,.\sin\omega.$$

Similarly, putting $B = B_0 + \beta$ we find that the formula for β differs from that for α only by having m in the denominator instead of n. Obviously, the difference between α and β will be very small over ordinary geodetic distances on an Earth-shaped spheroid. This result is expressed in Dalby's Theorem.

6.06 Dalby's Theorem. 'The convergence (as defined above) between plane section bearings at two points at geodetic distance apart on a spheroid of small excentricity is practically the same as if the points were at the same latitudes and longitudes on a sphere'. A practical use of this result is that if one bearing on the spheroid is known the other can be calculated by using the standard spherical formula

$$\tan \tfrac{1}{2}(B_0 - A_0) = \sin \tfrac{1}{2}(\psi + \phi) \cdot \sec \tfrac{1}{2}(\psi - \phi) \cdot \tan \tfrac{1}{2}\omega.$$

An immediate consequence of Dalby's result is that the sum of the three plane section angles of a spheroidal triangle will be practically the same as the sum of the corresponding three spherical angles.

A numerical example of spheroidal and spherical angles and bearings is shown graphically on Fig. 100.

The spheroid triangle is shown in full lines with exaggerated curvature intended to indicate that the sides are curves of alignment,

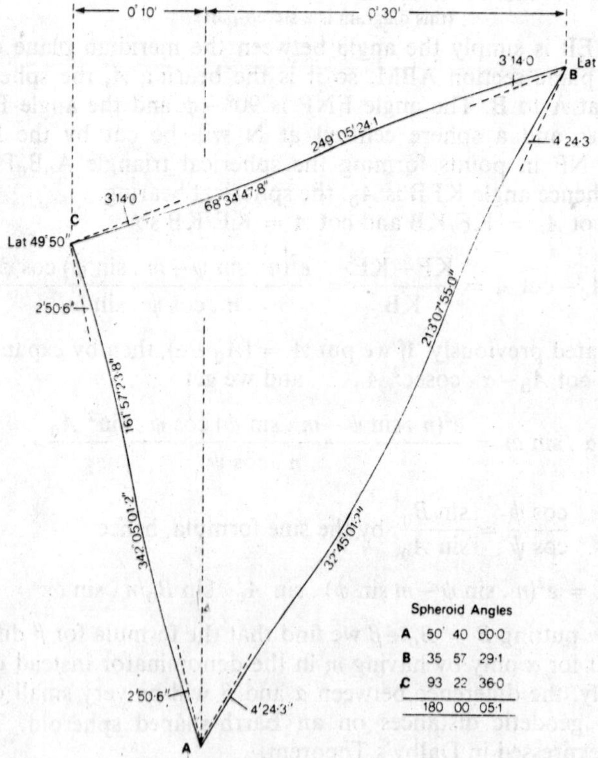

FIG. 100. SPHEROID TRIANGLE.

and the spheroid bearings are written along the lines. Bearings of the sides of the spherical triangle of points with the same latitudes and longitudes are indicated by the short broken lines and can be calculated by using the differences shown. The side AB is the same as the line described in Section 6.05. The spherical excess indicates an area of about 1000 km^2.

6.07 Spheroid Coordinates. A basic problem in geodetic computation, generally referred to by continental writers as The Principal Problem of Geodesy, is that of calculating the spheroid latitude and longitude of a point when its bearing and distance from a point of known coordinates are given. For instance, after a system of major triangulation has been adjusted to be internally consistent, positions of the points on a spheroid of reference need to be calculated, to provide a framework for further surveys or for mapping.

To locate the system on the spheroid, it is sufficient to specify the geodetic coordinates of one point and the bearing of one line; other positions follow by exact computation, using formulas appropriate to the geometry of the spheroid.

To be more specific, a point A has latitude ϕ and longitude L on a spheroid of reference: the bearing at A to another point B is A and the distance (along the line of plane section) from A to B is s. What are the latitude and longitude of B? There are several well-known sets of formulas for solving this problem.

It is necessary to consider what computational precision is required: obviously this must be related to the order of accuracy likely to be achieved in the adjusted survey. If the bearing of a line 40 km long is correct to $\frac{1}{2}''$, the relative positional accuracy between the two ends is 0·1 metre, or a fractional accuracy of 1/400 000. The accuracy of careful length measurement by EDM approaches this figure. The distance along a meridian equivalent to 1″ of latitude is about 31 m, so computation to 0·001″ of latitude corresponds with 0·03 m.

Thus, to avoid accumulation of error in extensive precise computations, lengths may be taken to 0·01 m, angles to 0·01″, latitudes and longitudes to 0·000 1, but there is no point in quoting final values closer than 0·1″ in angles and bearings, or 0·001″ for coordinates.

A distance of 40 km along a meridian represents about 1300 seconds of latitude; therefore we see that computation of differences of latitude (or longitude) can comfortably be done with 7-figure logarithms to the required accuracy of 0·001″. This is one reason why the formulas customarily used for geodetic computation have been arranged so as to give differences from point to point; if it were required to find a latitude directly from one of its trigonometrical functions, at least 9-figure tables would be needed, to give accuracy to 0·001″. With modern computing aids, however, other methods may be considered, see Section 6.10.

6.08 Clarke's Formulas. In a plane coordinate system, a line of length s and bearing A will have a northing component of $s \,.\, \cos A$ and an easting component of $s \sin A$. If we now think of the $s \,.\, \cos A$ as a distance along a meridian, then dividing by an appropriate radius ρ will give a difference of latitude; similarly, if we think of the $s \,.\, \sin A$ as being along the parallel of latitude ϕ then dividing by $\nu \,.\, \cos \phi$ will give a difference of longitude. The formulas $(s \,.\, \cos A)/\rho$ and $(s \,.\, \sin A)/(\nu \,.\, \cos \phi)$ are radian values: to give seconds of angle, they are expressed in the conventional forms $(s \,.\, \cos A)/(\rho \,.\, \sin 1'')$ and $(s \,.\, \sin A)/(\nu \,.\, \sin 1'' \,.\, \cos \phi)$.

At ordinary geodetic distances, these simple formulas give very close estimates of the differences of latitude and longitude: it is in the refinements required to provide the accuracy of 0·001″ that the various available formulas differ.

First consider the problem on a sphere of radius R, Fig. 101. The length s of AB has angle value s/R. BM is the perpendicular from B on to the meridian of A. Let the spherical excess of triangle

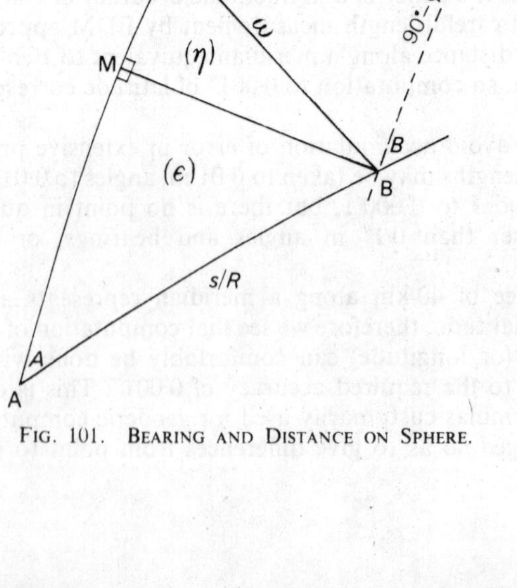

FIG. 101. BEARING AND DISTANCE ON SPHERE.

ABM be ε; then for computing ε it is quite sufficient to take the sides of the triangle as $s \cdot \cos A$ and $s \cdot \sin A$, its area as $\frac{1}{2}s^2 \sin A \cdot \cos A$ and the spherical excess therefore as $\frac{1}{2} \cdot \frac{s^2}{R^2} \cdot \sin A \cos A$ (radian). Since the angle at M is 90°, it is easy to see that the angle ABM must be $90° - A + \varepsilon$. The difference of longitude between A and B being denoted by ω, let BK be drawn so as to have also the angle value ω. Then the triangle BPK is peculiar in that, by the sine formulas, the sine of each of its angles is equal to the sine of the opposite side, that is, each angle is equal either to the opposite side or to its supplement: in this case obviously it is the angle BKM that is equal to $(90° - \psi)$. Then, denoting the spherical excess of triangle BMK by η, it follows that the third angle MBK must be $\psi + \eta$. Again, for calculating η, taking MB as $s \cdot \sin A$ and angle at B as ψ, the length of MK is approximately $s \cdot \sin A \cdot \tan \psi$. Thus the spherical excess n is $\frac{1}{2} \cdot \frac{s^2}{R^2} \cdot \sin^2 A \cdot \tan \psi$, which is also $\varepsilon \cdot \tan A \cdot \tan \psi$.

It is now necessary to calculate some lengths more precisely, and this is where Legendre's Theorem comes in. The linear length of AB is s, so to get the accurate linear lengths of AM and BM we use the sine formulas on the triangle A′B′M′ of Fig. 102. Thus, by

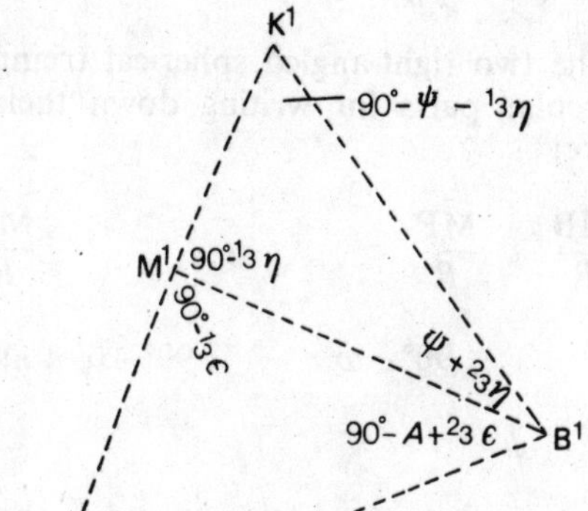

FIG. 102. LEGENDRE IRIANGLES.

Legendre's Theorem,

$$\text{AM} \doteqdot \text{A'M'} = \frac{s}{\sin(90° - \frac{1}{3}\varepsilon)} \cdot \sin(90° - A + \tfrac{2}{3}\varepsilon)$$

$$= \frac{s}{\cos\frac{1}{3}\varepsilon} \cos(A - \tfrac{2}{3}\varepsilon).$$

But we know that in practice ε will be only a few seconds, so that $\cos\frac{1}{3}\varepsilon$ can always be taken as equal to 1 (for example cos 1″ is about 0·999 999 999 988 . . . !!). Hence we have, to all needed accuracy, $\text{AM} = s \cdot \cos(A - \frac{2}{3}\varepsilon)$. By similar working we find that $\text{BM} = s \cdot \sin(A - \frac{1}{3}\varepsilon)$. Now proceed to the Legendre triangle B'M'K' to get the linear lengths of BK and MK. By the sine formula

$$\text{BK} \doteqdot \text{B'K'} = \frac{\text{BM}}{\sin(90° - \psi - \frac{1}{3}\eta)} \cdot \sin(90° - \tfrac{1}{3}\eta) = \frac{s \cdot \sin(A - \frac{1}{3}\varepsilon)}{\cos(\psi + \frac{1}{3}\eta)}.$$

But the linear length of BK is ωR, so we have

$$\omega = \frac{s \cdot \sin(A - \frac{1}{3}\varepsilon)}{R \cdot \cos(\psi + \frac{1}{3}\eta)} \text{ (radian)}.$$

This is the precise formula for difference of longitude.

It is found to be convenient to calculate the length of MK from BK. By Legendre and sine formula,

$$\text{MK} = \frac{\omega R}{\sin(90° - \frac{1}{3}\eta)} \cdot \sin(\psi + \tfrac{2}{3}\eta) = \omega R \sin(\psi + \tfrac{2}{3}\eta).$$

Now consider the two right-angled spherical triangles PMB and KMB: their circular parts for writing down their formulas by Napier's Rule are:

$\frac{\text{MB}}{R}$	$\frac{\text{MP}}{R}$	$\frac{\text{MB}}{R}$	$\frac{\text{MK}}{R}$
90° − (PBM)	90° − ω	90° − (ψ + n)	ψ
	ψ		90° − ω

Three elements are common to both sets and in the same relative positions except for being written opposite ways round. Hence the other corresponding elements must be equal, and so we get

(i) $\frac{\text{MP}}{R} = 90° - (\psi + \eta)$: (ii) $90° - (\text{PBM}) = \frac{\text{MK}}{R}$.

Now $\frac{MP}{R}$ is the co-latitude of the point M so from (i) the latitude of M is $(\psi+\eta)$. But the latitude of M is also

$$\phi+\frac{AM}{R} = \phi+\frac{s}{R} . \cos(A-\tfrac{2}{3}\varepsilon):$$

therefore

$$\psi = \phi+\frac{s}{R}\cos(A-\tfrac{2}{3}\varepsilon)-\eta.$$

This is the precise formula for latitude of B. It may be noted that η is the difference of latitude of B and M.

Consider the angles at point B. From (ii) we see that angle PBM is $90°-\omega . \sin(\psi+\tfrac{2}{3}\eta)$. The angle ABM is $90°-A+\varepsilon$, so to make up the 180° the angle B as shown must be

$$A+\omega . \sin(\psi+\tfrac{2}{3}\eta)-\varepsilon$$

and the reverse bearing is got by adding 180°. This is the precise formula for the reverse bearing, which of course is required for calculating the bearing to the next point to be computed, by adding or subtracting the appropriate adjusted angle at B in the survey figure.

The argument so far has provided formulas for working on a sphere: to transfer to the spheroid, we consider the sphere centred at N and of radius NB, the normal at B. This sphere will obviously touch the spheroid along the parallel of latitude of B. Elsewhere it will be above the spheroid as shown on Fig. 103. The normal through A cuts the sphere at A_1, and it can easily be proved that at ordinary geodetic distances the length of AA_1 will be at most only a metre or two. This being so, it is evident that the angle MA_1B and the distance A_1B cannot differ significantly from the given bearing A and distance s respectively. So now we have the figure A_1BP_1M on the sphere of radius NB ($= \nu_\psi$), and the formulas already developed can be applied where appropriate.

For the latitude difference, we have the length of A_1M equal to $s . \cos(A-\tfrac{2}{3}\varepsilon)$, and AM can be taken equal to this. But AM is on the spheroid and to convert it to a difference of latitude an average radius of curvature along AM must be used. It is easily shown that at ordinary geodetic distances the radius ρ_m for the mean of the latitudes of A and B is sufficiently accurate. Thus the formula for the spheroid is

$$\psi = \phi+\frac{s . \cos(A-\tfrac{2}{3}\varepsilon)}{\rho_m}-\eta.$$

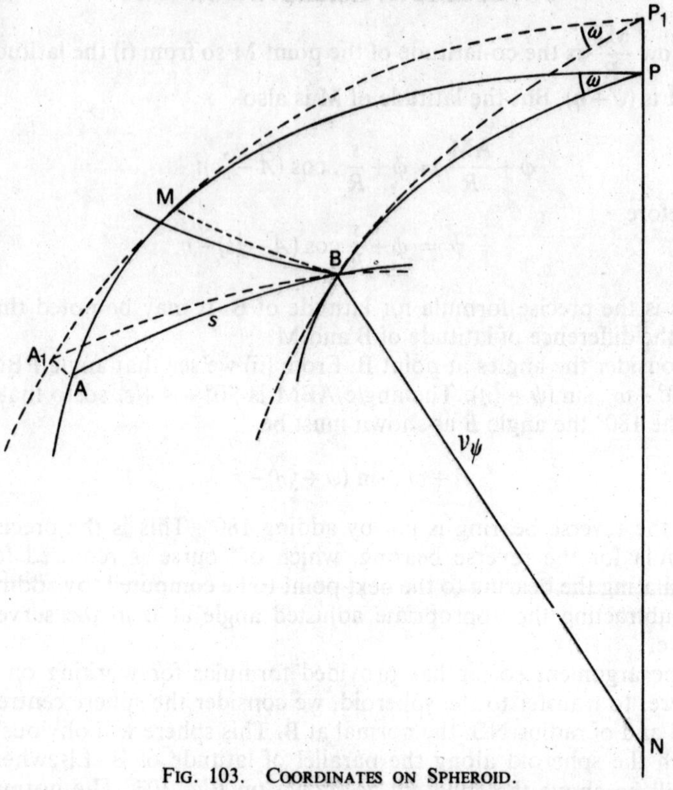

FIG. 103. COORDINATES ON SPHEROID.

As for the longitude, there is nothing more to do: it is the same on sphere as on spheroid and with the appropriate radius we have

$$\omega = \frac{s \cdot \sin\left(A - \frac{1}{3}\varepsilon\right)}{\nu_\psi \cdot \cos\left(\psi + \frac{1}{3}\eta\right)}.$$

Angles at B are also the same and we have the reverse bearing

$$A \pm 180° + \omega \cdot \sin\left(\psi + \tfrac{2}{3}\eta\right) - \varepsilon.$$

In actual computations of Clarke's Formulas, the steps are:

(a) calculate $s \cdot \cos A$ and $s \cdot \sin A$ and hence ε: an average curvature is introduced by using the formula

$$\tfrac{1}{2}s^2 \sin A \cdot \cos A / \rho\nu \sin 1''$$

where the values of ρ and ν are taken for an estimated mean latitude of the triangle ABM.

(b) from $s \cdot \cos \text{A}/(\rho \cdot \sin 1'')$ get an approximate latitude (ψ') of B.

(c) calculate η as $\varepsilon \,.\, \tan A \,.\, \tan \psi'$.
(d) proceed with the precise computations.

It is to be noted that the quantity ε, though shown on the diagrams as a spherical excess, must be given its sign according to the mathematical formula; it will be negative for instance when the bearing A is in the second quadrant. A computation example is given below. It is a computation by logarithms making use of prepared tables of

CLARKE'S FORMULAS

Data: $\phi = 50° \ 00' \ 00{\cdot}000''$N $s = 39\ 178{\cdot}29$ m
$L = \ 2 \ 00 \ 00{\cdot}000$ E $A = 128° \ 44' \ 17{\cdot}2''$

Spheroid: as for the Table in Section 6.04.

Preliminary Computations

log s	4·593 05	
log cos A	$\bar{1}$·796 41–	(1)
log $(1/\rho)$	$\bar{7}$·195 66	
log 206 265	5·314 43	
log $\Delta\phi''$	2·899 55–	(2)
log s	4·593 05	
log sin A	$\bar{1}$·892 10	(1)
log $(1/\nu)$	$\bar{7}$·194 44	
log $\frac{1}{2}$	$\bar{1}$·698 97	
log ε''	0·278 11–	(4)
log tan A	0·095 69–	(1)
log tan ψ'	0·072 79	(3)
log η''	0·446 59	(5)

(1) negative signs of the cosine and tangent are noted: the sine is positive.
(2) approximate change of latitude is $-794'' = -13'' \ 14''$; hence approximate ψ is $49° \ 46' \ 46''$ and, for the precise computation, ρ_m is taken at the mean latitude $49° \ 53' \ 23''$.
(3) use approximate ψ as given above.
(4) $\varepsilon'' = -1{\cdot}9''$; hence
$A - \frac{1}{3}\varepsilon = 128° \ 44' \ 17{\cdot}8''$,
$A - \frac{2}{3}\varepsilon = 128° \ 44' \ 18{\cdot}5''$.
(5) $\eta = 2{\cdot}796''$

Precise Computations

log s	4·593 045 5
log cos $(A - \frac{2}{3}\varepsilon)$	$\bar{1}$·796 412 3–
log $(1/\rho_m)$	$\bar{7}$·195 665 8
log 206 264·8	5·314 425 1
log $\Delta\phi''$	2·899 548 7–
antilog	−793·503″
=	−13′ 13·503″
$-\eta$	− 2·796
ϕ	50° 00 00·000
ψ	46 46 43·701
$\psi + \frac{1}{3}\eta$	49 46 44·63
$\psi + \frac{2}{3}\eta$	49 46 45·56
log s	4·593 045 5
log sin $(A - \frac{1}{3}\varepsilon)$	$\bar{1}$·892 101 5
log sec $(\psi + \frac{1}{3}\eta)$	0·189 944 4
log $(1/\nu_\psi)$	$\bar{7}$·194 455 4
log 206 264·8	5·314 425 1
log ω''	3·183 971 9
log sin $(\psi + \frac{2}{3}\eta)$	$\bar{1}$·882 844 9
	3·066 816 8
antilog	1 166·3″
=	+19′ 26·3″
$-\varepsilon$	+ 1·9
$A \pm 180°$	308 44 17·2
B	309 03 45·4
ω''	1 527·467″
=	+25′ 27·467″
L_ϕ	2 00 00·000 E
L_ψ	2 25 27·467 E

In practice, a computer would work to one more decimal place in coordinates and angles, to avoid accumulation of rounding errors.

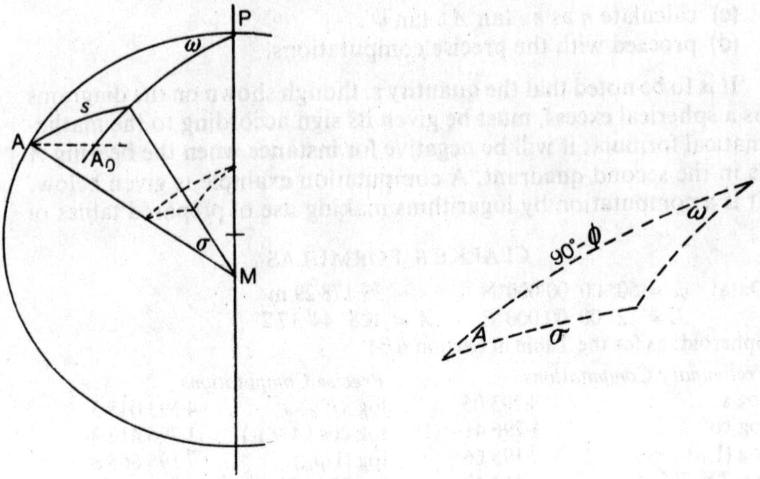

FIG. 104. GEOMETRY OF SERIES FORMULAS.

ρ, ν, etc. If an electronic computer of adequate size is used, all such quantities can be calculated from basic formulas by sub-routines.

6.09 Series Formulas: Puissant, etc. A different kind of approach to geodetic coordinate computation is that of converting the distance s to an angle equivalent, σ in Fig. 104, and generating a spherical triangle on a sphere centred at M on the normal at A.

Knowing σ, one has three elements of the triangle. It is then possible to calculate ω, the difference of longitude, directly from a standard spherical formula. The side opposite angle A is approximately the co-latitude of B, $(90° - \psi)$, but further computation is needed to get the exact value of ψ, since BM is not the normal at B.

The arc AB is part of the plane vertical section ellipse, and for given values of bearing A and latitude ϕ the dimensions of this ellipse can be found as described in Section 6.05. But first it is necessary to derive some general formulas of ellipse geometry, beginning with the expression of the ellipse in polar coordinates r and θ related to the tangent at A as indicated in Fig. 105. In relation to the (x, y) system shown, the coordinates of A are $(\nu . \cos \phi;$ $\nu(1-e^2) \sin \phi)$ and the chord r makes angle $(\phi+\theta)$ with the y axis. Hence the rectangular coordinates of the end of the chord are:

$$x = \nu . \cos \phi \qquad -r . \sin (\phi+\theta),$$
$$y = \nu(1-e^2) \sin \phi + r . \cos (\phi+\theta).$$

Substituting these into the ellipse formula given in Section 6.04

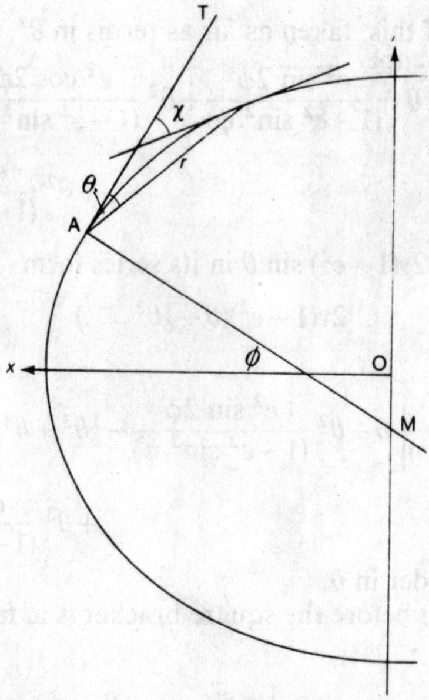

FIG. 105. POLAR COORDINATES.

gives, after some algebra, the polar equation

$$r(1-e^2 \sin^2(\phi+\theta)) = 2(1-e^2) \cdot \nu \cdot \sin\theta,$$

referred to AT as initial line. Incidentally, another way to get the formula for the whole chord AA′ as given in Section 6.05 is to put $\theta = 90°$ in this equation.

In dealing with a short arc length on the ellipse, the polar equation can be expanded in powers of the small angle θ, which is the vertical angle of depression at A to the other end of the arc. By the usual expansion procedure we have:

$$\sin^2(\phi+\theta) = \sin^2\phi + \theta \cdot \sin 2\phi + \tfrac{1}{2}\theta^2 \cdot 2\cos 2\phi + \tfrac{1}{6}\theta^3(-4\sin 2\phi)\ldots$$

in which the trigonometrical factors are successive differential coefficients of $\sin^2\phi$. Therefore:

$$1-e^2\sin^2(\phi+\theta)$$

$$= 1-e^2\sin^2\phi - \theta \cdot e^2 \sin 2\phi - \theta^2 \cdot e^2\cos 2\phi + \tfrac{2}{3}\theta^3 \cdot e^2 \sin 2\phi \ldots$$

$$= (1-e^2\sin^2\phi)\left[1-\theta \cdot \frac{e^2\sin 2\phi}{1-e^2\sin^2\phi} - \theta^2\frac{e^2\cos 2\phi}{(1-e^2\sin^2\phi)} + \tfrac{2}{3}\theta^3\frac{e^2\sin 2\phi}{(1-e^2\sin^2\phi)}\ldots\right].$$

The reciprocal of this, taken as far as terms in θ^2, is

$$\frac{1}{1-e^2\sin^2\phi}\left[1+\theta\cdot\frac{e^2\sin 2\phi}{(1-e^2\sin^2\phi)}+\theta^2\frac{e^2\cos 2\phi}{(1-e^2\sin^2\phi)}+\theta^2\frac{e^4\sin^2 2\phi}{(1-e^2\sin^2\phi)^2}\cdots\right]$$

Multiply this by $2\nu(1-e^2)\sin\theta$ in its series form

$$2\nu(1-e^2)(\theta-\tfrac{1}{6}\theta^3\ \ldots)$$

and we get:

$$r=\frac{2\nu(1-e^2)}{(1-e^2\sin^2\phi)}\left[\theta+\theta^2\frac{e^2\sin 2\phi}{(1-e^2\sin^2\phi)}-\tfrac{1}{6}\theta^3+\theta^3\frac{e^2\cos 2\phi}{(1-e^2\sin^2\phi)}+\theta^3\frac{e^4\sin^2 2\phi}{(1-e^2\sin^2\phi)^2}\cdots\right]$$

taken to third order in θ.

The expression before the square bracket is in fact 2ρ, because of the identity

$$\frac{1-e^2}{1-e^2\sin^2\phi}=\frac{\rho}{\nu}.$$

Now the angle θ is about half the subtended angle σ, so that it is unlikely that in practice θ will be greater than 1/200 radian, which corresponds with s being about 64 km. Even taking $\theta = 1/100$ radian, the value of $2\rho\theta^3 \cdot e^2$ is only about 8 cm. Clearly therefore, only the terms given below need be retained when ordinary geodetic survey distances are in consideration:

$$r=2\rho\left[\theta+\theta^2\cdot\frac{e^2\cdot\sin 2\phi}{1-e^2\cdot\sin^2\phi}-\tfrac{1}{6}\cdot\theta^3+\theta^3\cdot\frac{e^2\cdot\cos 2\phi}{(1-e^2\sin^2\phi)}\cdots\right.$$

The formula looks a bit neater if one uses the identity mentioned above and writes:

$$r \neq 2\rho\left[\theta+\theta^2\cdot\frac{\rho}{\nu}\cdot\frac{e^2}{1-e^2}\cdot\sin 2\phi-\tfrac{1}{6}\theta^3+\theta^3\cdot\frac{\rho}{\nu}\cdot\frac{e^2}{1-e^2}\cos 2\phi\ \ldots\right.$$

and it is quite sufficiently accurate to write the last term in the series as $\theta^3 \cdot e^2 \cdot \cos 2\phi$.

In computations of geodetic positions, it is the curve distance s that is known. As mentioned earlier, we require the angle σ, and it will be found that the angle χ will also be useful, because it is the difference of latitude from A. The required formulas can be obtained by standard differential expansion processes, as outlined below.

Let us assume a general series expression for r;

$$r = A \,.\, \theta + B \,.\, \theta^2 + C \,.\, \theta^3 + D \,.\, \theta^4 \ldots$$

as basis: the coefficients have linear dimension.

To obtain the curve length s we use the polar differential formula $(ds)^2 = (dr)^2 + r^2(d\theta)^2$. Now

$$dr = (d\theta)(A + 2B \,.\, \theta + 3C \,.\, \theta^2 + 4D \,.\, \theta^3 + \ldots).$$

Working to order θ^3, this expression is squared, $r^2(d\theta)^2$ is added, the square root is taken and the result is integrated, and the formula for s is found to be:

$$s = A \,.\, \theta + B \,.\, \theta^2 + (\tfrac{1}{6}A + C)\theta^3 + D \,.\, \theta^4 + \ldots.$$

Inverting this gives:

$$\theta = \frac{s}{A} - \frac{B}{A}\frac{s^2}{A^2} + \left(2\frac{B^2}{A^2} - \frac{1}{6} - \frac{C}{A}\right)\frac{s^2}{A^3}$$

$$- \left(5\frac{B^2}{A^2} - 5\frac{B}{A} - 5\frac{BC}{A^2} + \frac{D}{A}\right)\frac{s^4}{A^4} \ldots,$$

and substituting back into the series for r gives:

$$r = s - \tfrac{1}{6}A \,.\, \frac{s^3}{A^3} + \tfrac{1}{2}B \,.\, \frac{s^4}{A^4} \ldots.$$

Introducing temporarily the rectangular coordinate system $(\xi;\, \eta)$ shown on Fig. 106, the quantities we require are given by:

$$\tan \chi = d\eta/d\xi \quad \text{and} \quad \tan \sigma = \xi/(v - \eta).$$

Using the expansions for $\sin \theta$ and $\cos \theta$, we find that:

$$\xi = r \,.\, \cos \theta = A \,.\, \theta + B \,.\, \theta^2 + (C - \tfrac{1}{2}A) \,.\, \theta^3 + (D - \tfrac{1}{2}B) \,.\, \theta^4 \ldots$$

$$\eta = r \,.\, \sin \theta = A \,.\, \theta^2 + B \,.\, \theta^3 + (C - \tfrac{1}{6}A) \,.\, \theta^4 \ldots.$$

Then $\quad d\eta/d\theta = 2A \,.\, \theta + 3B \,.\, \theta^2 + 4(C - \tfrac{1}{6}A) \,.\, \theta^3 \ldots$

and $\quad d\xi/d\theta = A + 2B \,.\, \theta + 3(C - \tfrac{1}{2}A) \,.\, \theta^2 \ldots.$

By division and some algebraic working, we get:

$$\chi = 2 \,.\, \theta - \frac{B}{A} \,.\, \theta^2 + \left(2\frac{B^2}{A^2} - 2\frac{C}{A} - \frac{1}{3}\right)\theta^3$$

$$- \left(4\frac{B^3}{A^3} - 7\frac{BC}{A^2} - \frac{B}{A} + 3\frac{D}{A}\right)\theta^4 \ldots$$

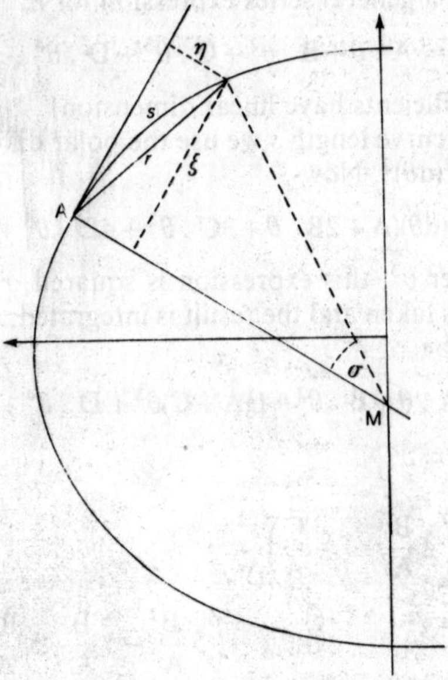

FIG. 106. GEOMETRY REFERRED TO ORIGIN AT A.

and by substituting the series for θ in terms of s:

$$\chi = 2\frac{s}{A} - 3\frac{B}{A}\frac{s^2}{A^2} + \left(8\frac{B^2}{A^2} - 4\frac{C}{A} - \frac{2}{3}\right)\frac{s^3}{A^3}$$

$$-\left(15\frac{B^3}{A^3} + 10\frac{B^2}{A^2} - 25\frac{BC}{A^2} - \frac{7}{3}\cdot\frac{B}{A} + 4\frac{D}{A}\right)\frac{s^4}{A^4}\dots$$

The series for σ comes out as:

$$\sigma = \frac{A}{v}\cdot\theta + \frac{B}{v}\theta^2 + \left(\frac{C}{v} - \frac{1}{2}\frac{A}{v} + \frac{A^2}{v^2} - \frac{1}{3}\frac{A^3}{v^3}\right)\theta^3$$

$$+\left(\frac{D}{v} - \frac{1}{2}\frac{B}{v} + 2\frac{AB}{v^2} - \frac{A^2B}{v^3}\right)\theta^4\dots$$

and when this is converted to a function in s the series is:

$$\sigma = \frac{s}{v} - \frac{1}{3}\left(2\frac{v}{A} - 1\right)\left(\frac{v}{A} - 1\right)\frac{s^3}{v^3} + \left(\frac{3}{2}\cdot\frac{B}{A} - \frac{B}{v}\right)\frac{s^4}{A^3 v}\dots$$

To complete the expressions in s, we find that:

$$\xi = s-\tfrac{2}{3}A\,.\,\frac{s^3}{A^3}+\tfrac{3}{2}B\,.\,\frac{s^4}{A^4}\ldots$$

$$\eta = A\,.\,\frac{s^2}{A^2}-B\frac{s^3}{A^3}+\left(2\frac{B^2}{A}-\tfrac{1}{2}A-C\right)\frac{s^4}{A^4}\ldots$$

By combination of the two series for σ and χ in terms of s, we get:

$$\chi = 2\frac{\nu}{A}\sigma-3\frac{B}{A}\,.\,\frac{\nu^2}{A^2}\,.\,\sigma^2-\frac{2}{3}\frac{\nu}{A}\left(2\frac{\nu}{A}-1\right)\left(1-\frac{\nu}{A}\right).\,\sigma^3$$
$$+\left(8\,.\,\frac{B^2}{A^2}-4\,.\,\frac{C}{A}-\frac{2}{3}\right)\frac{\nu^3}{A^3}\,.\,\sigma^3\ldots$$

to third order in σ.

The foregoing general formulas can now be applied to the computation on the spheroid. The distance s is along the plane section ellipse, the dimensions of which are given in Section 6.05, and the relevant expansion for r is

$$r = 2\rho_1\left[\theta+\frac{\rho_1}{\nu}\,.\,\frac{e_1^{\,2}\sin 2\phi_1}{1-e_1^{\,2}}\,.\,\theta^2-\tfrac{1}{6}\theta^3+\frac{\rho_1}{\nu}\,.\,\frac{e_1^{\,2}\cos 2\phi_1}{1-e_1^{\,2}}\ldots\right]$$

The coefficients of the powers of θ in this series are to be substituted for the 'A', 'B', and 'C' of the general formulas. The formula for converting s to r becomes very simple: the 'A' of the general formula is $2\rho_1$ and the 'B' is

$$\frac{2\rho_1^{\,2}}{\nu}\,.\,\frac{e_1^{\,2}\,.\,\sin 2\phi_1}{(1-e_1^{\,2})}$$

Using the relationships given in Section 6.05, 'B/A' can be written as

$$\frac{\rho_1}{\nu}\,.\,\frac{e^2\cos A\,.\,\sin 2\phi}{(1-e^2)},$$

so we get:

$$r = s-\frac{s^3}{24\rho_1^{\,2}}+\frac{e^2\,.\cos A\,\sin 2\phi}{(1-e^2)}\,.\,\frac{s^4}{16\nu\rho_1^{\,2}}\ldots$$

By putting some reasonable values into this formula, it is soon found that the term in s^4 will in practice be quite negligible, and even the term in s^3 is not likely to be more than a fraction of a metre. It is therefore not a critical question, what value to take for ρ_1; strictly it is the radius of curvature of the plane section and could be calculated exactly from the general formula of surface geometry:

$$\frac{1}{\rho_1} = \frac{\cos^2 A}{\rho} + \frac{\sin^2 A}{\nu}.$$

where the ρ and ν are of course the values for the spheroid at the latitude of A. Thus the working formula for chord length,

$$r = s - \frac{s^3}{24\rho_1{}^2},$$

is quite sufficiently accurate: for s = 100 km the second term is just over 1 m.

For the angle σ we have:

$$\sigma = \frac{s}{\nu} - \frac{1}{3}\left(\frac{\nu}{\rho_1} - 1\right)\left(\frac{\nu}{2\rho_1} - 1\right)\frac{s^3}{\nu^3}\ldots$$

and again it will be found that the next term, because it contains e^2 as a factor, represents an amount far smaller than can possibly be significant in practical cases. Substituting for ρ_1, this can be written:

$$\sigma = \frac{s}{\nu} + \tfrac{1}{6} \,.\, e^2 \,.\, \cos^2 A \,.\, \cos^2 \phi \,.\, \frac{s^3}{\nu^3} \ldots$$

quite accurately enough for practical use at ordinary geodetic distances.

With the value of s used in Section 6.08, and ν = 6 390 725·2 m, we find that (s/ν). 206 264·8 = 1264·505″ = 21′ 04·505″, and the other term is less than 10^{-5} second!

We now return to the spherical triangle of Fig. 104, and denote the side opposite angle A—the angle BMP—as $90° - (\phi + \delta)$. By standard spherical formula,

$$\sin(\phi + \delta) = \sin\phi \,.\, \cos\sigma + \cos\phi \,.\, \sin\sigma \,.\, \cos A.$$

By expansion, and use of trigonometrical series,

$$\sin\phi + \delta \,.\, \cos\phi - \tfrac{1}{2}\delta^2 \,.\, \sin\phi - \tfrac{1}{6}\delta^3 \,.\, \cos\phi \ldots$$
$$= \sin\phi + \sigma \,.\, \cos\phi \,.\, \cos A - \tfrac{1}{2}\sigma^2 \,.\, \sin\phi - \tfrac{1}{6}\sigma^3 \,.\, \cos\phi \,.\, \cos A \ldots$$

Then some algebraic work yields the explicit expression:

$$\delta = \cos \mathrm{A} \,.\, \sigma - \tfrac{1}{2}\tan\phi \,.\, \sin^2 A \,.\, \sigma^2 - \tfrac{1}{2}\cos A \,.\, \sin^2 A(\tan^2\phi + \tfrac{1}{3}) \,.\, \sigma^3 \ldots$$

Now the formula for σ in terms of s can be introduced, with the result:

$$\delta = \cos A \,.\, \frac{s}{\nu} - \tfrac{1}{2} \tan \phi \,.\, \sin^2 A \,.\, \frac{s^2}{\nu^2}$$
$$- \tfrac{1}{2} \cos A \,.\, \sin^2 A\,(\tan^2 \phi + \tfrac{1}{3}) \,.\, \frac{s^3}{\nu^3} \ldots$$

in which only terms up to third order (taking e^2 and $\frac{s}{\nu}$ as small) are included.

Now we consider the meridian of point B, with the point A_0, Fig. 104, at latitude ϕ; then to get the latitude of B we must add to ϕ the angle χ, with δ substituted for σ. The appropriate value of the coefficient 'A' is of course now 2ρ, since we are working on a spheroid meridian, and the 'B/A' is

$$\frac{\rho}{\nu} \,.\, \frac{e^2 \,.\, \sin 2\phi}{1 - e^2} \,.$$

It is easily seen that the coefficients of σ^3 in the formula for χ both have e^2 as factor, so if we confine our work to third order we are left with:

$$\chi = \frac{\nu}{\rho} \,.\, \delta - 3 \frac{\rho}{\nu} \frac{e^2 \sin 2\phi}{(1 - e^2)} \frac{\nu^2}{4\rho^2} \,.\, \delta^2 \ldots$$
$$= \frac{s}{\rho} \cos A - \tfrac{1}{2} \tan \phi \,.\, \sin^2 A \,.\, \frac{s^2}{\nu\rho} - \tfrac{1}{2} \cos A \,.\, \sin^2 A (\tan^2 \phi + \tfrac{1}{3}) \,.\, \frac{s^3}{\nu^2 \rho}$$
$$- \tfrac{3}{4} \,.\, e^2 \sin 2\phi \,.\, \cos^2 A \,.\, \frac{s^2}{\nu\rho} \ldots$$

This is the series formula for difference of latitude. If fourth-order terms are investigated, it is found that some of them might just be significant at the 0·001″ level for distances s over about 60 km.

The radius of curvature ρ is that at latitude ϕ: it seems reasonable that if the radius ρ_m for the mean latitude of A and B is used, the formula might become simpler; this is so. By considering the differential coefficient of $1/\rho$ with respect to ϕ, and using $s \,.\, \cos A/2\rho$ as the half-difference, we find that:

$$\frac{1}{\rho} = \frac{1}{\rho_m} + \frac{3}{4} \,.\, \frac{e^2 \sin 2\phi \,.\, \cos A}{(1 - e^2)} \,.\, \frac{s}{\nu} \ldots$$

and the formula for χ is now:

$$\chi = \frac{s \,.\, \cos A}{\rho_m} - \tfrac{1}{2} \tan \phi \,.\, \sin^2 A \,.\, \frac{s^2}{\nu \rho_m}$$
$$- \tfrac{1}{2} \cos A \,.\, \sin^2 A (\tan^2 \phi + \tfrac{1}{3}) \frac{s^3}{\nu^2 \rho_m} \ldots$$

This is essentially the formula published by Puissant in 1819. Incidentally it is, to this order of approximation, equal to $\frac{\nu}{\rho_m}\,.\,\delta$.

$s = 39\ 178{\cdot}29$ m $\quad A = 128°\ 44'\ 17{\cdot}2''$ $\quad \phi = 50°\ 00'\ 00{\cdot}000''$

$\nu = 6\ 390\ 725{\cdot}2$ m

$s/\nu = 0{\cdot}006\ 130\ 492$

$(s/\nu)206\ 264{\cdot}8 = 1264{\cdot}5047''$

$\sin A = 0{\cdot}780\ 014\ 3$

$\cos A = -0{\cdot}625\ 761\ 6$

$\cos\phi = 0{\cdot}642\ 787\ 6$

$\sin^2 A = 0{\cdot}608\ 422\ 4$

$\cos^2 A = 0{\cdot}391\ 6$

$\tan\phi = 1{\cdot}191\ 75$

The other term for σ is negligible: hence

$\sigma = 21'\ 04{\cdot}5047''$

$\cos A\,.\,\frac{s}{\nu}(206\ 264{\cdot}8)$	$-791{\cdot}2785$
$-\frac{1}{2}\tan\phi\,.\,\sin^2 A\,.\,\frac{s^2}{\nu^2}(206\ 264{\cdot}8)$	$-2{\cdot}8105$
$-\frac{1}{2}\cos A\,.\,\sin^2 A(\tan^2\phi+\frac{1}{3})\frac{s^3}{\nu^3}(206\ 264{\cdot}8)$	$+0{\cdot}0159$
$\delta =$	$-794{\cdot}0731''$
$=$	$-13'\ 14{\cdot}0731''$

$\sin\sigma = 0{\cdot}006\ 130\ 045\ 4$

$\sec(\phi+\delta) = 1{\cdot}548\ 630\ 3$

$\phi+\delta \quad 49°\ 46'\ 45{\cdot}9269''$

$\therefore\ \sin\omega = 0{\cdot}007\ 405\ 304\ 6$

$\omega = 1\ 527{\cdot}468''$

$= 0°\ 25'\ 27{\cdot}468''$

$\phi = 50°\ 00'\ 00''$

$\frac{1}{2}\Delta\phi \doteq -6\ 37$

$\phi_m = 49\ 53\ 23$

$\frac{\nu}{\rho_m}\,.\,\delta = -796{\cdot}299''$

$= -13'\ 16{\cdot}299''$

$\rho_m = 6\ 372\ 857{\cdot}3$

$\phi = 50\ 00\ 00{\cdot}000$

$\psi = 49\ 46\ 43{\cdot}701''$

Back-bearing by spherical formula:

$\phi_m = 49\ 53\ 21{\cdot}85''$

$\frac{1}{2}\Delta\phi \quad -6\ 38{\cdot}15$

$\frac{1}{2}\omega \quad 12\ 43{\cdot}734$

$\sin\phi_m \quad 0{\cdot}764\ 802\ 2$

$\sec\frac{1}{2}\Delta\phi \quad 1{\cdot}000\ 001\ 9$

$\tan\frac{1}{2}\omega \quad 0{\cdot}003\ 702\ 703$

$\therefore\ \tan\frac{1}{2}\gamma = 0{\cdot}002\ 831\ 840\ 8$

$\frac{1}{2}\gamma =$	$9'\ 44{\cdot}11''$
convergence $\gamma =$	$19\ 28{\cdot}2$
$A \pm 180°$	$308\ 44\ 17{\cdot}2$
B	$309\ 03\ 45{\cdot}4$

After calculating the difference of latitude as just described, there are several possible procedures. For example, referring to the spherical triangle, if σ and δ have been computed, then the difference of longitude ω is obtained directly from the sine formula: $\sin\omega = \sin\sigma \,.\, \sin A/\cos(\phi+\delta)$.

For the reverse bearing the spherical formula may be used, in accordance with Dalby's Theorem, as mentioned in Section 6.06: or the exact plane section bearing can be calculated as explained in Section 6.05. The numerical example on page 204 uses the same data as in Section 6.08, and the results are in agreement.

In the past, some ingenuity had to be exercised in choosing formulas and methods of computation to give the desired accuracy with available logarithmic aids—usually the familiar Chambers and Shortrede Tables.

As explained in previous Sections, spheroid coordinates had to be calculated by point-to-point differences, to get values correct to 0·001″. Advantage was taken of the smallness of certain quantities, and expansion in series was often helpful or even necessary. For instance, with 7-figure logarithms, a value of the radius of curvature ν can be found to 0·01 m accuracy from the series formula $a+\frac{1}{2}a \,.\, e^2 \sin^2\phi+\frac{3}{8}a \,.\, e^4 \sin^4\phi \ldots$ whereas the value computed from the closed formula $a(1-e^2 \,.\, \sin^2\phi)^{-\frac{1}{2}}$ could be wrong by a metre or two (and anyway this is a clumsy formula for logarithmic calculation).

Nowadays, even with desk-top electronic computers of quite modest size (and price), there is no trouble from restriction on the number of digits, and no need to search out small economies of computing labour, since the difference between one formula and another may be only a matter of a few milliseconds in the machine.

6.10 Computing in 3D Coordinates. Advantage can be taken of modern computer power to calculate geodetic work in the framework of a three-dimensional rectangular coordinate geometry.

In Fig. 107, which has much in common with Figs. 99 and 106, BK is the perpendicular from B on to the meridian plane of A: BE and KE are perpendicular to the normal AM: KH is perpendicular to the axis of the spheroid. In the notation of Section 6.09, $\xi = r \,.\, \cos\theta$, $\eta = r \,.\, \sin\theta$, and obviously $KE = \xi \,.\, \cos A$, $BK = \xi \,.\, \sin A$. Taking axes OX OY as shown in the meridian plane of A, the perpendicular BK may be denoted ζ in the third coordinate direction. See also Fig. 108.

The coordinates of A are $(\nu \,.\, \cos\phi: \nu(1-e^2) \,.\, \sin\phi: 0)$ (see Section 6.04), and it is easy to see that the coordinates of K are then:

$$x = \nu \,.\, \cos\phi-\eta \,.\, \cos\phi-\xi \,.\, \cos A \,.\, \sin\phi = HK,$$
$$y = \nu(1-e^2)\sin\phi-\eta \,.\, \sin\phi+\xi \,.\, \cos A \,.\, \cos\phi = HO,$$
$$z = 0.$$

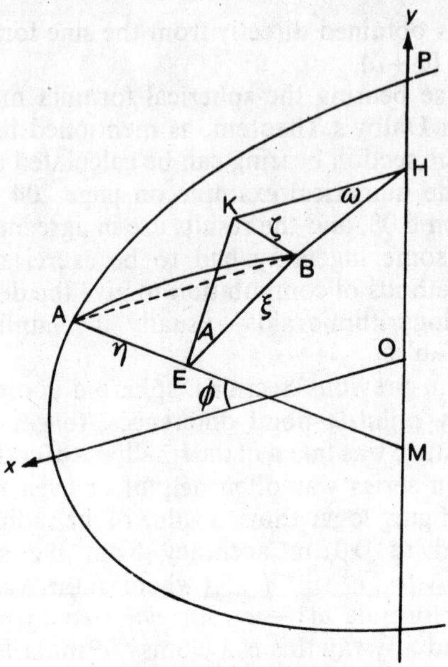

FIG. 107. 3D GEOMETRY.

Then we have immediately the longitude difference ω from

$$\tan \omega = \zeta/x = (r \,.\, \cos \theta, \sin A)/x.$$

Further, BH $= x \,.\, \sec \omega$, and the y coordinate of B is the same as that of K given above, so, if the latitude of B is denoted by ψ, we have;

$$\frac{y}{x \,.\, \sec \omega} = \frac{\nu_\psi \,.\, (1-e^2) \sin \psi}{\nu_\psi \,.\, \cos \psi}$$

whence the latitude formula:

$$\tan \psi = (y \,.\, \cos \omega)/x(1-e^2).$$

From this formula, ψ is obtainable to 0·001″ if the computing facility will operate to that order of precision: a little thought will indicate that calculations with 9 or 10 significant figures will be necessary.

As before, the reverse bearing can be calculated by precise formula of Section 6.06.

Thus, once the values of r and θ are known, the rest of the calculation follows by exact and simple formulas. Instead of calculating r and θ, the values of ξ and η can be derived more directly from s, using the formulas of Section 6.09; taking account of the dimensions of the plane section AB we find that, to sufficient accuracy,

$$\xi = s - \tfrac{1}{6} \,.\, s \,.\, \frac{s^2}{\rho_1{}^2} \ldots$$

$$\eta = \tfrac{1}{2}s \,.\, \frac{s}{\rho_1} - \tfrac{1}{4}s \,.\, \frac{e^2 \cos A \,.\, \sin 2\phi}{(1-e^2)} \,.\, \frac{s^2}{\nu\rho_1} \ldots$$

The computations below use the same data as in the previous examples. To get the value of s/ρ_1, one can use the general formula for curvature of a skew line in the form:

$$\frac{\nu}{\rho_1} = \frac{\nu}{\rho}\cos^2 A + \sin^2 A = \left(1 + \frac{e^2 \,.\, \cos^2 \phi}{1-e^2}\right)\cos^2 A + \sin^2 A$$

$$= 1 + \frac{e^2 \,.\, \cos^2 \phi \,.\, \cos^2 A}{1-e^2}$$

Then $$\frac{s}{\rho_1} = \frac{s}{\nu} \,.\, \frac{\nu}{\rho_1} = \frac{s}{\nu}\left(1 + \frac{e^2 \,.\, \cos^2 \phi \,.\, \cos^2 A}{1-e^2}\right)$$

$s = 39\ 178{\cdot}29$ m $\quad A = 128° \ 44' \ 17{\cdot}2''$ $\quad \phi = 50° \ 00' \ 00{\cdot}000''$

$e^2 = 0{\cdot}006\ 694\ 43$ $\quad e^2/(1-e^2) = 0{\cdot}006\ 739\ 547$

$\sin\phi$	0·766 044 443 1	$\nu =$	6 390 725·2 m
$\cos\phi$	0·642 787 609 7	$\nu(1-e^2) =$	6 347 942·9 m
$\sin A$	0·780 014 3	s/ν	0·006 130 492
$\cos A$	−0·625 761 6		6 685
$\sin 2\phi$	0·984 8	s/ρ_1	0·006 137 177

$s = 39\ 178{\cdot}29$ m $\quad \tfrac{1}{2}s\dfrac{s}{\rho_1} = 120{\cdot}222$ m

$\tfrac{1}{6}s \,.\, \dfrac{s^2}{\rho_1{}^2} = 0{\cdot}25$ $\quad -\tfrac{1}{4}s \,.\, \dfrac{e^2 \,.\, \cos A \,.\, \sin 2\phi}{1-e^2} \,.\, \dfrac{s^2}{\nu\rho_1} = +0{\cdot}002$

ξ 39 178·04 m $\quad \eta$ 120·224 m

Coordinates of K:

	x		y
$\nu \,.\, \cos\phi$	4 107 878·96	$\nu(1-e^2)\sin\phi$	4 862 806·40
$-\eta \,.\, \cos\phi$	− 77·28	$-\eta \,.\, \sin\phi$	−92·10
$-\xi \,.\, \cos A \sin\phi$	+ 18 780·44	$+\xi \cos A \,.\, \cos\phi$	− 15 758·66
x	4 126 582·44	y	4 846 955·64

$\tan\omega = \xi \sin A/x = 0{\cdot}007\ 405\ 509$ $\quad \omega = 25' \ 27{\cdot}468''$

$\cos\omega = 0{\cdot}999\ 972\ 730\ 6$

$$\tan\psi = \frac{y \,.\, \cos\omega}{x(1-e^2)} = \frac{4\ 846\ 822{\cdot}74}{4\ 098\ 957{\cdot}01} = 1{\cdot}182\ 452\ 69$$

$\psi = 49° \ 46' \ 43{\cdot}700''$.

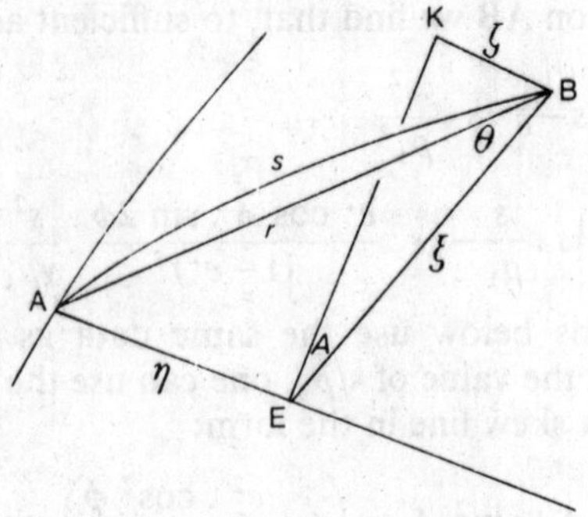

FIG. 108. 3D GEOMETRY.

6.11 Inverse Computations. The calculation of distances and bearings between two given points on the spheroid is usually called the Inverse Problem. Viewed in the three-dimensional system of the previous Section, this is a simple exercise. If ϕ and ψ are the latitudes and ω the difference of longitude, we see immediately that the coordinates of A and B are

A: $(\nu_\phi \,.\, \cos\phi: \quad \nu_\phi(1-e^2)\sin\phi: 0)$

B: $(\nu_\psi \,.\, \cos\psi \,.\, \cos\omega: \nu_\psi(1-e^2)\sin\psi: \nu_\psi \,.\, \cos\psi \,.\, \sin\omega)$

These values are calculated, the three differences pair by pair are squared and added, and the square root of the sum is the chord r. Coordinates must be calculated to 0·01 m precision, which means dealing in 9-digit numbers. The surface distance s is then obtained by adding $r^3/24\rho_1{}^2$ to r. Bearings can of course be found by using the exact formulas of Section 6.06.

Clarke's Formulas can also be worked in the inverse sense. Referring to Section 6.08, we see that the latitude and longitude formulas can be written:

$$s \,.\, \cos(A - \tfrac{2}{3}\varepsilon) = (\psi - \phi + \eta)\rho_m$$

$$s \,.\, \sin(A - \tfrac{1}{3}\varepsilon) = \nu_\psi \,.\, \cos(\psi + \tfrac{1}{3}\eta).$$

On the right-hand side of each of these, everything is known except η. But η and ε are both very small angles, so for computing them we can use the approximate quantities:

$$s \,.\, \cos A \doteqdot (\psi - \phi)\rho_m: \qquad s \,.\, \sin A \doteqdot \nu_\psi \,.\, \cos\psi,$$

whence η and ε are got from the Clarke Formulas:

$$\varepsilon = 206\,265 \,.\, (s \,.\, \sin A)(s \,.\, \cos A)/2\rho\nu \qquad \text{angle seconds,}$$

$$\eta = \varepsilon \,.\, \tan\psi \,.\, (s \,.\, \sin A)/(s \,.\, \cos A).$$

Now if we divide the exact longitude formula by the exact latitude formula, of which the right-hand sides are now completely known, we have

$$\frac{\sin(A-\frac{1}{3}\varepsilon)}{\sin(A-\frac{2}{3}\varepsilon)} = \frac{\nu_\psi \cdot \cos(\psi+\frac{1}{3}\eta)}{(\psi-\phi+\eta)\rho_m}$$

and this ratio must be very nearly tan A; calling it tan A', we can look up the angle A' in the tables and it only needs a small correction to get the true A. A little mathematical working shows that $A = A'+\frac{1}{3}\varepsilon(1+\sin^2 A')$. The distance s can now be obtained from either of the exact formulas, or preferably from both as a check, and the reverse bearing from the Clarke formula.

6.12 Plane Coordinates. The computation methods that have been described in this Chapter are applicable to geometrical figures and positions of points on a spheroidal reference surface. Extensive surveys are normally calculated in terms of spheroid coordinates, since this provides a basic description of the horizontal geometry of the survey. For mapping purposes, there must be a second stage of conversion of the spheroid coordinates to a plane system through the formulas of a chosen map projection, as described in detail in Chapter IX.

Indeed, for most purposes other than geodetic studies concerned with the shape of the Earth, plane rectangular coordinates will be the usual way of expressing the positions determined by surveying. Moreover, surveyors and engineers will not want to have to calculate spheroid coordinates, if they can derive the desired plane coordinates directly from the survey measurements. In order to do this correctly, account must be taken of the properties of the projection by either (i) using appropriate formulas, or (ii) modifying the angles and lengths of the survey geometry so that they become 'projection values' and can be used in standard plane trigonometrical formulas. These methods are described in Chapter IX.

An isolated survey may be calculated in an arbitrary plane rectangular coordinate system if the area covered is small, but any survey, of whatever extent, that is connected to an existing mapping system, must be calculated with due regard to the structure of that system, so that the plane coordinates obtained for the points are those that would have been obtained if the survey had been included in the main system in the first place.

In brief, the geometry of an extensive survey framework can be described either by angles lengths and spheroid coordinates on a reference surface, or by plane rectangular coordinates in a specified projection system. In the latter case, the angles and lengths of the plane figure will differ slightly from those of the spheroid figure; these differences are sometimes called projection errors, a rather misleading term since they can be calculated exactly from formulas appropriate to the projection!

EXAMPLES

1. On a sphere of radius 6 370 km, what are the bearings at the ends of a portion of great circle of length 20 km, both ends of which are exactly at latitude 50°?

2. On a sphere of radius 6 370 km the lengths of the sides of a triangle are 85·167 39, 94·003 24, and 71·925 83 km. Calculate the angles of the triangle.

3. Show that if the length of the ellipse arc between latitudes $(\phi+\frac{1}{2}\delta)$ and $(\phi-\frac{1}{2}\delta)$ is calculated as $\rho \,.\, \delta$, where ρ is the radius of curvature at latitude ϕ, the error is approximately $\frac{3}{4}ae^2 \,.\, \cos 2\phi \,.\, (\delta-\sin\delta)$, which for an arc of 1° is not more than 3 cm.

4. Calculate the differences of latitude and longitude due to a line of length exactly 50 km starting at bearing 45°, at a point at latitude 40° on the Krassovsky ellipsoid—a = 6 378 245 m, f = 1/298·3.

REFERENCES

BLAIKIE A. G. 'Report on the application of electronic desk top calculators in surveying'. *New Zealand Surveyor*, **5**, 239, 1971.

JACKSON J. E. 'A note on Clarke's Formulae'. *Empire Survey Review* No. 90, October 1953.

KING C. W. B. 'Computation of simple figures on sphere and spheroid'. *Survey Review* No. 162, October 1971.

SMITH J. R. 'The Hewlett-Packard 9810 desk calculator and its predecessors'. *Survey Review* No. 163, January 1972.

LEGENDRE A. M. *Eléments de géométrie*. Paris, 1813.

PUISSANT L. *Traité de géodésie*, 3rd Edn., Paris, 1842.

THOMAS J. L. 'Desk computers in surveying'. *Chartered Surveyor*. Vol 104 No. 5, 1971.

CHAPTER VII

GEODETIC LEVELLING AND HEIGHTS

7.01 The Third Dimension. The three methods commonly used in the third dimension of surveying, that is for finding relative heights of points, are Levelling, Vertical Angles, and Barometric Observations. A fourth method, Hydrostatic Levelling by means of a tube filled with water, has been used occasionally for transferring level across sea channels and in other special situations. The first and second of these four methods are described in this chapter.

Heights are expressed in linear units, metres or feet. The zero or datum for height is usually mean sea level as determined from a series of tidal observations. For the survey of a restricted area, it may be sufficient to use an arbitrary datum by assigning a value to one point in the survey. Nowadays, however, it is unlikely that a surveyor will have to establish a datum, because he will be able to connect his work to benchmarks provided by the national surveying organisation in the country in which he is working. These benchmarks have been established by systematic levelling operations of the highest precision, as described below.

Mean Sea Level. In order to determine the average elevation of the sea surface, a continuous record of its varying position in relation to a fixed point on land must be kept for as long a time as possible. The period of observation should cover an integral number of lunations, and unless it extends to at least a year the effects of the solar components will not be averaged out. The employment of a self-registering tide gauge is almost essential, and its zero must be connected and periodically checked by precise levelling to a permanent benchmark near at hand. Tide gauges are discussed in Vol. I and questions relating to mean sea level as a datum surface for heights in Section 7.09 of this volume. When a datum is defined in terms of mean sea level the particular period covered by the observations should be stated in the records and on maps. The adoption of a datum is effected by specifying the height of a benchmark near the tide gauge, and a datum should be adhered to as long as possible, to avoid confusion of records.

7.02 Precise Levelling. Levelling of the highest precision is used in establishing networks of permanent benchmarks for future reference. This sort of work is mostly done by national surveying organisations, but the surveyor engaged on other extensive work, especially if it is concerned with the flow of water, may decide that high-precision levelling is necessary to control the setting out of engineering structures and surfaces. The field method is in principle the same as ordinary levelling, but special instruments are used and certain refinements are applied in the treatment of the observations. The accuracy is indicated by the fact that the staff readings are booked to 0·1 mm or finer.

In ordinary levelling, when readings are taken by direct estimation of the position of the cross line against the graduations on the staff, the smallest interval of reading at normal sight distances is about $\frac{1}{2}$ mm. Precise levelling may be defined as levelling done with equipment designed so that a finer precision of reading is possible. This has been achieved, in most of the work done to date, by use of the parallel-sided-plate micrometer fitted to a telescope of magnification about 30× and provided with a tubular bubble of high sensitivity.

FIG. 109. PRECISE LEVEL WILD N3.
(photograph supplied by Wild Heerbrugg (U.K.) Limited)

A typical precise level is illustrated in Fig. 109. Apart from the parallel plate and its associated mechanism, the instrument is of the familiar tilting design, as described in Vol. I, Section 107. The tripod for a precise level must be of robust and very stable construction. In some instruments, the bubble can be seen through the telescope eyepiece, alongside the staff image, with the advantage that the

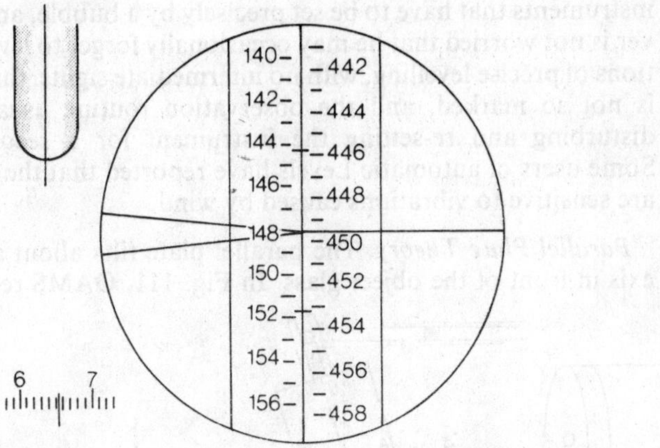

FIG. 110. EXAMPLE OF READING WITH WILD N3 PRECISION LEVEL
READING: 148,653 CM.

observer can check it while making the cross-line settings on the staff. The bubble must be a very sensitive one if the reading precision mentioned above is to be actually achieved. At a sighting distance of 40 metres, reading to 0·1 mm corresponds with a tilt of $\frac{1}{2}$ second angle: this is about the limit of sensitivity that is practicable in an instrument that is to be used on a portable tripod in variable weather conditions.

Automatic Levels. There are now available several types of Level in which the precise setting of the line of sight occurs automatically as soon as the instrument is approximately levelled by reference to a comparatively insensitive circular bubble. This setting is done by the effect of gravity on a component suspended inside the instrument on fine threads. The result is that if the levelling is slightly disturbed, as by a small movement of a levelling screw for instance, the cross line appears to move along the staff but immediately comes back to the original reading. The movement of the suspended component must of course be heavily damped, and this is usually done by surrounding the component or part of it by a loose-fitting container that provides an air-cushion effect. Some of these automatic instruments look like conventional Levels, but some, such as those produced by the firm Salmoiraghi, look like short periscopes. In many automatic Levels, the view of the staff is not inverted. There is not much published information on the sensitivity of automatic levels in comparison with those using high-quality bubbles.

In ordinary levelling work, automatic Levels are quicker than instruments that have to be set precisely by a bubble, and the observer is not worried that he may occasionally forget to level. In conditions of precise levelling, with no intermediate sights, this advantage is not so marked, and the observation routine usually requires disturbing and re-setting the instrument for a second reading. Some users of automatic Levels have reported that the instruments are sensitive to vibrations caused by wind.

Parallel Plate Theory. The parallel plate tilts about a horizontal axis in front of the object glass. In Fig. 111, OAMS represents the

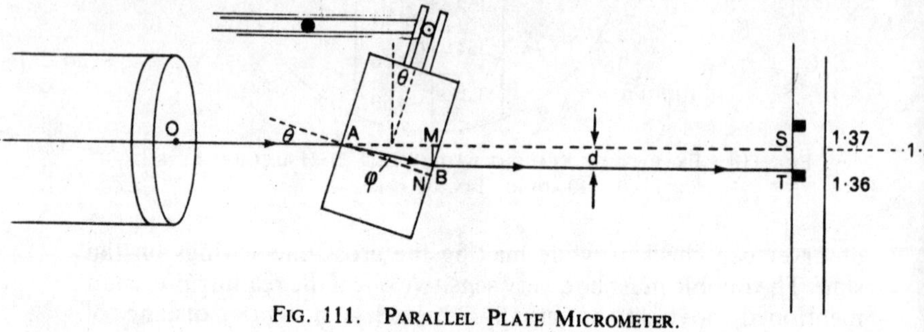

FIG. 111. PARALLEL PLATE MICROMETER.

centre line of the telescope, so the staff reading without the plate would be at S. The observer tilts the plate so that the cross-line appears to coincide with a specific graduation of the staff, e.g. 1·36 m as indicated. The small additional amount ST is read on a scale associated with the tilting mechanism. Let the plate be of thickness a, refractive index μ, and suppose that it is tilted at angle θ from the vertical. The refracted path of the line of sight is along AB and it emerges from the other face of the plate in a direction exactly parallel to AS, provided that the faces of the plate are optically parallel. If angle BAN is called ϕ, the refraction law gives $\mu \,.\, \sin \phi = \sin \theta$; then $AB = a \,.\, \sec \phi$, $BM = ST = d = a \,.\, \sec \phi \,.\, \sin (\theta - \phi)$. As a reasonable example, take $a = 25$ mm, $\mu = 3/2$: some corresponding values of θ and d are given in the table below.

θ	ϕ	d(mm)	$d/\tan \theta$
5°	3° 20′	0·7293	8·336
10	6 39	1·4718	8·347
15	9 56	2·2404	8·361
20	13 11	3·0490	8·377
25	16 22	3·9122	8·390
30	19 28	4·8454	8·392
35	22 29	5·8647	8·376
40	25 22	6·9863	8·326

The graduations on the scale for reading values of the displacement ST must of course be placed to correspond with equal increments of d, and movement over the total length of the scale must displace the line of sight by exactly one staff graduation interval—usually either 5 or 10 mm. These are design problems, but reference to the last column of the table shows an almost constant ratio of d to $\tan\theta$: therefore, if the tilting of the plate is done by horizontal movement of a rod in a guide, as shown schematically in the figure, the movement of the rod can be taken as proportional to d, and uniform graduation of the scale will be quite accurate enough.

It will be seen from the table that, with a plate 25 mm thick, a range of tilting of just beyond $\pm 30°$ will provide a displacement of 10 mm. If the push rod is 40 mm from the axis of tilt of the plate, a total movement of about 47 mm will be required. The push rod may be operated by a cog wheel and ratchet, and the fine reading scale may be marked round the wheel by which the cog is turned: it is easily seen that a scale divided to represent intervals of 0·1 mm of d will be quite open and estimation to 0·02 mm will be possible (if hardly justifiable). As described above, the micrometer scale readings would be from $+5$ to -5 mm; to avoid possible confusion over signs, it is convenient to number the scale from zero at one end, thus introducing a constant addition to all recorded staff readings but having no effect on calculated differences of level.

7.03 Precise Levelling Staves. In conformity with the precision of setting of the Level, it is usual to employ staves with graduations marked on strips of invar: moreover, the staff is in one piece. The commonly used length is 3 metres (or 10 feet in the British system). There are various designs of staff, but the general principle of construction is a wooden or metal frame with grooves in which an invar strip some 20 or 30 mm wide is an easy fit. The strip is fixed only at the bottom end of the frame; it may be sufficiently thick, 2 or 3 mm, to stand up rigidly in its grooves, or it may be a thin tape kept taut by a spring at its upper end. The graduation marks may be painted on the invar, or preferably the paint may be filled into shallow recesses cut in the metal on a dividing machine and then smoothed off flush with the surrounding metal. The numbers and marks other than the graduations themselves are on the frame. Part of a typical metric staff is seen in Fig. 110: the short graduation lines are 1 mm thick and spaced at 10 mm intervals, and there are two sets of graduations offset from each other, so that the observer can take two readings with different micrometer settings. The setting is made by getting a graduation mark symmetrically between the slightly inclined lines as shown. On the staff illustrated, the scale of smaller numbers indicates centimetres from the bottom of the staff. In another system, the cross-line is set in symmetrical relation to two staff divisions, as in Fig. 112.

In British practice, much levelling has been done with graduations at 0·02 foot spacing, and the fine reading taken by setting the central horizontal cross-line on one edge of a graduation mark. Micrometer readings have been estimated to 0·0001 ft.

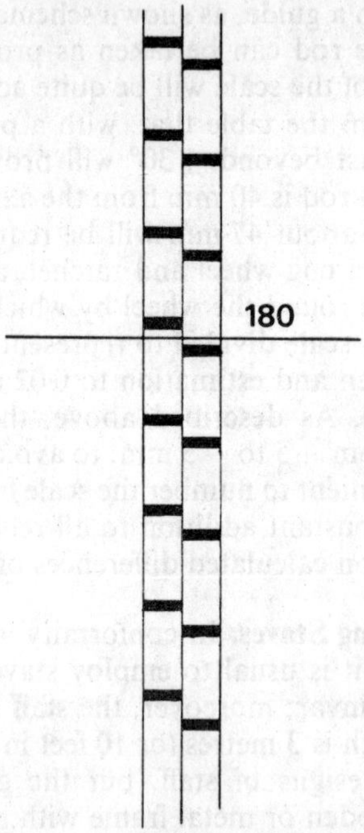

FIG. 112. STAFF GRADUATIONS.

The base of the staff is usually a thick metal plate machined accurately flat, and the invar strip is fixed to this. The staff, when not on a benchmark, must never stand on the ground: precise levelling accuracy can only be maintained by use of firmly placed supports such as heavy footplates with rounded heads or large round-headed spikes driven into the ground. The base surfaces of staves must be kept clean.

The staff must be fitted with a circular bubble adjusted to indicate verticality, and the staff-holder should have a pair of light poles about 2 m long which he grips with the staff to form a tripod and hold the staff steadily in the vertical position while observations are being taken.

Staff Calibration. Graduation of invar strips is done as accurately as possible by the manufacturers, but it is always necessary to keep a check on graduation accuracy, and often in fact there will be graduation errors appreciable at the standard of precision of the observations. Staff graduation may be checked by comparing it with a calibrated invar tape that has been tested by a Standards Authority such as the National Physical Laboratory. Small differences between staff and tape graduations may be measured by use of a microscope, or by moving the staff (or the tape) so that corresponding graduations coincide, and recording the movements by a suitable micrometer arrangement.

Staff calibration should be checked regularly, say once a year, because invar is liable to undergo physical changes and become unstable after much use in field conditions.

7.04 Benchmarks. Precise levelling is done to provide a basic reference system for height, and the end product of all the work is a network of lines of benchmarks in recorded positions and with adjusted heights. Much thought and preparatory work must be done on a large levelling scheme before any observations are actually taken. A suitable network of lines must be arranged and these will usually run along roads, not only for ease of access but chiefly because these are the places where height values are most likely to be required for subsequent use. The layout of a primary levelling system will depend very much on the type of country and the availability of suitable routes. A typical portion of a primary network is shown in Fig. 113. Along the lines of levelling permanent benchmarks must be constructed; they should be closer together in built-up areas, say two or three per kilometre, and perhaps as much as 2 or 3 km apart in unoccupied territory. In any case, it is convenient to establish the benchmarks sufficiently close so that the levelling between consecutive benchmarks will not take more than a morning's or an afternoon's work; moreover, if a mistake is made, the shorter section means less time spent on re-levelling.

It is usual to construct specially stable benchmarks, called Fundamental Benchmarks, at junctions of levelling routes and spaced at distances of 20 to 50 km.

The ordinary benchmarks along the lines may be of various forms, according to circumstances, but every benchmark must have a definite upper surface of reference such as is provided by a round-headed bolt of non-corrosive metal, or a smooth hard round stone like flint. If solid rock is available at or near the ground surface, metal bolts may be cemented into drilled holes; bolts may be fixed on permanent buildings or bridge abutments; concrete

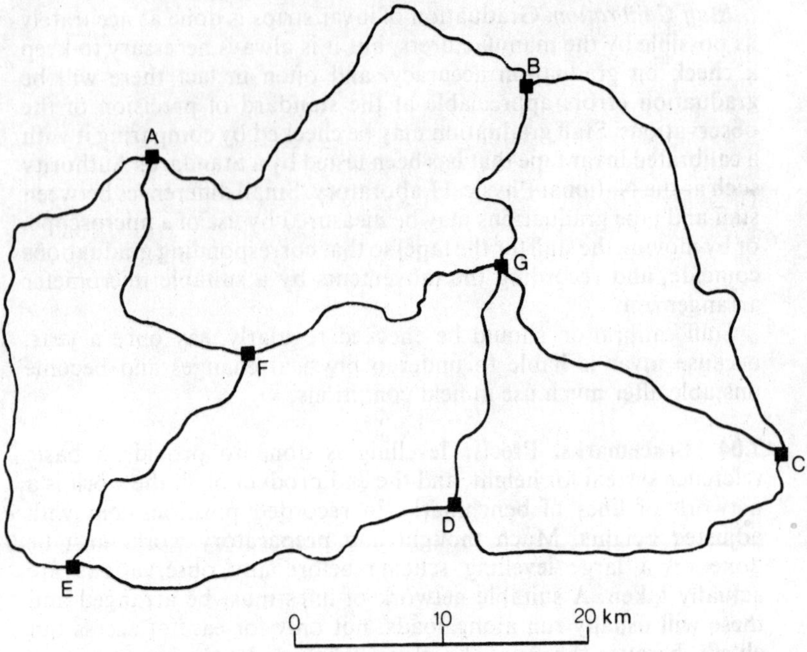

FIG. 113. LEVELLING NETWORK.

blocks about 0·3 metre cube may be cast in shallow holes so that the metal bolt on top is just below ground level. Some designs for ordinary benchmarks are illustrated in Fig. 114. The height of the crow's foot can be found by having a staff station immediately below it and reading the height of the groove on the staff. The flush brackets are used by the Ordnance Survey: a movable attachment can be hung from them in such a way as to provide a staff support at exactly the same height as the permanent reference surface.

Benchmarks must of course be located in positions where a 3 m staff can be placed on them or very close to them. The choice of positions for benchmarks is often one of some difficulty; a benchmark too far from a road may be disturbed by agricultural operations but too near the road it may be removed or covered over by widening or reconstruction work.

The Fundamental benchmarks are much more elaborate constructions, with one or more underground marks as well as one above surface.

Constructed benchmarks should be allowed to settle for some months, even a year. before levelling is done on them.

An essential part of the preliminary work of a levelling scheme is the preparation of location sketches for all benchmarks constructed; for each benchmark, measurements to 4 or 5 visible objects such as corners of buildings, telegraph posts, road signs, road centre, identifiable trees, should be recorded. Positions should also be marked on any available large-scale maps or air photographs. The relocation of a buried benchmark may be made easy by covering it with pieces of charred wood, which will remain conspicuous for a very long time.

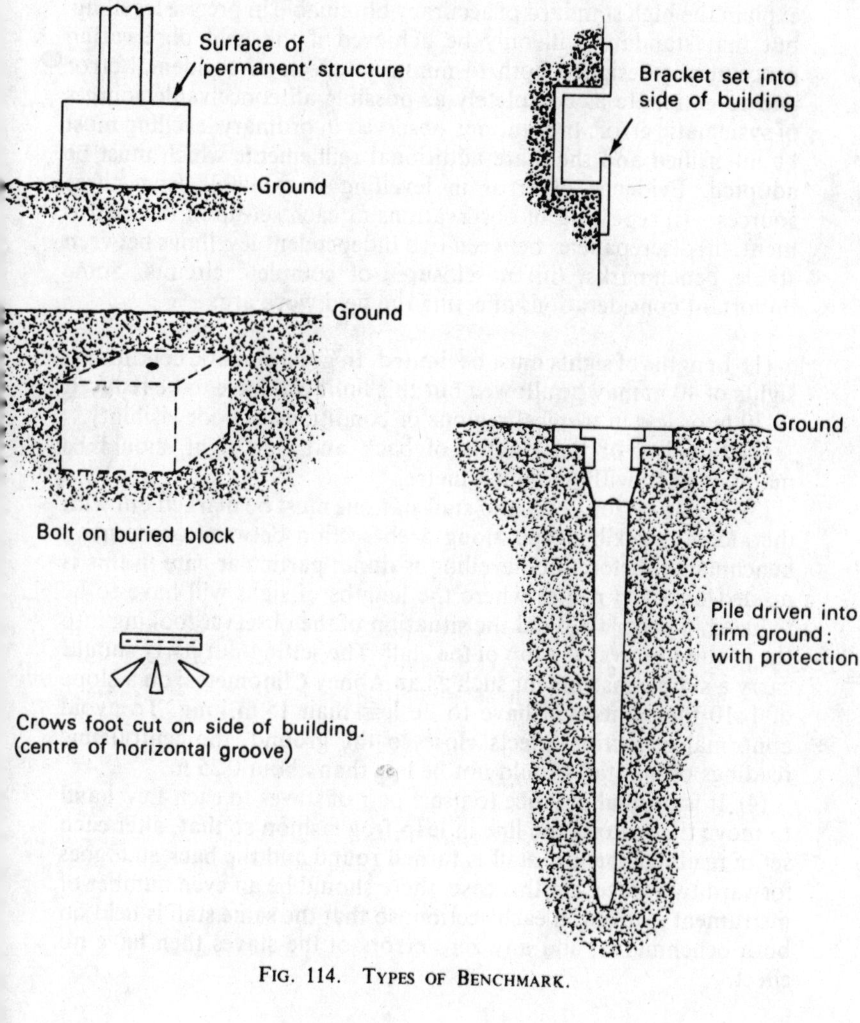

FIG. 114. TYPES OF BENCHMARK.

It must be recognised that a proportion of benchmarks in a large national system will inevitably be lost, and for this reason, if not for others also, most such levelling systems have to be re-established at intervals of the order of 20 years. In countries subject to tectonic and seismic disturbance (Finland, Japan, etc.) and in areas of mining subsidence, the interval will have to be shorter.

7.05 Methods for Accuracy. Levelling is perhaps the simplest operation in surveying, but it is repeated very many times in the course of establishing an extensive network. This simplicity may explain the high standard of accuracy obtainable in precise levelling, but that standard will only be achieved if the field observation procedure is designed both to minimise effects of accidental error and to eliminate as completely as possible all conceivable sources of systematic error; precautions observed in ordinary levelling must be intensified and there are additional refinements which must be adopted. Evidence of error in levelling is available from three sources—(i) repeating of observations at each set-up of the instrument, (ii) discrepancies between two independent levellings between stable benchmarks, (iii) misclosures of complete circuits. Some important considerations affecting the field work are:

(1) Lengths of sights must be limited. In good climatic conditions, sights of 40 m may be allowed but this limit will have to be reduced to 30 m or less in tropical regions or conditions of poor visibility,

(2) Equality of the lengths of back and fore sight should be maintained to within about $\frac{1}{2}$ metre,

(3) Positions for Level and staff stations must be marked out with the aid of a flexible tape along each section between consecutive benchmarks, before any levelling is done; particular care in this is needed on steep routes where the lengths of sight will have to be reduced in order to avoid the situation of the observer looking into the ground or over the top of the staff. The setting-out party should carry a simple instrument such as an Abney Clinometer: on a slope of 1:10 the sights will have to be less than 15 m long. To avoid abnormal refraction effects close to the ground, the central line readings of the staff should not be less than about 0·25 m.

(4) It is normal practice to use a pair of staves to each Level and to move them along the line in leap-frog fashion so that, after each set of readings the fore staff is turned round and the back staff goes forward two places; in this case, there should be an even number of instrument stations in each section, so that the same staff is held on both benchmarks, and any zero errors of the staves then have no effect.

Observation Procedure. The staff readings taken at each set-up should include stadia readings on each staff and two pairs of level readings: since the level readings are the definitive ones, they should be consecutive readings as between the two staves. The stadia readings provide immediate arithmetical checks and verify equality of sight lengths. A suitable programme for readings would be:

Back Staff—stadia lines
Back Staff—level line
Fore Staff—level line
Fore Staff—stadia lines

Disturb the instrument and re-set and take readings:

Fore Staff—level line
Back Staff—level line

A suitable form for bookings is shown in Fig. 115: the numbers in brackets in the first two sections show the order in which the observations are taken.

In precise levelling it is most desirable to employ booking assistants so that the surveyors can take readings as rapidly as possible, while the bookers have time to do the arithmetic necessary for maintaining the observational standards required in the work.

It is also important to reverse the order of readings as between the back and fore staffs at alternate set-ups: if two staves are in use, this means that first readings are always taken to the same staff, as is indicated on the specimen form. The necessity for this procedure has been proved in practice and is thought to be due to the natural steady change of refraction which would cause a systematic error if the back staff was always observed first. This is, however, still somewhat of a mystery, because two pairs of readings in reverse order are taken anyway.

Observation Checks. Before a Level station is vacated, the necessary arithmetic must be done so that certain discrepancies can be checked. Limiting values for these discrepancies must be specified, and if they are exceeded a complete new set of observations must be taken, preferably after the instrument has been fully disturbed and re-levelled:

(i) for the difference between the two readings of the levelling line on the same staff a reasonable limit would be 0·4 mm,

(ii) for the difference between the sum of the two stadia readings and the sum of the two level readings on the same staff, a reasonable limit would be 1 mm,

(iii) for the difference between stadia intercepts fore and back, a reasonable limit would be 5 mm, representing $\frac{1}{2}$ m in sight lengths.

PRECISE LEVELLING — Page . . . .

From benchmark . . . . to benchmark . . . . Date . . . .

Observer . . . .

Set Up No.	Fore Staff No.		Back Stadia	Back Levelling	Fore Levelling	Fore Stadia	Notes
			(1)	(3)	(4)	(5)	
			(2)	(8)	(7)	(6)	
		Sum					
		Diff.					
			(13)	(12)	(11)	(9)	
			(14)	(15)	(16)	(10)	
		Sum					
		Diff.					
		Sum					
		Diff.					
15	38		2·0638	1·9276	2·5658	2·7013	
			1·7916	1·9279	2·5658	2·4304	
		Sum	3·8554	3·8555	5·1316	5·1317	
		Diff.	0·2722	3	0	0·2709	
16	37		0·9844	0·8370	1·9680	2·1153	
			0·6890	0·8369	1·9678	1·8198	
		Sum	1·6734	1·6739	3·9358	3·9351	
		Diff.	0·2955	1	2	0·2955	
17	38						

FIG. 115. PRECISE LEVEL BOOKING.

Double Levelling. All levelling should be done twice, preferably by different observers at different times using different staff and instrument positions. It is usual practice also for the two observers to proceed in opposite directions, but this arrangement may be applied only section by section, that is the observers may proceed in the same direction along the line as a whole.

Office Work. Field observation records should be sent to a supervising surveyor or local office at frequent intervals, or as soon as the levelling of 2 or 3 sections is completed. All arithmetic must be checked and staff calibration corrections worked out. The calibration corrections need not be actually applied to the individual staff readings: it is only necessary that the corrections appropriate to the staff readings be added together and the difference between total correction for fore and back readings along a whole section be found. If, as usual, two readings of the central cross-line are taken on each staff and added together, the crude difference back-fore is calculated and divided by 2, and then the differential calibration correction can be applied. As soon as possible, the two corrected levellings of a section must be compared, so that if any re-levelling is needed the observers can be informed before they have travelled far from the section to be re-levelled. A limit of discrepancy is required, and a reasonable value might be 2·5 mm for a 1 km section, and proportional to the square root for other lengths.

The rate of progress of precise levelling is in the region of 2 to 5 kilometres a day, depending on conditions.

7.06 Orthometric Height. Consider a section of the Earth's surface along a meridian as shown in Fig. 116. Owing to the ellipsoidal shape of the Earth, level surfaces at different heights are not parallel. This can perhaps be visualised by supposing that a layer of water covers an ellipsoidal Earth, then the water surface will be level but the depth of the water will decrease from Equator to Pole. If A_0B_0 represents the physical surface of the Earth and is level, and AB is mean sea level, then for the same reason these will not be parallel. Our natural concept and definition of 'height' is 'vertical distance above a datum surface such as sea level': evidently, the orthometric heights of A_0 and B_0 are not equal, but a leveller proceeding from A_0 to B_0 will record no difference of back and fore readings. Thus, straightforward levelling as described so far does not measure differences of orthometric height; this phenomenon may be described by saying that the staff is a geometrical instrument but the Level is a gravitational instrument, and the simple process of levelling mixes up the two kinds of quantity. Anyway, to get the

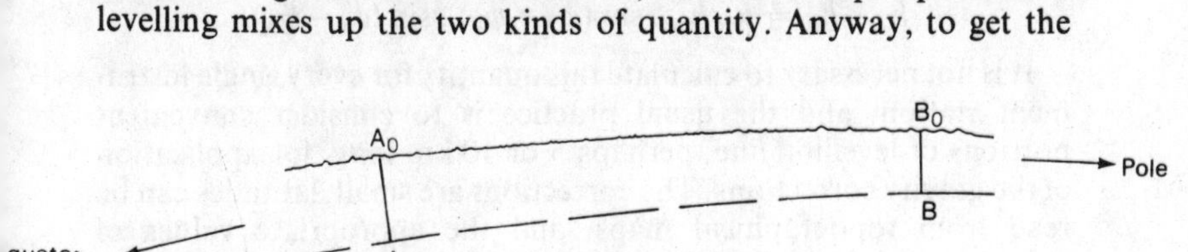

FIG. 116. LEVEL SURFACES.

difference of orthometric height by levelling from A_0 to B_0 a correction is needed, equal to the difference between AA_0 and BB_0. This is calculated from the fact that AB and A_0B_0 are both level surfaces and their separation depends on the strength of gravity in such a way that the mechanical work used in taking an object from A up to A_0 is exactly the same as that used in taking the same object from B up to B_0.

7.07 Gravity Formula Corrections. If g_A and g_B are the strengths of gravity at the two places, then the equality is expressed as $g_A \,.\, h_A = g_B \,.\, h_B$, where h_A and h_B are the heights of the two points. We are concerned here with the relationship between the latitude of a place and the strength of gravity there; as the result of extensive gravity surveys it has been found that this relationship can be expressed by a formula:

$$g = g_0(1+\alpha \,.\, \sin^2 \phi)$$

where g_0 is the strength of gravity at the Equator, ϕ is the latitude, and α has the value 0·005 30. Thus, if ϕ_A and ϕ_B are the latitudes of the two places, the requirement of equality of work leads to:

$$h_A(1+\alpha \,.\, \sin^2 \phi_A) = h_B(1+\alpha \,.\, \sin^2 \phi_B),$$

the factor g_0 having cancelled out. We then get:

$$\frac{h_A}{h_B} = \frac{1+\alpha \,.\, \sin^2 \phi_B}{1+\alpha \,.\, \sin^2 \phi_A},$$

leading to:

$$\frac{h_A-h_B}{h_A+h_B} = \frac{\alpha(\sin^2 \phi_B - \sin^2 \phi_A)}{2},$$

omitting terms in the denominator which are negligible in comparison with 2. Then:

$$h_A - h_B = \tfrac{1}{2}(h_A+h_B)\alpha(\sin^2 \phi_B - \sin^2 \phi_A).$$

We write h_m for $\frac{1}{2}(h_A+h_B)$, the mean height, and transform the trigonometrical expression, and get:

$$h_A - h_B = \alpha \,.\, h_m \,.\, \sin(\phi_A+\phi_B) \,.\, \sin(\phi_B-\phi_A).$$

It is not necessary to calculate this quantity for every single instrument station, and the usual practice is to consider convenient portions of levelling line, perhaps 5 or 10 km long, for application of the gravity corrections. The corrections are small, latitudes can be read from topographical maps, and the appropriate values of h_m can be roughly estimated from the levelling itself or perhaps also estimated from maps. The formula is usually transformed further

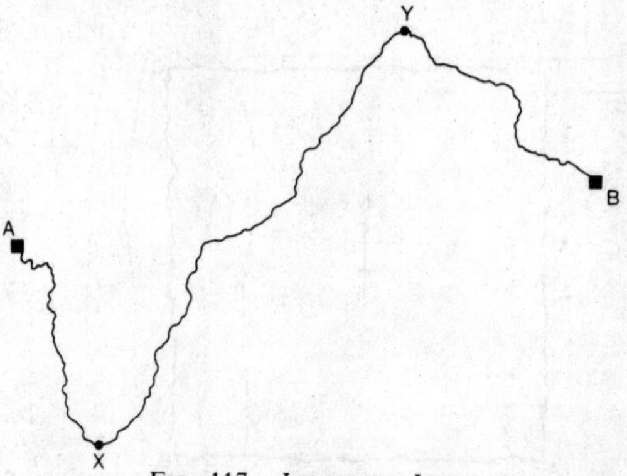

FIG. 117. LEVELLING LINE.

by writing $2\phi_m$ for $(\phi_A+\phi_B)$ with ϕ_m as the mean latitude; and because $(\phi_B-\phi_A)$ is small its radian measure can be used and written as M/R where R is a mean radius for the Earth and M is the north–south extent of the levelling route expressed in the same units as for R: so the working formula is $\alpha \,.\, h_m \,.\, \sin 2\phi_m \,.\, M/R$.

A part of a levelling line is shown in Fig. 117, and the application of the corrections to three portions is illustrated in the table below

Section	*Level Book* metres	h_m metres	ϕ_m	*M* km	*Correction*	*Orthometric Height difference* metres
AX	+214·0273	435	51° 21′	−10·8	+0·0038	+214·0311
XY	+ 61·8515	570	51° 24′	+23·1	−0·0107	+ 61·8408
YB	−216·0178	495	51° 28′	− 8·0	+0·0032	−216·0146
						+ 59·8573

The gravity corrections are not negligible in relation to the accuracy achieved in precise levelling: moreover, they do not necessarily cancel out round a closed circuit. Consider the simplified circuit shown in Fig. 118, composed of four more or less straight lines each 20 km long situated at about latitude 45°N. We suppose that the points A and D are exactly at datum level, points B and C are at exactly 500·000 m orthometric height. Then point C is above the level surface through B by 0·0083 m as calculated from the formula. A leveller starting from A and making error-less observations, will get the following differences of back-minus-fore staff readings:

AB	+500·0000 m	There is no gravity
BC	+ 0·0083	correction for lines
CD	−500·0000	AB and CD because they
DA	0·0000	run east–west.
	+ 0·0083 misclosure	

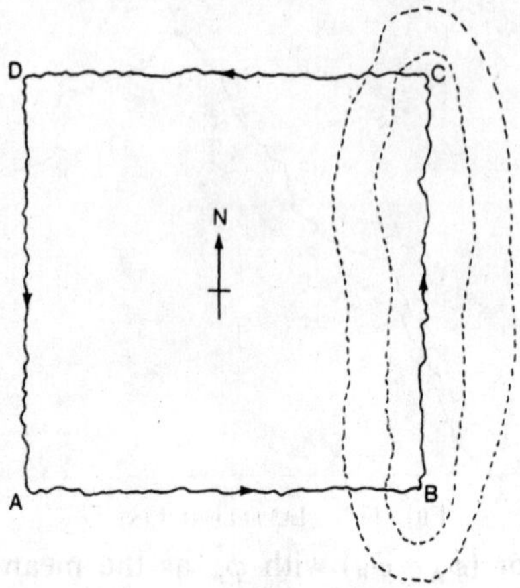

FIG. 118. LEVELLING CIRCUIT.

This misclosure is of the same order of magnitude as the probable error of 80 km of precise levelling, so the corrections are necessary.

7.08 Level Circuit Adjustment. After the levelling lines have been corrected as detailed above, to give observed differences of orthometric height, there remains an adjustment problem: the observed differences between junction points must be amended so that there is exact closure round each circuit. In the small network of Fig. 113 there are 11 lines and 5 closed circuits—11 adjustments to be found to satisfy 5 conditions if the condition method is used: or 6 variations to be calculated, since one junction point may be held fixed and given an arbitrary value for the adjustment by variation of approximate values of the other 6 junction points. Details of the adjustment are not given here, but since the level lines will be of different lengths it will be necessary to adopt weights, and it is usual to take weights inversely proportional to the lengths of the lines. See Section 5.30.

7.09 Height Datum. When a new levelling scheme is connected to existing benchmarks the published values of these must be held and the adjustment must include conditions to preserve the given differences between them if there are more than one. Otherwise, a datum must be adopted and normally this will be done by connecting the levelling system to a tide-gauge reference mark whose height above mean sea level has been determined. Using the adjusted differences of orthometric height along the level lines, the definitive heights of the junction points are first calculated, then the lines of benchmarks are adjusted to these. For the geodetic levelling of Great Britain, the datum now adopted is mean sea level as determined at Newlyn in Cornwall from tide-gauge observations in the period 1915–1921.

7.10 Dynamic Height. Though our natural concept of height is orthometric, the fact that the points A_0 and B_0 in Fig. 116 are on the same level surface is not without interest and may even be of practical significance in any extensive engineering works concerned with water surfaces and movements.

Points A_0 and B_0 are at the same level, hydrostatically considered, and a system in which these points have the same numerical values will embody this concept. This has been done by the convention illustrated in Fig. 119. For a national system of levelling, a reference

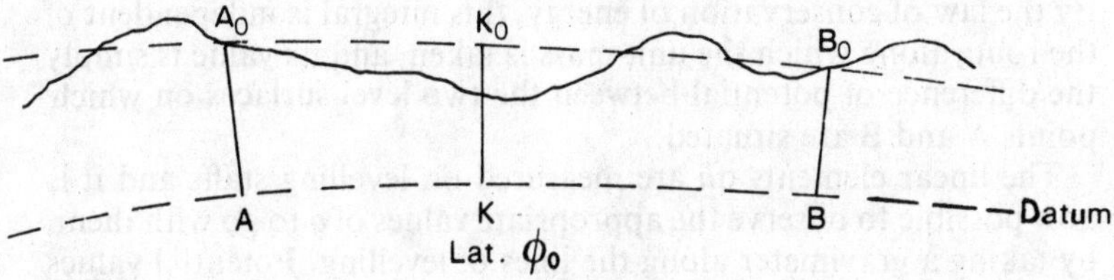

FIG. 119. DYNAMIC HEIGHT.

latitude ϕ_0 is adopted, at about the mean latitude of the area covered. For England this is 53°N. The *dynamic height* of a benchmark is then defined as the orthometric height at latitude ϕ_0 of the level surface through the benchmark. Thus, the dynamic heights of points A_0 and B_0 are both equal to the linear distance KK_0 which is the orthometric height of K_0. The orthometric heights AA_0 etc. will have been determined by the levelling and adjustment processes already described: to find the dynamic height of A_0, the difference between AA_0 and KK_0 is calculated from the formula already given, using the height of A_0, the mean latitude of A and K, and the north–south distance AK. It is usual to publish both the orthometric and dynamic heights of the principal benchmarks of a levelling system. The differences in most cases do not exceed a few centimetres: for instance, the fundamental benchmark at Hawick has orthometric height 202·33 m and dynamic height 202·29 m.

7.11 Geopotential. A theoretical formula has been used in defining the quantity called *dynamic height*, to take account of the non-parallelism of level surfaces, and in principle to assign to any particular level surface a particular numerical value. But actual values of gravity may differ considerably from values calculated from the formula, and actual level surfaces have undulations which follow the gravity field as it is locally or regionally modified by topographical and geological irregularities. Thus, dynamic height, as defined above, depends on an adopted formula of reference and to that extent it is not a real quantity.

The development of gravimeters in the 1930s has made it possible to use a unique natural unit for labelling the level surfaces of the actual gravity field. These surfaces are surfaces of equal values of potential, and are orthogonal to the directions of gravity. Differences of potential are measured in terms of mechanical work, the product of force and distance. In mathematical terms, the integral $\int_A^B g \cdot dh$ is the work done in taking a unit mass from point A to point B, dh being an element of linear measure along the vertical and g being the average value of gravity acceleration on that element. By the law of conservation of energy, this integral is independent of the route along which the unit mass is taken, and its value is simply the difference of potential between the two level surfaces on which points A and B are situated.

The linear elements dh are measured on levelling staffs and it is now possible to observe the appropriate values of g to go with them, by taking a gravimeter along the lines of levelling. Potential values are thereby derived entirely from quantities measured on the ground.

The basic process of potential measurement is illustrated in Fig. 120, which shows a single set-up on a levelling line. The back staff reading is dh_1, and g_1 is the value of gravity at the mean reading $\frac{1}{2} dh_1$: similarly for dh_2 and g_2 at the fore staff. Then the difference of potential from A to B is strictly $(+g_1 \cdot dh_1 - g_2 \cdot dh_2)$. In practice,

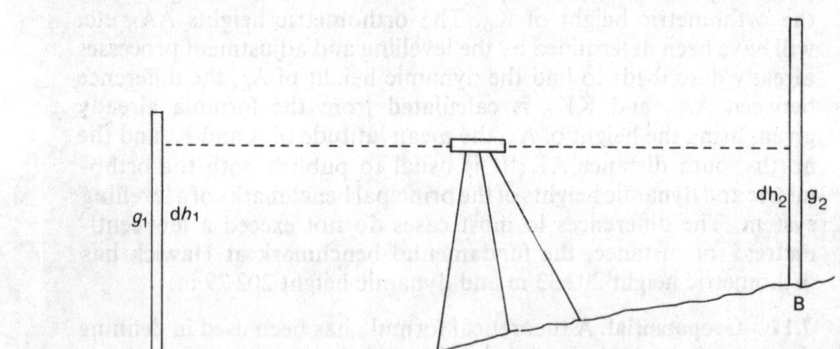

FIG. 120. GEOPOTENTIAL DIFFERENCES.

it is accurate enough to assume that $g_1 = g_2$, and it is unnecessary to measure gravity at every position of the Level: most of the gravity values can be interpolated from observations taken much further apart—up to several kilometres in flat country. Permanent benchmarks are usually placed at intervals of $\frac{1}{2}$ to 2 km, and it will generally be convenient to use these, or some of them, as gravity stations, for identification purposes.

Geopotential Units. Geopotential differences are thus found by multiplying the observed height difference of each levelling section by the appropriate mean observed value of gravity for the section. It has been recommended that for this purpose the linear measurements—i.e. corrected staff readings—should be expressed in metres and the gravity values should be in kilogals (see Section 7.12). Since the values of gravity on the Earth's surface are always approximately 0·98 kilogal, the recommendation means that geopotential differences will be numerically about 2% less than ordinary height differences expressed in metres. These quantities expressed in *geopotential units* (gpu) are true measures of level which give to each actual level surface a unique number, as soon as a particular level surface has been assigned the value zero. Naturally, the observed position of mean sea level at some tide gauge station is taken as datum.

For the European Unified Levelling System the geopotential of a certain benchmark at Amsterdam has been adopted as datum. In Great Britain, the Newlyn Datum is used, and the geopotential at a benchmark near the tide gauge there has the value 4·66 gpu. Many countries have now computed their principal levelling networks in terms of geopotential, as well as in orthometric and dynamic height.

It is of course not necessary to observe gravity values at the same time as the levelling, indeed it will often be convenient to arrange the observation programmes at different times, and in many countries the gravity survey has been done along lines levelled some years previously.

7.12 Gravity Measurement. Dimensionally, gravity is expressed as an acceleration—the acceleration of an object falling freely in a vacuum. In the Metric System, the proper unit of acceleration is one centimetre per second per second, called a *gal*, and gravity on the Earth's surface has values around 980 in this unit. In SI Units the values would be around 9·8 metres per second per second. For geopotential calculation, the unit is taken ten times greater, as explained in Section 7.11. The acceleration of gravity has been directly measured at only a few places in the world, because, until very recently, such a measurement, if it is to achieve the desired accuracy, has needed very elaborate apparatus and much time. The first reasonably accurate absolute determination was made in 1905 at the Geodetic Institute in Potsdam. All other published values have been derived from the Potsdam value by differential methods using pendulums or gravimeters. Reference stations have been established in most countries, and from these as bases the gravity values at many places can readily be found by use of gravimeters. It is now known that the Potsdam datum value is about 0·014 gal too great, but this datum is being retained in order to avoid confusion; anyone requiring the actual value should subtract the above amount from any published value unless there is an indication that the correction has already been made. A recent international conference recommended the adoption of a new "International Reference System 1967", which included the 14 mgal correction; this could cause some confusion if published values of gravity and geopotential are not carefully described.

Gravimeters. A gravimeter operates essentially by the principle of the spring balance, in which the weight of a suspended part of the instrument is supported by elastic stress of a thread, spring, or similar device. A feature of gravimeters is that they do not directly measure anything; readings are obtained on a scale which must be calibrated by taking the instrument to places where the gravity values are already known. The relation between scale readings and gravity values can thus be established, but the calibration relationship is liable to drift. Gravity surveys are therefore made by methods similar to the use of barometers in height determination. A gravimeter is taken to a place where the gravity value is known, a reading of the scale is recorded, the instrument is then taken to other places where gravity values are required, and the series is terminated (the same day if possible) by returning to the starting point or to another place where the gravity value is known. Misclosures of reading are adjusted in any reasonable manner. Gravity values at benchmarks could be found by such procedures, depending in detail upon the availability of reference stations in the country where the levelling is done.

For practical purposes, surveyors will probably continue to use orthometric heights, except that dynamic heights might be taken into consideration on very extensive works involving flow and levels of water. The differences between the two systems are hardly appreciable at ordinary levelling accuracies. Geopotential values could be used in place of dynamic heights but their primary merit is that they are basic material for the study of the Earth's gravity field, since they are independent of any arbitrary reference system or formula.

7.13 Heights from Vertical Angles. In the days before air photographs became the principal source material for the preparation of topographical maps, it was the practice for observers of triangulation systems to measure vertical angles so that the heights of trigonometrical stations and other observed points could be calculated. The stations themselves were necessarily on tops of hills, high buildings or observation towers; moreover the topographers with their plane-tables required control points, both for mapping and for contouring, to be in positions of prominence and easy visibility.

Nowadays, the photogrammetrist setting up photographs on the plotting machine requires height control points to be easily identifiable on the pictures, and at points of low as well as high altitude. The modern surveyor establishing such control positions must select suitable points by reference to the photographs, and he does not have to consider their intervisibility on the ground. See Vol. I, Chapter XIII.

The determination of heights of air-survey control points is now largely done by simple levelling or by careful systematic observations with aneroid altimeters, but the use of vertical angles at short distances is not excluded.

The limitations to the use of vertical angles are those which arise from the well known but only roughly calculable effects of atmospheric refraction. In early triangulation works, vertical angles were observed over long distances in various meteorological conditions, and the heights calculated from them could well be in error by 10 or 20 metres: nevertheless, such results were better than having no heights at all. The fact that Mt. Everest is the highest point on the Earth was discovered by use of vertical angles measured at points more than 100 km from the mountain!

7.14 Atmospheric Terrestrial Refraction. Suppose there is a source of light at point S in Fig. 121. If the atmosphere were of uniform density, the light would spread out from S in concentric spherical wave-fronts travelling at the speed appropriate to the density of the air. In fact, the density decreases with increasing height, so the light travels faster upwards than horizontally, and the wave fronts are non-concentric surfaces as shown in a very exaggerated manner on the diagram. A surveyor at A, looking at the light source, will set his telescope perpendicular to the wave fronts, and will see the source apparently in the direction indicated by the arrow. The broken line is the conventional ray which is everywhere perpendicular to the wave fronts. The angle of refraction at A, angle v, is the angle between the apparent direction of the light and the true direction AS.

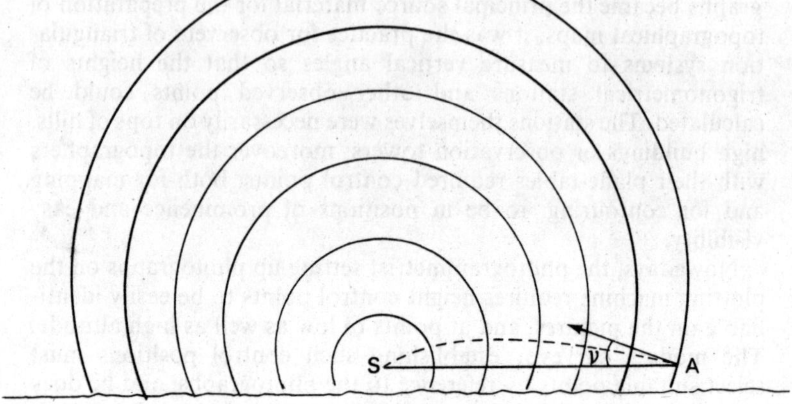

FIG. 121. ATMOSPHERIC REFRACTION.

Obviously, the refraction angle will depend on the physical condition of the atmosphere between S and A, especially on the upwards rate of decrease of the density. In the circumstances of practical surveying, it is impossible to find out enough about the atmosphere to be able to calculate the value of v, and the best that can be done is to estimate its value and to use methods that minimise its effects on the final results.

(The refraction angle is more accurately calculable if the light comes from outside the atmosphere, as in astronomy.)

The situation of a terrestrial surveyor is illustrated very crudely in Fig. 122. His 'line of sight' from A to B is the curve shown, which will normally be above the straight chord. The actual route of the curve will vary with the atmospheric conditions. The refraction angles v_1 and v_2 directly affect the observed vertical angles.

The true zenith angles at the points are VAB and WBA, but the observed angles will be (VAB-v_1) and (WBA-v_2). If we suppose that the verticals meet at O, then OA and OB will both be roughly 6370 km, since the surface variations of the Earth are very small compared with its radius. If A and B are points in a computed survey system, the angle θ can be calculated; it is about 1° for 111 km distance between the points, so in practice θ will usually be much less than 1°.

If the observed zenith angles are Z_1 and Z_2, then the zenith angles of the chord are Z_1+v_1 and Z_2+v_2. The three angles of the plane triangle OAB are thus θ, $(180° - Z_1 - v_1)$ and $(180° - Z_2 - v_2)$.

Setting the sum of these to 180°, we find that

$$v_1+v_2 = 180°-Z_1-Z_2+\theta.$$

Thus, if the two zenith angles are observed, and the line of sight is the same for both observations, the total curvature of the line can be immediately calculated.

It is usually assumed that $v_1 = v_2$, and if this is so the two vertical angles can be corrected and the difference of height OB–OA can be calculated by suitable application of plane trigonometry to the triangle OAB. If we write OA $= R+h_1$, OB $= R+h_2, v_1 = v_2 = v$,

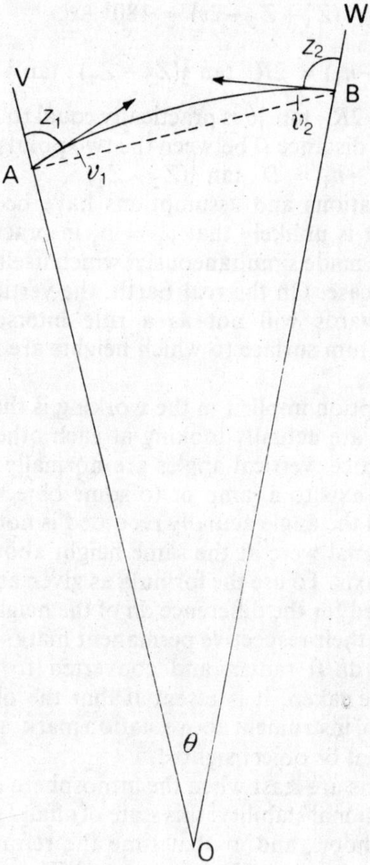

FIG. 122. VERTICAL ANGLES.

then

$$\frac{\text{OB}}{\text{OA}} = \frac{R+h_2}{R+h_1} = \frac{\sin(Z_1+v)}{\sin(Z_2+v)},$$

hence

$$\frac{h_2-h_1}{2R+h_1+h_2} = \frac{\sin(Z_1+v)-\sin(Z_2+v)}{\sin(Z_1+v)+\sin(Z_2+v)}$$

$$= \frac{\cos\frac{1}{2}(Z_1+Z_2+2v)\,.\,\sin\frac{1}{2}(Z_1-Z_2)}{\sin\frac{1}{2}(Z_1+Z_2+2v)\,.\,\cos\frac{1}{2}(Z_1-Z_2)}$$

$$= \tan\tfrac{1}{2}(Z_1-Z_2)\,.\,\cot\tfrac{1}{2}(Z_1+Z_1+2v).$$

But $$(Z_1+Z_2+2v) = 180^\circ+\theta,$$

so $$(h_2-h_1) = 2R\,.\,\tan\tfrac{1}{2}(Z_2-Z_1)\,.\,\tan\tfrac{1}{2}\theta.$$

But θ is small, and $2R\,.\,\tan\frac{1}{2}\theta$ is practically equal to $R\theta$, which is the known horizontal distance D between the two points, and we get the simple formula $h_2-h_1 = D\,.\,\tan\frac{1}{2}(Z_2-Z_1)$.

Many simplifications and assumptions have been made in the above working. It is unlikely that $v_1 = v_2$ in practice, even if the observations were made simultaneously, which itself is very unlikely to have been the case. On the real Earth, the verticals at A and B continued downwards will not as a rule intersect each other. Sections of the datum surface to which heights are referred are not generally circular.

Another assumption implicit in the working is that the observers at the two points are actually looking at each other's theodolites! Of course in practice, vertical angles are normally measured from theodolite transit axis to a lamp or to some object erected on the sighted point, and the angle actually recorded is not what would be obtained if the signal were at the same height above station mark as the theodolite axis. To use the formula as given above, each angle should be corrected for the difference dh of the heights of signal and instrument above their respective permanent marks. The correction can be taken as dh/D radian and converted to seconds. When vertical angles are taken, it is essential that the observers should record the height of instrument above station mark, and also describe precisely each point or object sighted.

Refraction effects are least when the atmosphere is in a condition nearest to gravitational stability: this state of affairs generally occurs in the early afternoon, and at that time the refraction angle v is usually found to be about 1/15 of the angle θ. The ratio v/θ is called the Coefficient of Refraction, not to be confused with the Refractive

Index mentioned in Chapter IV. The coefficient may vary enormously in special conditions: values at least double the above fraction can occur, and negative values—line of sight below the chord, sometimes producing a mirage—are not unknown. Vertical angles observed in such abnormal conditions are useless.

7.15 Single Angles. If vertical angles are not observed reciprocally, the geometry of a one-way, or single, observation must be used. In considering this problem, altitudes, positive or negative, will be used instead of zenith angles. Point A in Fig. 123 is at height h_1 above

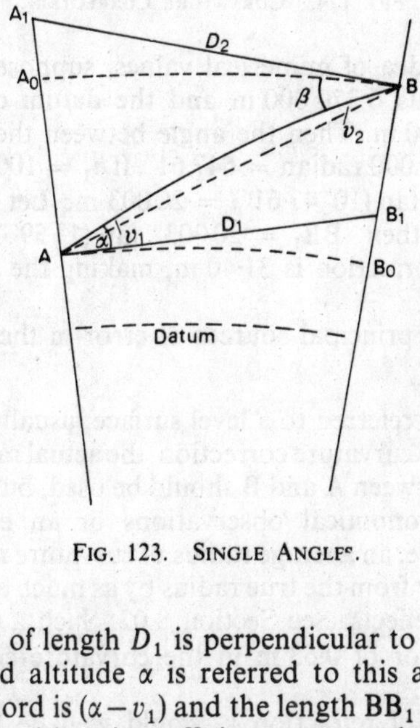

FIG. 123. SINGLE ANGLES.

datum; line AB_1 of length D_1 is perpendicular to the vertical at A, and the observed altitude α is referred to this as zero. The true altitude of the chord is $(\alpha - v_1)$ and the length BB_1 can be calculated as $D_1 \tan(\alpha - v_1)$ with quite sufficient accuracy because in practice the angle AB_1B will be so little different from a right angle.

The point B_0 is at the same height as A above datum, so BB_0 is the required difference of height between A and B: to obtain this, the curvature correction B_0B_1 must be added to the formula given above. If the curved line AB_0 is parallel to the datum surface and is taken to be part of a circle of radius R_1, then B_0B_1 can be calculated as $D_1{}^2/2R_1$: referring to Fig. 124, we have:

$$R_1{}^2 + D_1{}^2 = (R_1 + (B_0B_1))^2 = R_1{}^2 + 2R_1(B_0B_1) + (B_0B_1)^2.$$

Cancelling the $R_1{}^2$, we get $D_1{}^2/2R_1 = (B_0B_1) + (B_0B_1)^2/2R_1$, and the last term is negligible in practice.

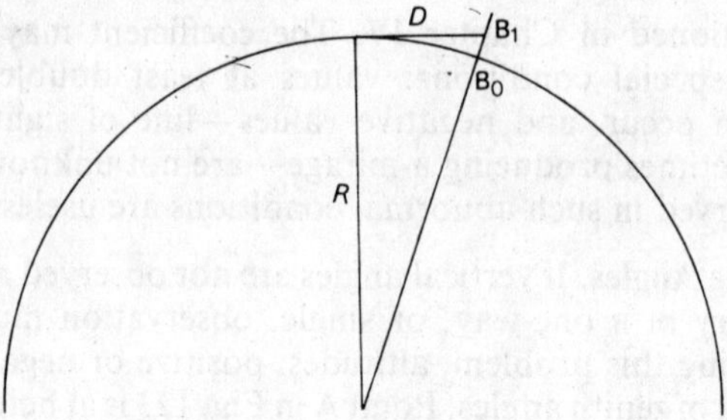

FIG. 124. CURVATURE CORRECTION.

To get some idea of numerical values, suppose that the datum surface has radius 6 370 000 m and the datum distance between A and B is 20000 m. Then the angle between the two verticals is $\theta = 20\,000/6\,370\,000$ radian $= 647{\cdot}61''$. If $h_1 = 1000$ m, then strictly $D_1 = 6371000 \,.\, \tan(10'\,47{\cdot}61'') = 20003$ m. Let us take $\alpha = 2°$ and $v_1 = 43''$: then $BB_1 = 20003 \,.\, \tan(1°\,59'\,17'') = 694{\cdot}35$ m. The curvature correction is 31·40 m, making the height difference 725·75 m.

There are two principal sources of error in the single height as derived above:

(i) Heights are referred to a level surface, usually mean sea level, as datum. For the curvature correction, the actual radius of curvature of this surface between A and B should be used, but this can only be found from astronomical observations or an elaborate gravity survey. In practice, an average radius of curvature must be assumed, and this can differ from the true radius by as much as 2% on account of local gravity effects. See Section 8.02. Such a difference would introduce an error of 0·63 m in the curvature correction, in the worked example.

(ii) The angle of refraction v_1 could well be in error by 10″: this obviously has an effect proportional to itself and to the distance, and in the above example would lead to an error of about 1 m.

It is obvious that the consideration of refinements in the geometrical theory is pointless in view of the magnitude of errors that can occur when the observation is not reciprocal.

With reference to Fig. 123, the reciprocal observation would be expressed by the *depression* angle β to which the refraction angle v_2 must be *added*: then the length of AA_1 would be calculated and the curvature correction *subtracted* from it to give the difference of height AA_0. Thus the refraction and curvature terms come into the two computations with opposite signs and will practically cancel out in the mean. This illustrates the great advantage of two-way observations, and the fact that the simple formula $D \cdot \tan \frac{1}{2}(\alpha+\beta)$ can then be used.

The assumption that $v_1 \equiv v_2$ will hardly ever be exactly true over distances as long as 20 km even if the observations are simultaneous. To arrange for simultaneous observations in a triangulation scheme is usually impracticable. The next best thing to do is to instruct the observers to take vertical angles at the time of minimum refraction. In traversing by EDM however, there should not be much difficulty in providing a second theodolite: if only for the purpose of the slope corrections, the observation of reciprocal vertical angles is advised if the best results are to be obtained where the lines are comparatively steep.

Where the points are much closer together, the use of vertical angles begins to approach other methods as regards accuracy obtainable. At a distance of one kilometre, for instance, the refraction angle will normally be only 2 or 3 seconds; the curvature correction is only 8 centimetres, and any possible error in either of these is not likely to be significant. Reciprocal vertical angles at such a distance will certainly give results comparable in accuracy with that of simple levelling.

Thus, the use of vertical angles is a method available to the surveyor for obtaining heights where the necessary sighting distances are greater than is practicable in levelling or tachymetry, but not so great that the possible errors of the method would be unacceptable. The accuracy over a kilometre or two will be good enough for air survey control and some kinds of engineering survey work. The method can be used for crossing water, and in situations where other methods would be difficult. In earlier times, there would often be a problem in obtaining the necessary distance measure, but now there is EDM.

REFERENCES

ANGUS-LEPPAN P. V. *Surface Effects on Refraction in Precise Levelling.* University of New South Wales.

ANGUS-LEPPAN P. V. 'Diurnal seasonal variations in the coefficient of refraction'. *Canadian Surveyor*, June 1967.

ANGUS-LEPPAN P. V. 'The three-peg adjustment for levelling instruments'. *Survey Review* No. 159, January 1971.

BERTHON-JONES P. 'An investigation of the instrumental sources of error in levelling of high precision by means of automatic Levels'. *Survey Review* No. 132, April 1964.

BROWN M. S. 'The variation of refraction with elevation in an arid climate'. *Survey Review* No. 158, October 1970.

CORNISH R. J. 'Trigonometrical levelling'. *Empire Survey Review* No. 119, January 1961.

CORNISH R. J. 'Trigonometrical levelling'. *Survey Review* No. 155, January 1970.

DICKSON W. L. 'The calculation of the height of Kilimanjaro'. *Empire Survey Review* No. 91, January 1954.

DYUS G. R. 'Primary levelling in Tanganyika'. *Survey Review* No. 136, April 1965.

GODSON W. L. 'Atmospheric refractions and trigonometric levelling'. *Canadian Surveyor*, June 1970.

GORDON D. L. & SUTHONS C. T. 'Mean sea level in the British Isles'. *Bulletin Géodésique* No. 77, 1965.

GRIFFIN W. L., JONES B. G., & MCALINDEN J. M. 'Establishing tidal datum lines for sea boundaries'. *Surveying and Mapping*, September 1968.

HALLIDAY E. X. 'An experiment in hydrostatic levelling'. *Survey Review* No. 152, April 1969.

HOGG F. B. R. & ARMSTRONG J. A. 'Two new self-aligning Levels'. *Empire Survey Review* No. 111, January 1959.

HRADILEK L. 'Trigonometric levelling and spatial triangulation in mountain regions'. *Bulletin Géodésique* No. 87, 1968.

HRADILEK L. 'Refraction in trigonometric and three-dimensional terrestrial networks'. *Canadian Surveyor*, March 1972.

JACKSON J. E. 'Height and geopotential'. *Survey Review* No. 128, April 1963.

JESSEN A. 'Experiments in hydrostatic levelling'. *Bulletin Géodésique* No. 89, 1968.

LENNON G. W. 'Instrumentation concernant le niveau moyen, ses limitations et l'optimalisation du fonctionnement des marégraphes conventionnels en Grande-Bretagne'. *International Hydrographic Review* (1971) 2, p. 133.

LEVALLOIS J. J. 'Sur la frequence des measures de pesanteur dans les nivellements'. *Bulletin Géodésique* No. 74, 1964.

MAKOWSKA A. 'Vertical angles and refraction in mountain areas'. *Bulletin Géodésique* No. 91, 1969.

NIBLOCK J. 'Graduations on geodetic levelling staves'. *Survey Review* No. 138, October 1965.

PUGH K. T. 'The height determination of Mt. Kilimanjaro, Tanganyika, E. Africa: August–September 1952'. *Empire Survey Review* No. 91, January 1954.

RAINSFORD H. F. 'Trigonometrical heights and refraction'. *Empire Survey Review* No. 98, October 1955.

RAINSFORD H. F. 'The weighting of differences of trigonometrical heights'. *Survey Review* No. 131, January 1964.

RAMSAYER K. 'The admissible distances of the gravity points for the determination of geopotential numbers in high mountains, medium high mountains and flat countries'. *Bulletin Géodésique* No. 76, 1965.

ROWLAND J. B. 'Atmospheric refraction on long-line vertical angles'. *Surveying and Mapping*, March 1969.

SNEDDON J. 'Hydrostatic levelling—a review'. *Australian Surveyor*, June 1972.

STEYNBERG C. A. 'Levelling from island to island in the South China Sea'. *Survey Review* No. 127, January 1963.

THOM A. 'An empirical investigation of atmospheric refraction.' *Empire Survey Review* No. 108, April 1958.

WALLEWIJN A. 'Hydrostatic levelling in the Netherlands'. *Survey Review* No. 131, January 1964.

CHAPTER VIII

GEODESY

8.01 Figure of the Earth. It was Isaac Newton who suggested that the movement of objects in space might be explained, in terms of known mechanical rules, by the assumption that there was a force of attraction acting between the objects. One consequence of this idea was that the Earth should be bulging round the equator, or flattened at the poles, on account of its rotation. The rate of rotation was quite accurately known, and Newton himself made an estimate of the flattening, based on the assumption then quite generally accepted that the interior of the Earth was entirely liquid, with just a thin crust on top.

Newton's deduction gave the surveyors a new problem because it meant that the curvature of the Earth would vary with the latitude. Theoretically, the shape and size of an ellipse can be determined by finding the radius of curvature at two different latitudes: this was put to the test in grand style by the French Academy of Science which sent out two expeditions, one to Peru and one to Lappland in the early eighteenth century. The principle of the method was the same as that used by Eratosthenes twenty centuries earlier: north–south distances were measured by surveying chains of triangulation, and the differences of latitude covered by the surveys were found by astronomical observation. It was then a very simple matter to estimate the radius of a circle along which the measured distance would correspond with the measured angle. The results of the two enterprises verified the flattening of the Earth, making it not quite so much as Newton's estimate.

Subsequently, several other similar works were conducted for the scientific purpose of determining the Figure of the Earth, and then the great national surveys made in European countries, in India, and in the United States contributed further to the information on the subject. All the results differed slightly among themselves, for the reason that there is nothing about the real Earth that is exactly elliptical: each determination of the figure was an exercise in curve fitting, in fact the fitting of an ellipse to a section of a real surface, the geoid or mean sea level. The primary requirement for each national survey was a spheroidal surface of reference for its mapping system.

In working out a reference surface suitable for an area covered by a survey system, there are two distinct problems. One is to decide on the geometrical shape and size of the surface; this is defined by specifying two parameters; formerly, the two semi-axes a and b of the meridian ellipse would be given, but nowadays it is the custom to give a with the flattening f. The other problem is to locate the spheroid in relation to the surface of the Earth or, more explicitly, to the points constituting the survey system.

8.02 Spheroid Fitting. Figure 125 illustrates the situation in a simple way. The full line of irregular curvature represents the geoid, except that the variations of curvature are not in reality anything like so

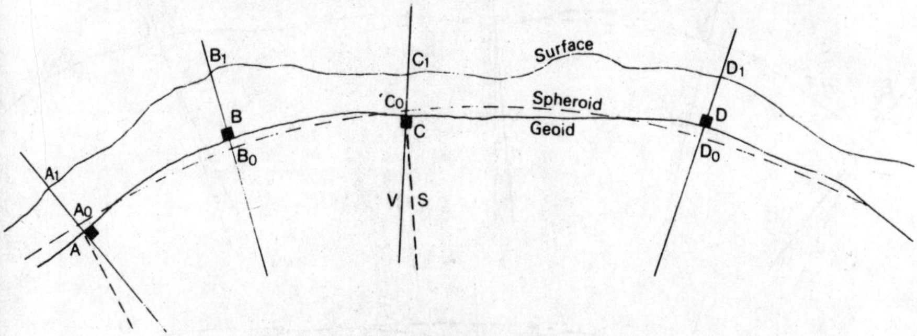

FIG. 125. SPHEROID AND GEOID.

violent as shown. A_1 B_1 C_1 D_1 are points in a survey system on the Earth's physical surface, and the straight lines through them are the verticals, or directions of gravity at the points. The geoid, being a level surface, is necessarily perpendicular to the verticals at the points A B C D. The broken curve represents a reasonable position for a spheroid reference surface.

The angles between the vertical lines can be found by astronomical observation, and by no other method. To make the illustration as simple as possible, suppose the diagram represents a north–south section of the Earth's surface; then the angles between the verticals are the differences of astronomical latitude between the points. The distances along the geoid, AB BC CD, are derived from the ground survey; if the measured lengths in the survey work are reduced to sea level, all computed distances will also be referred to that surface.

When a new survey is to be computed, it is necessary to adopt a preliminary spheroid and a datum point. That is, values of a and f must be assumed, and a spheroid with these dimensions must be placed in a defined position in relation to the survey points, so that spheroid coordinates of all the points can be calculated by the formulas explained in Chapter VI. It is usual to assume at first that the spheroid coordinates at a particular point are the same as the astronomical latitude and longitude, and that the spheroid at that point is at sea level. Taking a numerical example, suppose that the astronomical latitude of B is 60° 00′ 00″; adopting this point as datum is equivalent to placing the spheroid so that it passes through B, and its normal at a point of spheroid latitude 60° 00′ 00″ coincides with the gravity vertical there. This is illustrated in Fig. 126(a).

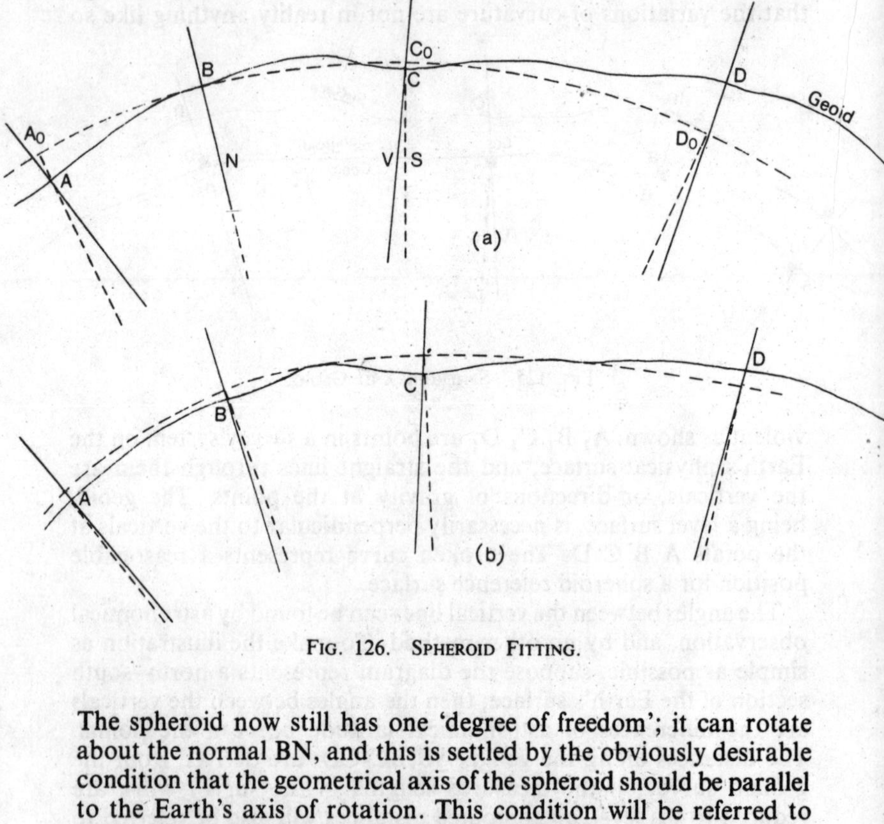

FIG. 126. SPHEROID FITTING.

The spheroid now still has one 'degree of freedom'; it can rotate about the normal BN, and this is settled by the obviously desirable condition that the geometrical axis of the spheroid should be parallel to the Earth's axis of rotation. This condition will be referred to again. The condition will be satisfied if the geometrical meridian on the spheroid through B lies in the direction of true north, that is the astronomical and spheroid meridians at B coincide.

The next stage is to calculate the spheroid coordinates of A_0 C_0 D_0. Let us suppose that the spheroid latitude of C_0 is found to be 60° 27′ 19·205″; this is the angle made by the normal CS with the spheroid equatorial plane. The astronomical latitude of C is determined by the direction of the gravity vertical at that point, and if it is found by astronomical observation it might be, say, 60° 27′ 25″; it is the angle made by CV with the spheroid equatorial plane. The angle VCS, about 6″, is a *deviation of the vertical* in the meridian direction. The precision to which deviations can be quoted is of course limited to the accuracy of observation of astronomical latitude. There will be deviations at any other points where the astronomical latitude is observed.

Study of the deviations indicates how the spheroid could be modified so as to be a better fit to the geoid. In the simple case illustrated, some ways of improvement are immediately suggested by looking at the geometry. The trial spheroid is probably too small. More obviously, it should be more sharply curved in the region of point B and less curved in the stretch CD; this indicates an increase of the flattening. Some improvement would be made simply by bodily tilting the reference surface about point B or about C in an anticlockwise direction; this will introduce a deviation at B and needs further consideration because if the whole spheroid is tilted the condition that its axis is parallel to the Earth's axis of rotation will be disturbed. Suppose the tilt desired is 5″; then what must be done is not to tilt the whole spheroid but simply to shift it so that a point on it having latitude 59° 59′ 55″ is at B. The improved spheroid might be located somewhat as shown in Fig. 126(b).

The foregoing description of spheroid fitting is geometrical and qualitative, and can be put into more mathematical form. When the computations of spheroid coordinates have been made, deviations will be revealed at all points where astronomical values have been observed. The computer can then change the reference surface by altering the semi-axis a, by altering the flattening f and by introducing a deviation at the datum point as described above; furthermore, he can move the spheroid 'vertically' so that it is not exactly at sea level at B, if this will help to get a better overall fit with the geoid.

On the spheroid of modified dimensions and position, computations using the same survey data will result in slightly different spheroid latitudes, hence changes in the deviations. Since the changes are small, simple differential formulas can be used, and the equations connecting the changes of spheroid with the changes of deviations can be used to derive a better fitting spheroid, on some such basis as minimising the sum of the squares of the residual deviations.

Before the days of telegraphy, the observation of astronomical longitude was a tedious and difficult operation, and the accuracy of the results did not approach that of latitude observation. Therefore, longitude observations did not play much part in determining

Figure of the Earth, but this was of little concern because it is of course in the variation of curvature with latitude that the elliptical shape has most effect. Radio time signals have made it possible, with the aid of impersonal methods of recording observations, to determine longitude with accuracy about the same as that of latitudes. A spheroid fitting process can now include longitude observations. A deviation of longitude is the difference between a spheroid longitude and the astronomical longitude at the same place, and modification of an initial spheroid must aim at reducing these deviations as well as the deviations of latitude. The best-fitting spheroid for an area is the spheroid of such dimensions and in such position that the differences of astronomical and spheroid latitudes and longitudes at all points where astronomical values have been observed are randomly distributed and as small as possible on some statistical criterion such as the principle of least squares.

8.03 Laplace Point. There is an important geometrical property of longitude deviations that has significance in adjustment processes. As explained in Section 1.06, the astronomical or real meridian plane of a point is the plane determined by the gravity vertical there and the line through the point parallel to the Earth's axis of rotation. In Fig. 127, the line ZAV is the vertical and AP is parallel to the

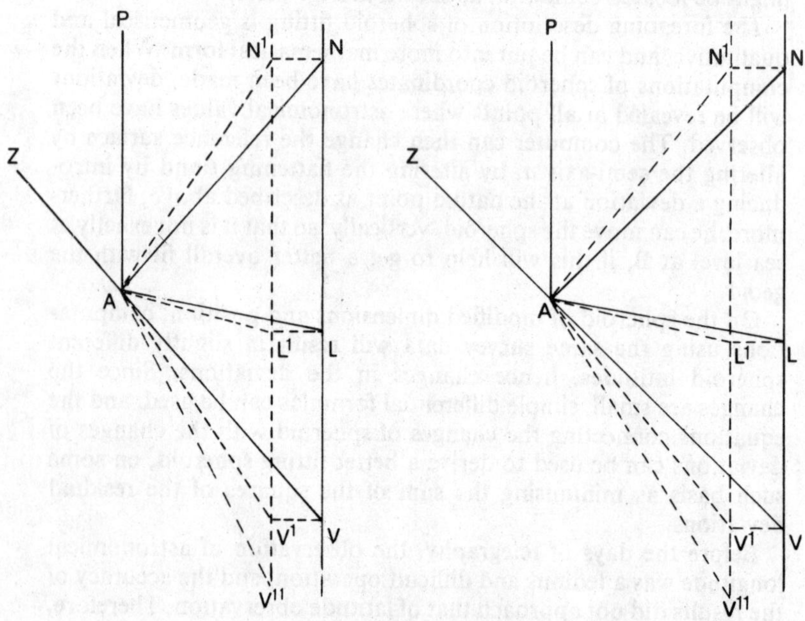

FIG. 127. LAPLACE POINT.
(this diagram is a stereo-pair)

Earth's axis. AN is the line through A perpendicular to AZ and by definition this is the direction of true or astronomical North at A; a horizontal angle measured clockwise from AN is a true bearing or azimuth. The angle made by the plane containing the lines AP, AN, AV with the Greenwich meridian plane is the Greenwich longitude of point A. Now if a spheroid of reference has been established with its axis parallel to the Earth's axis, the spheroid meridian plane will contain the line AP, but if there is a longitude deviation the spheroid meridian plane will not contain the lines AN and AV. In that case, the direction of the spheroid meridian at A, that is the spheroid North, will not lie along AN and is indicated as AN′; the lines AP and AN′ are in the spheroid meridian plane.

Let AL and AL′ be perpendicular to AP, therefore perpendicular to the Earth's axis. The angle LAL′ is the longitude deviation, being the angle between the spheroid and true meridian planes. Spheroid bearings are of course referred to the direction AN′: thus, in the situation represented by Fig. 127, spheroid bearings will be greater than true bearings by the angle NAN′. This angle can therefore be called the 'azimuth deviation'.

There is a simple relationship between the longitude deviation and the azimuth deviation. Angle ZAP is the angle between zenith and pole, and is therefore the co-latitude of A; the angle PAN is the latitude ϕ, as is also the angle ANL. Thus AL/AN = sin ϕ. The lines NLV and N′ L′ V′ are drawn parallel to AP and equidistant from it, hence NN′ = LL′ = VV′. The longitude deviation $\delta\lambda$ is LL′/AL and the azimuth deviation δA is NN′/AN; hence by using the ratio AL/AN given above we find that $\delta A = \delta\lambda \,.\, \sin\phi$.

The relationship given at the end of the previous Section is called the Laplace Equation. It is of considerable practical importance, because $\delta\lambda$ can be found by observing longitude and then δA can be calculated. It follows that if the astronomical azimuth is observed the addition or subtraction of δA will give the spheroid bearing. Now when coordinates are calculated on the spheroid, the computed bearings will be affected by any errors in the angles used to carry the bearings forward, and this can only be checked by independent determination of *spheroid* bearings, as is provided by the procedure just described. Thus, Laplace Points distributed through a survey system provide checks against accumulation of directional error, and the mathematical conditions for closure between these check bearings can be included in the adjustment of the system.

Clearly, the use of the Laplace correction is the more necessary the greater the latitude, and there is no correction at a point on the equator anyway. Otherwise, the number of Laplace Points desirable in a system depends on the strength of its geometry. The principle is that the separation between Laplace Points should be such that the accumulated error of direction between them is likely to be appreciably greater than the probable error of the Laplace bearing itself. In a strong triangulation system the reasonable separation might be of the order of 100 to 200 km, but in a network of traverses the Laplace Points are the only checks on the angle measurement, so they should be much closer. In the traverse network covering Australia there are about 533 Laplace Points, and in the re-triangulation of Great Britain 15 were used.

The introduction of a latitude deviation at a datum point may help to improve the fit of a spheroid, and so may a longitude deviation. Evidence for the desirability of a longitude deviation may come from the corrections at Laplace Points. If these show a strong systematic tendency to be of the same sign, it is obviously an indication that the initial spheroid is 'twisted' away from most of the true meridians by being established at a point where the true meridian is out of sympathy with most of the rest of the survey. This situation is illustrated in Fig. 128, which represents a 'view' of the Earth from a position far above the Pole. A B C D are points in a survey system and the full lines through them are sections of their meridian planes.

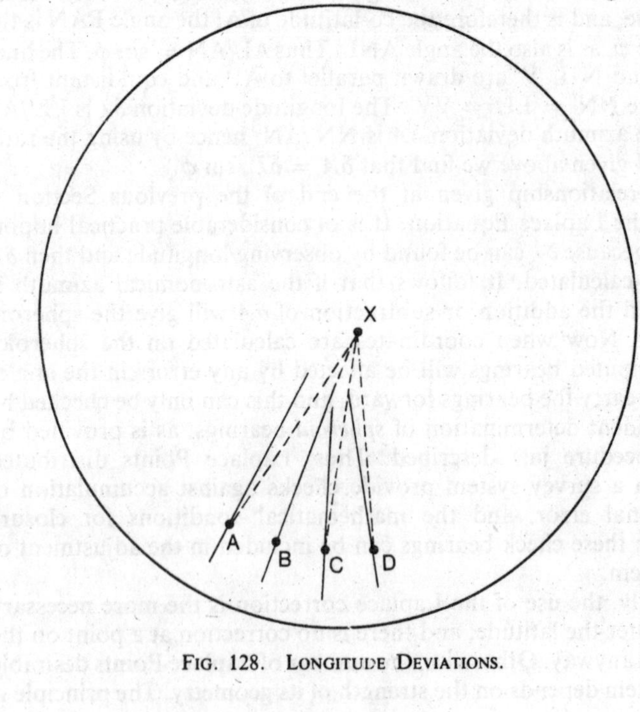

FIG. 128. LONGITUDE DEVIATIONS.

If an initial spheroid is placed at B with its meridian BX coinciding with the true meridian, there will be longitude deviations and consequent azimuth deviations at all the other points as shown by the angles between the full and broken lines. Obviously, a suitable rotation of the spheroid at B will remove the longitude deviations at the other points and introduce one at B: the rotation must of course be made about the line through B parallel to the Earth's axis.

A consequential effect on bearings must not be overlooked. The initial spheroid bearing at B for starting off the computations of coordinates was also the true bearing: when the spheroid is rotated as suggested by the situation in Fig. 128, the initial spheroid bearing will be greater by the amount given by the Laplace Equation. This change of initial bearing will go through the whole computation process, and must be allowed for in calculating the various changes of deviations referred to in Section 8.03.

In Fig. 127, the angle AVN is the angle between the gravity vertical at A and the direction of the Earth's axis of rotation, and it is therefore the astronomical co-latitude of A: it is also equal to the angle AV'N' which is the spheroid co-latitude of A if AV' is the spheroid normal. In that case there is no latitude deviation at A. But the spheroid normal might be in some slightly different direction in the spheroid meridian plane, such as AV", in which case there is a latitude deviation equal to the angle V'AV"; the spheroid latitude of A is greater than the astronomical latitude. Thus, the angle VAV" is the total deviation, the angle between the gravity vertical and the spheroid normal. It is customary to describe deviations by their north–south and east–west components: the angle V' AV" is the latitude deviation usually denoted by ξ, and the angle VAV' is the 'prime-vertical' deviation usually denoted by η. It is easy to show that $\eta = \delta\lambda \, . \cos \phi$.

8.04 Black's Method for Spheroid Bearing. Spheroid bearing can be found without direct observation of longitude, by an ingenious method published by A. N. Black in 1951. Observationally, the method is that of azimuth by hour angle, as described in Section 2.17. Fig. 129 is the spherical triangle of directions at the survey station. P is in the direction of the Pole and Z is in the direction of the zenith or gravity vertical. Z' is in the direction of the normal of the spheroid; ZZ' is therefore the total deviation of the vertical but its direction and amount are unknown. PZ' is the spheroid meridian, assuming that the direction of the spheroid axis is also that of the Pole. The spheroid coordinates of the survey station have been computed, and PZ' is thus the spheroid co-latitude. The observation consists of

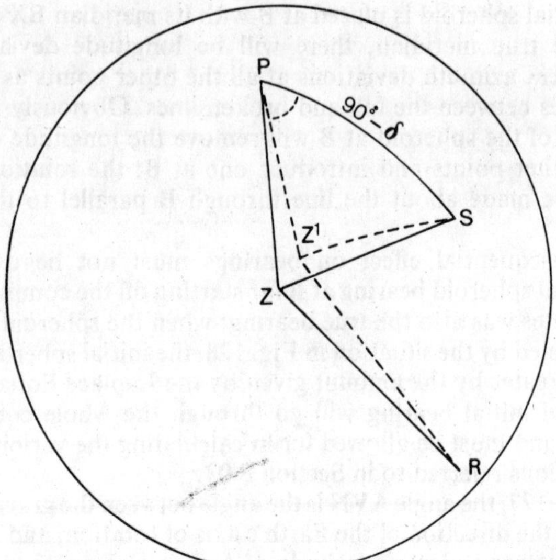

FIG. 129. BLACK'S METHOD FOR SPHEROIL AZIMUTH.

a recording of the moment of setting the vertical crossline on a star, and measurement of the horizontal angle between star and a fixed reference signal R on the ground.

Since the theodolite is necessarily levelled about the gravity vertical, the horizontal angle measured will be SZR.

Now, if the *spheroid* longitude is used in calculating the hour angle of the star from the chronometer reading, the spheroid hour angle SPZ′ will be obtained. Two sides PS, PZ′ and included angle are therefore known and the *spheroid* bearing of the star, angle PZ′S, can be calculated. What is now wanted is the angle SZ′R, to get the spheroid bearing of R, but actually it was the angle SZR that was observed.

The relevant geometry is exactly the same as that of the effect of a dislevelment of the theodolite on observed horizontal angles, discussed in Section 2.31. The reference signal will in practice be at almost zero altitude, so, as explained in that Section, the addition of $\varepsilon \,.\, \tan h \,.\, \sin \theta$ to the observed angle SZR would give the value of SZ′R: ε is the deviation, h is the altitude of the star, and θ is the direction of the star referred to the direction ZZ′ of the total deviation. Of course numerical values of the formula cannot be calculated because ε and θ are not known. The essence of Black's method is the use of a number of stars in directions well distributed round the horizon. Strictly, if all the stars were observed at the same altitude in directions evenly spaced round the circle, the mean of all the values of the correction formula would be zero, so the mean of all the azimuths of R obtained by using horizontal angles measured round Z would be the correct *spheroid* azimuth.

In practice, if the directions are fairly evenly spaced and altitudes approximately the same, any residual error will be small. Moreover, advantage can be taken of the hour angle method by setting on stars at quite low altitudes, thus keeping the factor $\tan h$ to small fractional values. More precisely, if vectors of lengths proportional to $\tan h$ drawn in the directions of the stars are vectorially added, the mean azimuth obtained will be the correct spheroid azimuth if the vector diagram closes up.

The differences of the individual values of the azimuth from their mean will show a cyclic relation to the directions of the stars, and thus can be analysed to give the amount and direction of the deviation of the vertical. If A_0 is the azimuth of the deviation and A the azimuth of a star, then $\theta = A - A_0$ and the formula can be written

$$\varepsilon \,.\, \cos A_0 \,.\, \tan h \,.\, \sin A - \varepsilon \,.\, \sin A_0 \,.\, \tan h \,.\, \cos A$$

then $\varepsilon \,.\, \cos A_0$ and $\varepsilon \,.\, \sin A_0$ are the components ξ and η of the deviation of the vertical as defined in Section 8.03.

Suppose for example that the computed spheroid azimuth of a star was 68° 19′ 12·9″ and the measured horizontal angle from star to signal was 74° 33′ 20·2″: the altitude of the star was 14° 46′ (this can be calculated from the triangle, or noted on the theodolite; it is not required precisely). After calculating the values of $\tan h \,.\, \sin A$ and $\tan h \,.\, \cos A$, the result of this observation is

$$142^\circ\ 52'\ 33{\cdot}1'' + 0{\cdot}245\xi - 0{\cdot}097\eta$$

for the spheroid azimuth of R. Assuming the azimuth of R to be $142^\circ\ 52'\ 40'' + \alpha''$ this observation provides a residual of (observed-adopted value) $= 0{\cdot}245\xi - 0{\cdot}097\eta - \alpha - 6{\cdot}9''$. Each observation gives a residual with different coefficients of ξ and η, and different numerical terms. If there are more than 3 residuals, the usual least-squares process yields the 'most probable' values of ξ, η and α. It may be noted that the definitions of ξ and η here used are such that spheroid coordinates are north and east of the astronomical quantities.

8.05 Deviations and Gravity. Deviations of the vertical are of course related to the topographical and geological irregularities of the Earth's surface form and crustal structure.

It is natural to think of a mountain as attracting the plumb-bob of an instrument away from the position in which it would hang if the mountain were not there. To put it the other way round, if the mountain could be spirited away, the plumb-bob would swing back on being released from the sideways pull of the mountain. It is a simple matter in theory to calculate what the swing would be; one needs Newton's formula, a reasonably good topographical map to give the geometrical dimensions of the mountain, an intelligent guess of the density of the rocks of which the mountain is made, and enough enthusiasm to do an enormous amount of simple arithmetical computation. If t is the horizontal pull of the mountain the plumb-bob would swing away by the angle t/g radian, a very small angle, g being the strength of gravity in the same units as t.

In the early days of geodetic studies, there was an idea that if the Earth could be smoothed down by removal of topographical irregularities and filling up the sea with rocks, then not only would the physical surface be a simple geometrical figure, but the fitting of a spheroid to the verticals would be a non-problem. This idea was soon found to be wrong.

Studies of deviations in places where they might be expected to be large, particularly in North India, showed that they were very much smaller than the results of calculations done on the basis suggested above. These discoveries led to theories of isostasy or compensation associated especially with the names of Archdeacon Pratt of Calcutta and Sir George Airy the Astronomer Royal. The idea in both systems was that the weights of columns of the same cross section down to a certain depth were equal, so that the material of a mountain or the low density of sea water were compensated respectively by reduction or increase of material below.

In Figs. 130 and 131 there are indicated three columns of the

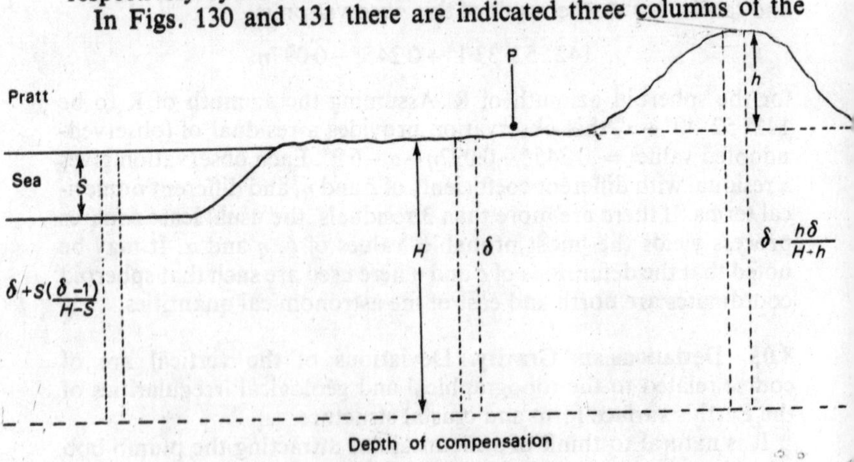

FIG. 130. PRATT'S ISOSTASY.

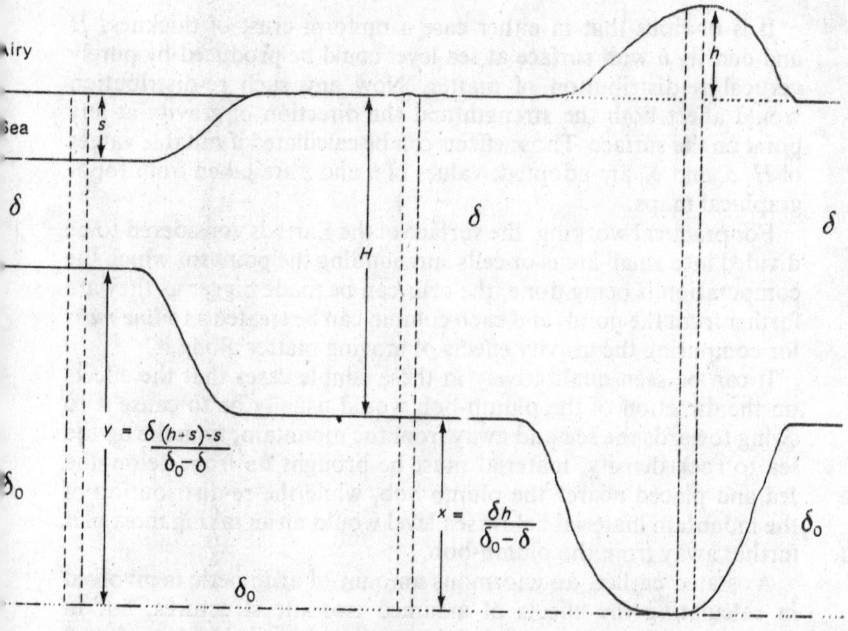

FIG. 131. AIRY'S ROOTS OF MOUNTAINS.

Earth's crust of equal cross-sections, one under a mountain, one at sea and one in a lowland area not much above sea level. In both cases, the 'normal' crust below the lowland area has depth H of rock of density δ grams per cubic centimetre.

On the Pratt (Isostasy) hypothesis, the column below the mountain of height h is uniformly less dense than δ down the depth H below sea level, and the column below the sea of depth s is uniformly more dense than δ, so as to give equal masses of material in all the columns. The crust densities shown on the diagram are easily calculated, and the density of sea water has been taken as 1 for simplicity. In any case, reasonable guesses have to be made about rock densities.

On the Airy (Roots of Mountains) hypothesis, there is a crust of uniform density everywhere, but variable thickness, thinner than H below the sea and thicker than H below the mountain. This crust floats on material of higher density δ_0. Again, the formulas on the diagram are easily calculated. and it is to be noticed that in this case the depth H does not appear in them; this is understandable from the mechanics of the situation.

It is obvious that in either case a uniform crust of thickness H and density δ with surface at sea level could be produced by purely vertical re-distribution of matter. Now any such re-distribution would affect both the strength and the direction of gravity at any point on the surface. Those effects can be calculated if suitable values of H, δ, and δ_0 are adopted; values of h and s are taken from topographical maps.

For practical working, the surface of the Earth is considered to be divided into small areas or cells surrounding the point for which the computation is being done; the cells can be made bigger as they are further from the point, and each column can be treated as a *line mass* for computing the gravity effects of moving matter along it.

It can be seen qualitatively in these simple cases that the effects on the direction of the plumb-bob would usually be to cause it to swing towards the sea and away from the mountain; to make up the sea to rock density, material must be brought up from below the sea and placed nearer the plumb-bob, while the re-distribution of the mountain material below sea level would mean taking most of it further away from the plumb-bob.

As stated earlier, an enormous amount of arithmetic is involved in calculating the effects of assumed isostatic structures, but in most areas where the work has been carried out it has been found that deviations of the vertical would actually be largely eliminated. This makes spheroid fitting easier, but perhaps the principal result of these studies has been to show that most of the Earth's topographical irregularities are in fact isostatically compensated to a considerable degree.

8.06 Global Geodesy. At the present time, methods of geodetic investigation have moved over to approach the problem from the other side, by imagining 'models' of the Earth. Seismological and other geophysical research has provided a great amount of information about densities of rocks and the depths at which layers of different kinds of rocks are situated. Guided by this knowledge, and aided by great computer power, it is now possible to set up quite complicated models of crustal structure, and to modify them simply by feeding different parameters into the machine. In this way, a model can be generated which closely simulates the deviations and gravity values actually observed, and then of course values can be calculated at any other points.

Since gravimeters came into use about 40 years ago, study of the detailed shape of the geoid has mostly been based on gravity measurement, because it is far easier to cover an area with gravimeter stations than with astronomical stations. Gravity methods are not discussed in this book.

A more recent development in geodetic instruments is the artificial satellite. This provides both gravitational and geometrical information. Analysis of the movements of satellites as observed at stations on the Earth gives information about the gravity field above the Earth, and locates the major 'undulations' of the level surfaces. In the geometrical application, illuminated satellites are photographed by large cameras at fixed stations, and the directions of the satellite positions are found by comparing them with the stars in the background. With simultaneous photographs from several stations, the satellite is simply a trig signal high enough to be seen from places several thousands of kilometres apart. The process is triangulation in three dimensions and has provided geodesists for the first time with a method for encircling the whole Earth in a system of geometrical measurement.

In the early days of geodesy, many 'Figures of the Earth' were calculated from surveys of comparatively small portions of the Earth's surface, and were used as the bases of national mapping systems. Very slowly, contiguous survey systems were joined up; it was not until after the second world war that the project of joining up the geodetic surveys of Europe was put into action. Extensions of this system into South Africa and S.E. Asia have now been connected.

Meanwhile, the geodesist Hayford had calculated what the shape of the geoid in the United States would be if the Earth's topography were smoothed away by distributing the material of the mountains down into the crust in accordance with the theory of isostasy as mentioned in Section 8.05. The spheroid derived by Hayford was recommended by the international association in 1924 to be adopted as an International Spheroid of Reference. Its dimensions were $a = 6\,378\,388$ m, $f = 1/297$: along with these dimensions there goes an appropriate theoretical formula for the relation of gravity to latitude. The gravity formula has been useful as a model with which to compare actual measurements of gravity: very few countries have adopted the spheroid as a basis of mapping, because they are already mapped and there is no sense in re-drawing a set of national maps just to put them on another spheroid which may or may not be a better fit to the local piece of the geoid.

There is now a new recommendation for an international reference spheroid. Improved measurements of gravity in the 1930s and later showed that the model g formula was too big by about 14 milligals. Much evidence shows that the Hayford spheroid is dimensionally too big also. And now the artificial satellites, after thousands of girdles of the Earth, have indicated a slight reduction of the flattening. The recommended *Geodetic Reference System* 1967 has $a = 6\,378\,160$ m, $f = 1/298{\cdot}247$. The extent to which this, with its appropriate gravity formula, is adopted remains to be seen.

Meanwhile, established national mapping and coordinate systems will go on, and the surveyor who has to connect his work to one of them must take care to put the correct numbers into his geodetic calculations and projection formulas.

REFERENCES

BLACK A. N. 'Laplace Points in moderate and high latitudes'. *Empire Survey Review* No. 82, October 1951.

BLACK A. N. 'A note on azimuth determination'. *Empire Survey Review* No. 89, July 1953.

BOMFORD A. G. 'The geodetic adjustment of Australia 1963–66'. *Survey Review* No. 144, April 1967.

COOK A. H. *Gravity and the Earth*. Wykeham Publications, London, 1969.

FISCHER I. 'The development of the South American Datum'. *Survey Review* No. 158, October 1970.

HAYFORD J. F. 'The Figure of the Earth and Isostasy from measurements in the United States'. *U.S. Coast and Geodetic Survey* 1909.

HIRVONEN R. A. 'The reformation of Geodesy'. *Journal of Geophysical Research* Vol 66, p. 1471, 1961.

HUNTER W. H. 'Satellite Geodesy'. *Australian Surveyor*, September 1969.

JACKSON J. E. 'The quantity of a degree in our English Measure'. *Empire Survey Review* No. 86, October 1952.

JONES H. E. 'Geodetic ties between continents by means of radio telescopes'. *Canadian Surveyor*, September 1969.

LEVALLOIS J. J. 'Le systeme géodésique de reference 1967 et le calcul des anomalies de la pesanteur'. *Bulletin Géodésique* No. 103, March 1972.

MATHER R. S. AND FRYER J. G. 'Orientation of the Australian geodetic datum'. *Australian Surveyor*, March 1970.

MATHISON O. 'Determination of deflection of vertical with a small instrument'. *Bulletin Géodésique* No. 93, 1969.

ROBBINS A. R. 'A geoidal section through Great Britain'. *Survey Review* No. 128, April 1963.

SCHMID H. H. 'A world survey control system and its implications for national control networks'. *Canadian Surveyor*, June 1971.

VEIS G. 'The determination of the radius of the Earth and other geodetic parameters as derived from optical satellite data'. *Bulletin Géodésique* No. 89, 1968.

CHAPTER IX

MAP CONSTRUCTION

9.01 Map Projection. The ultimate object of most surveying operations is mapping. A map is a plane representation of the curved surface of the spheroid, on which positions are defined as latitudes and longitudes. The construction of a map therefore requires the transformation of spheroid coordinates into a plane system: such a transformation is a *map projection*.

It is however unnecessary to find spheroid coordinates of every point surveyed and then transform them to the plane system: the usual procedure is to calculate spheroid latitudes and longitudes of the main framework of a survey; some or all of these values may also be used, in conjunction with astronomical observations, in studies of the shape and size of the Earth (Chapter VIII). The rest of the points are needed only for control of mapping, engineering surveys, etc. Nevertheless, since all the survey measurements must be made on the Earth, it is necessary to ensure that all positions are correctly expressed in the plane system.

Two problems have now to be considered: (i) the transformation of spheroid coordinates expressed as latitudes and longitudes, into a system of rectangular coordinates for plotting on a plane surface, by using the geometry and mathematics of a map projection; and (ii) the methods a surveyor may adopt for obtaining correct plane coordinates of surveyed points directly from his observations on the Earth, thus bypassing, as it were, their positions on the surface of reference, but of course taking into account the structure of the projection in which the points have to be located.

There are innumerable constructions which can be used for representing the spheroid surface on a plane, and many of these can be seen in atlases. Because an atlas map must comprise a large portion, or perhaps the whole, of the Earth, the distortion of shape and variations of scale unavoidable in any plane representation are quite conspicuous. The surveyor and engineer, on the other hand, are concerned with maps on much larger scales, typically 1/50 000 for topographical maps, or 1/2000 for cadastral and engineering plans. These will usually be based on a regional or national projection system limited to a very small fraction of the spheroid, such an area, for instance, as that of Great Britain.

If the area covered on one coordinate system is sufficiently small, it is easy to devise a projection structure such that the distortions mentioned above are imperceptible, indeed practically unmeasurable on a map printed on ordinary paper. The projection properties are graphically invisible, and are only retained in the mathematical formulas and computation methods appropriate to the projection.

Consider a small circle of angular radius θ on a sphere of radius R. Its diameter in its own plane is $2R \, . \sin \theta$, but the distance across the circle along the surface of the sphere is $2R\theta$ (θ in radian). If the circle is mapped so that its circumference is of correct length at a certain scale, its diameter measured on the map will come out too small, to the same scale, in the ratio $(\sin \theta)/\theta$, that is approximately $1-\frac{1}{6}\theta^2 \ldots$. Now suppose we require the map to have no scale variation greater than 1/2000, then θ must be less than $(6/2000)^{\frac{1}{2}}$, approximately 1/18 radian, about 3°. This indicates the limitation of extent that can be covered on a single projection system, given that a limit is set to the acceptable scale variation.

Scale. The foregoing remarks do give rise to the question of what is the exact meaning of 'scale of a map'. As stated earlier, it is impossible for all distances on a plane map to be correct at any one scale; hence, any scale quoted on a map must be a nominal figure. A sensible meaning may be given to the idea of scale by considering a true scale model of the spheroid, about which there is no doubt as to the meaning. The surface of the model can then be transformed to a plane through some defined geometrical relationship keeping closely to dimensions, and the scale of the model can be quoted as the nominal scale of the map. Alternatively, one could think of the full-size spheroid being mapped at nominal scale 1/1, and then the notional map so produced being reduced to, say, 1/20 000 to give the actual map.

9.02 Projection Coordinates. In the most general terms, a map projection may be described as a pair of formulas:

$$N = f_1(\phi, L): \qquad E = f_2(\phi, L)$$

for expressing rectangular coordinates, Northings and Eastings, as functions of latitude and longitude. In some literature on surveying and map projections, the coordinates are denoted x and y, and many surveyors have been accustomed to the x coordinate being in the north–south direction, but there can be confusion, even between different surveyors, and it is best to use the N and E notation, as is now done in the Ordnance Survey, for instance.

Geometrically, a map projection can be visualised by means of the pattern or 'graticule' of the spheroid meridians and parallels as they appear in the rectangular coordinate system of the map. This is illustrated in Fig. 132: the broken lines are the coordinate grid

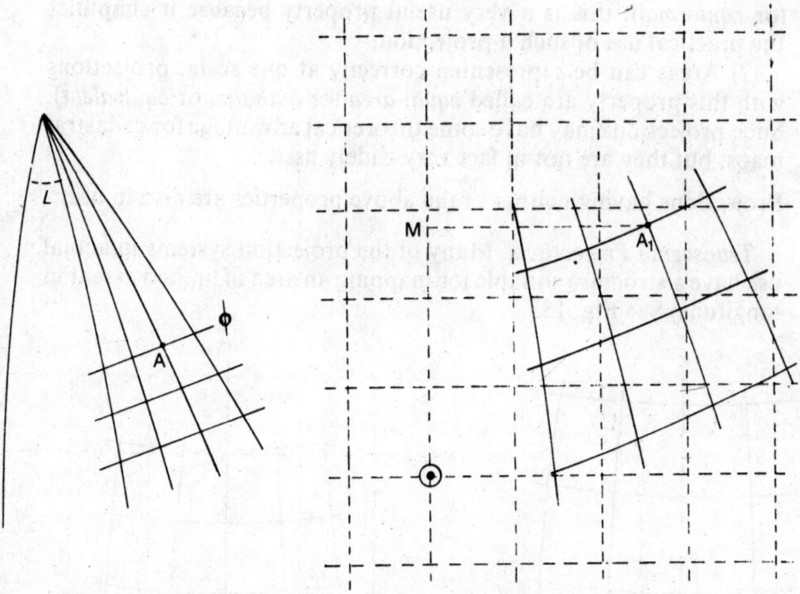

FIG. 132. SPHEROID AND MAP.

system with origin at O, and then the Northing and Easting of point A_1 are respectively OM and MA_1.

It is now customary to print topographical maps covering areas the edges of which are lines of constant value of one of the plane coordinates, so that the sheet edges are rectangles. Alternatively, maps may be printed with portions of parallels and meridians as limits, and then the sheet edges will be (theoretically) curved in most cases.

The number of different projections in actual use for mapping is not great, and nowadays is tending to become even more restricted, on account of the widespread adoption of a particular projection, the Transverse Mercator. There is no point in going through a general classification of projections in this volume, but two important properties must be mentioned. It has already been emphasised that no projection system can show all distances correctly at one scale. Two other geometrical features, however, can be preserved, though not simultaneously.

(1) Angles can be represented correctly throughout a projection; that is, if two lines on the spheroid intersect at a certain angle, then the lines representing them on the projection will intersect at that same angle. Projections having this property are called *orthomorphic*

(or *conformal*): this is a very useful property because it simplifies the practical use of such a projection.

(2) Areas can be represented correctly at one scale: projections with this property are called *equal-area* (or *authalic*, or *equivalent*). Such projections may have some theoretical advantage for cadastral maps, but they are not in fact very widely used.

Projections having neither of the above properties are also in use.

Transverse Projections. Many of the projection systems in actual use have a structure suitable for mapping an area of limited extent in longitude. See Fig. 133.

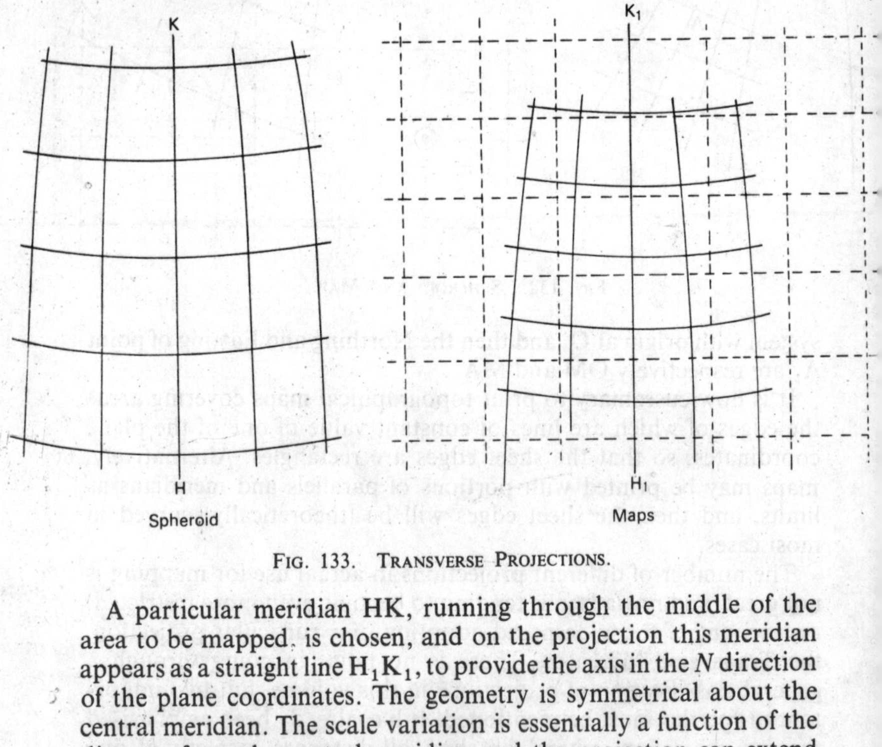

FIG. 133. TRANSVERSE PROJECTIONS.

A particular meridian HK, running through the middle of the area to be mapped, is chosen, and on the projection this meridian appears as a straight line H_1K_1, to provide the axis in the N direction of the plane coordinates. The geometry is symmetrical about the central meridian. The scale variation is essentially a function of the distance from the central meridian and the projection can extend in latitude to the borders of the area mapped. There are three well-known projections in this class:

(i) Polyconic, (ii) Cassini, and (iii) Transverse Mercator.

In the descriptions of the construction of projections, the spheroid will be considered to be reduced to the nominal scale (which need not be specified) and the methods and formulas for the construction will be related to this model. The expression *true to scale* will then mean that a line on the spheroid model is represented on the projection by a line of exactly the same length. Latitude of a typical point will be denoted ϕ, and for the *transverse* projections the longitude will be denoted λ and reckoned positive eastwards from the central meridian.

9.03 Polyconic Projection. This projection is rarely used for mapping now, and presumably it came into use, when it did, because its geometry is as simple for a spheroid as for a sphere.

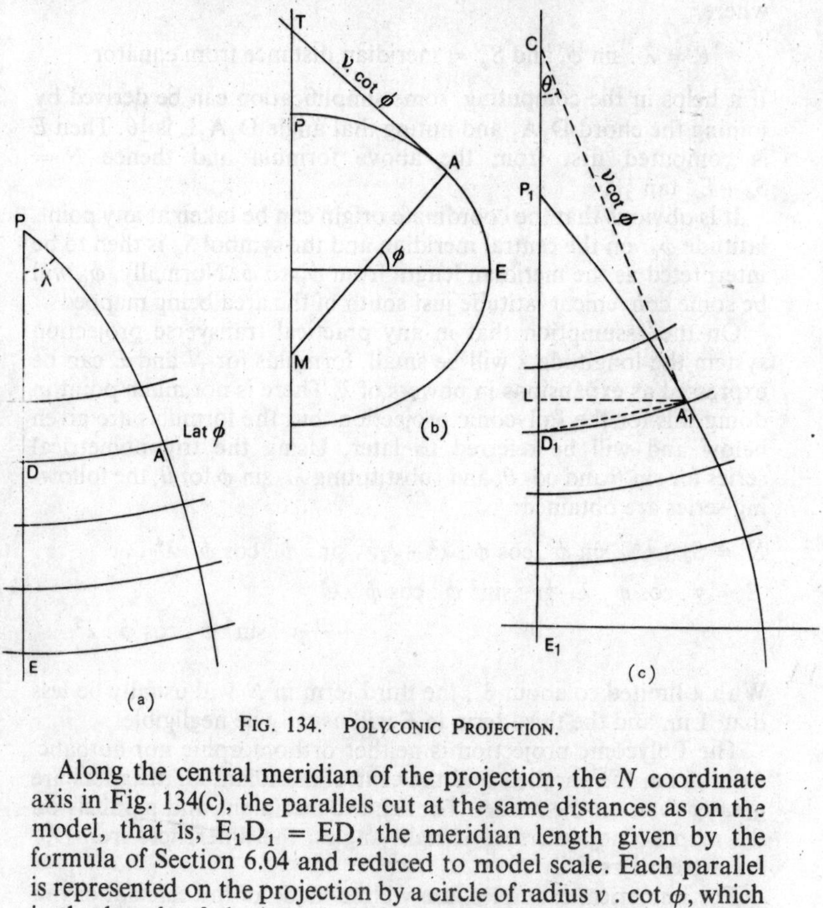

FIG. 134. POLYCONIC PROJECTION.

Along the central meridian of the projection, the N coordinate axis in Fig. 134(c), the parallels cut at the same distances as on the model, that is, $E_1D_1 = ED$, the meridian length given by the formula of Section 6.04 and reduced to model scale. Each parallel is represented on the projection by a circle of radius $\nu \cdot \cot \phi$, which is the length of the tangent AT shown on Fig. 134(b): thus the parallels are not concentric and in fact they spread out as indicated in an exaggerated way on Fig. 134(c). The Equator has infinite radius and is therefore a straight line. Scale is correct along all the parallels: hence arc D_1A_1 = arc DA = $(\nu \,.\, \cos\phi) \,.\, \lambda$, ($\lambda$ in radian). The angle θ at the centre of the circle representing parallel ϕ is therefore $(\nu \,.\, \cos\phi) \,.\, \lambda/(\nu \,.\, \cot\phi) = \lambda \,.\, \sin\phi$. The coordinates of A_1 can now be written down by use of elementary trigonometry. Taking the origin at E_1 on the map equator, we have:

$$N = E_1D_1 + D_1C - CL = S_\phi + \nu \,.\, \cot\phi - \nu \,.\, \cot\phi \,.\, \cos\theta,$$
$$E = LA_1 = \nu \,.\, \cot\phi \,.\, \sin\theta,$$

where

$$\theta = \lambda \,.\, \sin\phi, \text{ and } S_\phi = \text{meridian distance from equator.}$$

If it helps in the computing, some simplification can be derived by joining the chord D_1A_1 and noting that angle D_1A_1L is $\frac{1}{2}\theta$. Then E is computed first from the above formula and thence $N = S_\phi + E \,.\, \tan\frac{1}{2}\theta$.

It is obvious that the coordinate origin can be taken at any point, latitude ϕ_C on the central meridian and the symbol S_ϕ is then to be interpreted as the meridian length from ϕ_0 to ϕ. Normally, ϕ_0 will be some convenient latitude just south of the area being mapped.

On the assumption that in any practical transverse projection system the longitude λ will be small, formulas for N and E can be expressed as expansions in powers of λ. There is not much point in doing this for the Polyconic projection, but the formulas are given below and will be referred to later. Using the trigonometrical series for $\sin\theta$ and $\cos\theta$, and substituting $\lambda \,.\, \sin\phi$ for θ, the following series are obtained:

$$N = S_\phi + \tfrac{1}{2}\nu \,.\, \sin\phi \,.\, \cos\phi \,.\, \lambda^2 - \tfrac{1}{24}\nu \,.\, \sin^3\phi \,.\, \cos\phi \,.\, \lambda^4 \ldots$$
$$E = \nu \,.\, \cos\phi \,.\, \lambda - \tfrac{1}{6}\nu \,.\, \sin^2\phi \,.\, \cos\phi \,.\, \lambda^3 + \tfrac{1}{120}\nu \,.\, \sin^4\phi \,.\, \cos\phi \,.\, \lambda^5 \ldots$$

With λ limited to about 3°, the third term in N will usually be less than 1 m, and the third term in E will usually be negligible.

The Polyconic projection is neither orthomorphic nor authalic. Apart from the parallels and the central meridian, all distances are greater than on the model. Though the meridians and parallels on the map do not intersect at right angles, the differences from 90° are extremely small.

For numerical examples, use will be made of the calculations in Chapter VI, Section 6.08. Taking the meridian 2° west of point A as the centre meridian for the projection, the positions to be converted to projection coordinates are as shown on Fig. 135. Thus,

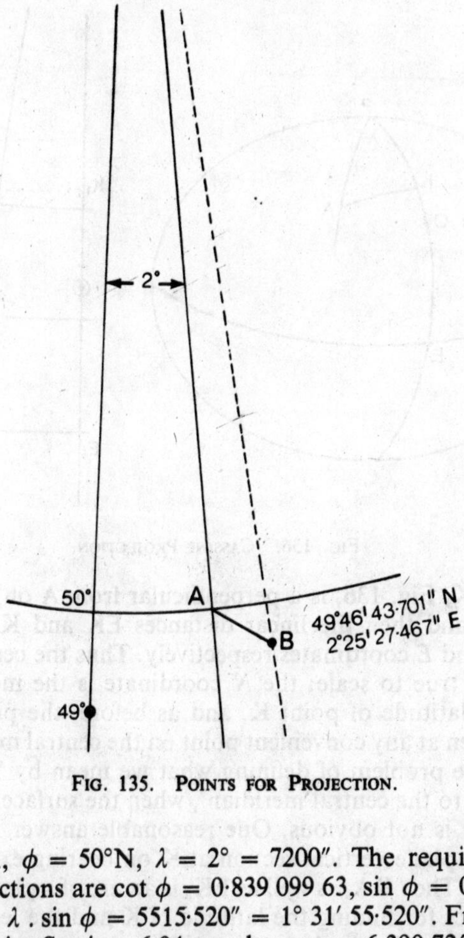

FIG. 135. POINTS FOR PROJECTION.

for point A, $\phi = 50°\text{N}$, $\lambda = 2° = 7200''$. The required trigonometrical functions are $\cot \phi = 0{\cdot}839\,099\,63$, $\sin \phi = 0{\cdot}766\,044\,44$, whence $\theta = \lambda \,.\, \sin \phi = 5515{\cdot}520'' = 1°\,31'\,55{\cdot}520''$. From the geodetic table in Section 6.04 we have $\nu = 6\,390\,725{\cdot}2$ m, hence $\nu \,.\, \cot \phi = 5\,362\,455{\cdot}1$ m and $E = \nu \,.\, \cot \phi \,.\, \sin \theta = 143\,374{\cdot}94$ m. We take the point at latitude 49° on the centre meridian as origin: then $S_{50} - S_{49} = 111\,219{\cdot}81$ m. $E \,.\, \tan \tfrac{1}{2}\theta$ is 1917·04 m, so the coordinate $N = 113\,136{\cdot}85$ m.

The corresponding computations for point B give:

$$N = 89\,437{\cdot}32 \text{ m}, \qquad E = 174\,578{\cdot}17 \text{ m}.$$

9.04 Cassini Projection. This projection was used for many of the early mapping systems in European countries, including England. In some writings it is given the curious title 'Spherical Rectangular Coordinates'. The basic idea is simple, in that on the model spheroid

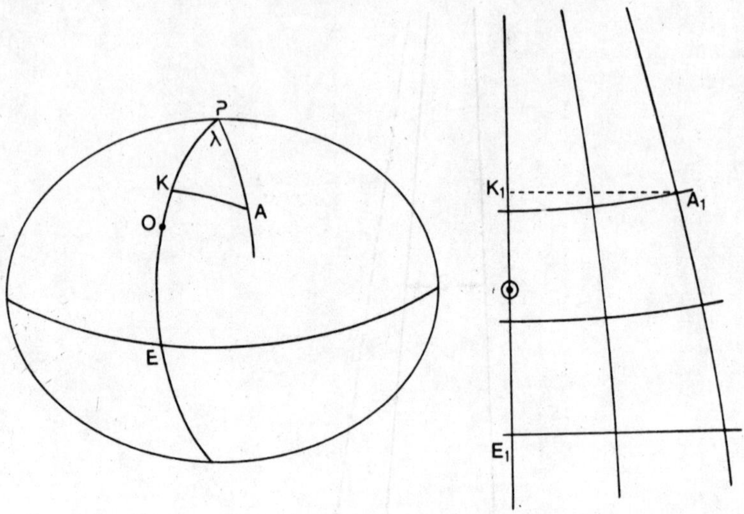

FIG. 136. CASSINI PROJECTION.

the line AK, Fig. 136, is a perpendicular from A on to the central meridian, and then the linear distances EK and KA are plotted as the *N* and *E* coordinates respectively. Thus the central meridian comes out true to scale; the *N* coordinate is the meridian length up to the latitude of point K, and as before, the practical origin can be taken at any convenient point on the central meridian. There remains the problem of defining what we mean by 'perpendicular from A on to the central meridian', when the surface is a spheroid; the answer is not obvious. One reasonable answer is to locate K so that the plane vertical section at K on bearing exactly 90° goes through A. Then KA, as well as EK, is an arc of a plane ellipse. The mathematics for finding the latitude of K and the length of KA is very tedious, and only the results will be given: the formulas are:

$$N = S_\phi + \tfrac{1}{2}\nu \,.\, \sin\phi \,.\, \cos\phi \,.\, \lambda^2 + \tfrac{1}{24}\nu \,.\, \sin\phi \,.\, \cos^3\phi\left(6\frac{\nu}{\rho} - \sec^2\phi\right).\,\lambda^4 \ldots$$

$$E = \nu \,.\, \cos\phi \,.\, \lambda - \tfrac{1}{6}\nu \,.\, \sin^2\phi \,.\, \cos\phi \,.\, \lambda^3 - \tfrac{1}{120}\nu \,.\, \sin^2\phi \,.\, \cos^3\phi\left(8\frac{\nu}{\rho} - \tan^2\phi\right).\,\lambda^5 \ldots$$

and in practice the value of $\frac{\nu}{\rho}$ can be taken to be 1. For the points

A and B of Fig. 135, the terms in the Cassini projection series are given below:

Point A		Point B	
N	*E*	*N*	*E*
111 219·81	143 392·03	86 616·97	174 608·55
+1 917·15	−17·09	+2 820·59	−30·38
+ 0·29	—	± 0·63	—
113 137·25 m	143 374·94 m	89 438·19 m	174 578·17 m

This projection also is neither orthomorphic nor authalic: the scale is correct along the *E* coordinate lines, and is too large elsewhere except on the central meridian. The projection is the spheroid transverse version of the simple cylindrical equidistant projection of the sphere, illustrated in Fig. 137. On that projection, every

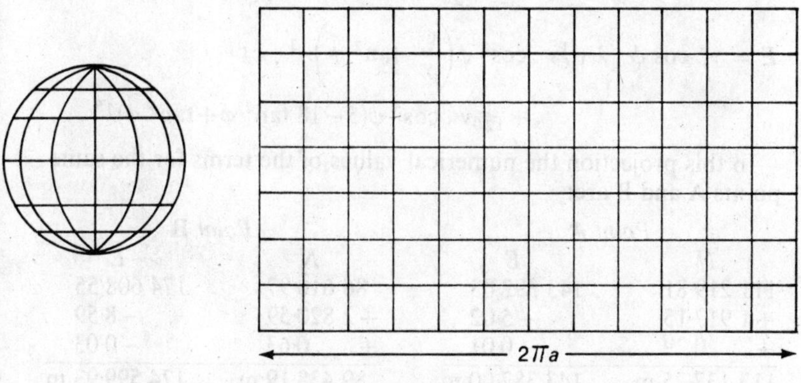

FIG. 137. SIMPLE CYLINDRICAL PROJECTION OF SPHERE.

parallel has length $2\pi a$, compared with the model length $2\pi a \cdot \cos\phi$, so the scale along parallel ϕ is exaggerated in the factor sec ϕ, which for small values of ϕ is $1+\frac{1}{2}\phi^2 \ldots$. This is $1+\frac{1}{2000}$ when ϕ is about 1/32 radian, representing roughly 200 km on the Earth. Applied to the Cassini projection, this gives an idea of how to estimate the scale exaggeration along the *N* coordinate lines. Formulas for scale factors are quoted later.

9.05 Transverse Mercator Projection. This is the orthomorphic transverse projection now so widely used for map coordinate systems. Its formulas are obtained by standard mathematical processes dealing with orthomorphic (conformal) transformation. Taking note of the form of the series expressions of the two projections already described, we write:

$$N = S_\phi + A \,.\, \lambda^2 + B \,.\, \lambda^4 \ldots$$
$$E = P \,.\, \lambda + Q \,.\, \lambda^3 \ldots$$

in which the coefficients A, B . . . P, Q, . . . are functions of ϕ only. Orthomorphism is ensured if the coefficients satisfy the conditions:

i) $\dfrac{\partial N}{\partial \phi} \cdot \dfrac{\partial N}{\partial \lambda} + \dfrac{\partial E}{\partial \phi} \cdot \dfrac{\partial E}{\partial \lambda} = 0$

ii) $\nu^2 \,.\, \cos^2 \phi \left[\left(\dfrac{\partial N}{\partial \phi}\right)^2 + \left(\dfrac{\partial E}{\partial \phi}\right)^2\right] = \rho^2 \,.\, \left[\left(\dfrac{\partial N}{\partial \lambda}\right)^2 + \left(\dfrac{\partial E}{\partial \lambda}\right)^2\right].$

A succession of equations is obtained, and the resulting series are:

$$N = S_\phi + \tfrac{1}{2}\nu \,.\, \cos\phi \,.\, \sin\phi \,.\, \lambda^2 + \tfrac{1}{24}\nu \,.\, \cos^3\phi \,.\, \sin\phi(5 - \tan^2\phi) \,.\, \lambda^4 \ldots$$

$$E = \nu \,.\, \cos\phi \,.\, \lambda + \tfrac{1}{6}\nu \,.\, \cos^3\phi\left(\frac{\nu}{\rho} - \tan^2\phi\right)\lambda^3 + \tfrac{1}{120}\nu \,.\, \cos^5\phi(5 - 18\tan^2\phi + \tan^4\phi)\lambda^5 \ldots$$

In this projection the numerical values of the terms for the same points A and B are:

Point A		*Point B*	
N	*E*	*N*	*E*
111 219·81	143 392·03	86 616·97	174 608·55
+1 917·15	−5·02	+2 820·59	−8·59
+ 0·29	−0·01	+ 0·63	−0·03
113 137·25 m	143 387·00 m	89 438·19 m	174 599·93 m

The greater values of the eastings, compared with the other projections, is of course the result of the necessity to increase scale in this direction in order to achieve orthomorphism by matching the inevitable increases in the northing direction.

9.06 Definition of Actual Scale. A clear conception of actual scale on a projection can only be realised by considering the relation between infinitesimally small lengths. If a line of length ds on the spheroid model is represented by a length ds_1 on the map, then the limiting value of the ratio ds_1/ds, as the lengths are made smaller and smaller, will differ from unity, except perhaps along certain lines on which by construction the projection is 'true to scale'. In general, this ratio, or scale factor, will depend not only on the position of the element in the coordinate system, but also on its direction. It is an important characteristic of orthomorphic projections that the scale factor at any one point is the same in all directions. It follows that any infinitesimally small shape on the spheroid model will come out on the map as exactly the same shape, though generally larger or smaller in size. This is why orthomorphic projections are sometimes described as showing shape correctly, as the basic meaning of the word implies, but this is not true of figures of finite size. In the Polyconic and Cassini projections as described above, the scale is true in a certain direction at each point in the system, and in fact the scale factor is greater than 1 in other directions, except on the central meridian. In the Transverse Mercator projection, the scale factor is greater than 1 in all directions, at every point that is not on the central meridian. There is considerable advantage, however, from the factor being the same in all directions and having a value depending only on the position in the system.

9.07 Spheroid-Map Relationships. As stated earlier, the calculation of geodetic latitudes and longitudes would be superfluous for the great majority of the points of a survey system: only projection coordinates are needed. Thus, a situation will often arise such as the one illustrated in Fig. 138. Points A, Q, and Z could be primary trigonometrical stations of which geodetic and projection coordinates have been computed. Line AB, BC, . . . could be elements in a secondary triangulation survey, or, more likely nowadays, the route of a traverse. The data for the computation will be the

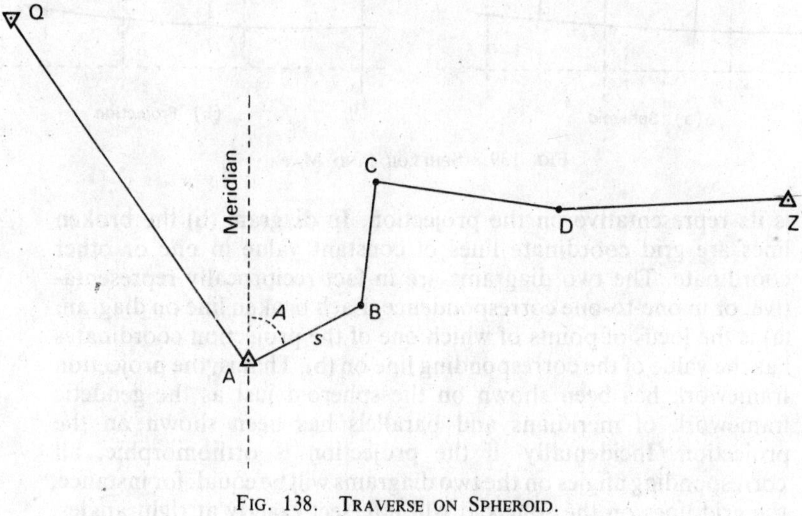

FIG. 138. TRAVERSE ON SPHEROID.

lengths of the lines and the angles at the points. Since point A is in the geodetic system, the knowledge of an angle such as QAB will make it possible to calculate the true bearing *A* of line AB. Thus, one might expect to be able to calculate the projection coordinates of B from formulas depending on *s* and *A*. It is true that this can be done, but in practice it is more convenient to use another angle instead of *A*, as explained below. For this purpose, a further look at the relationships between the geometries on the spheroid and plane will help. In Fig. 139, HK is the central meridian and H_1K_1

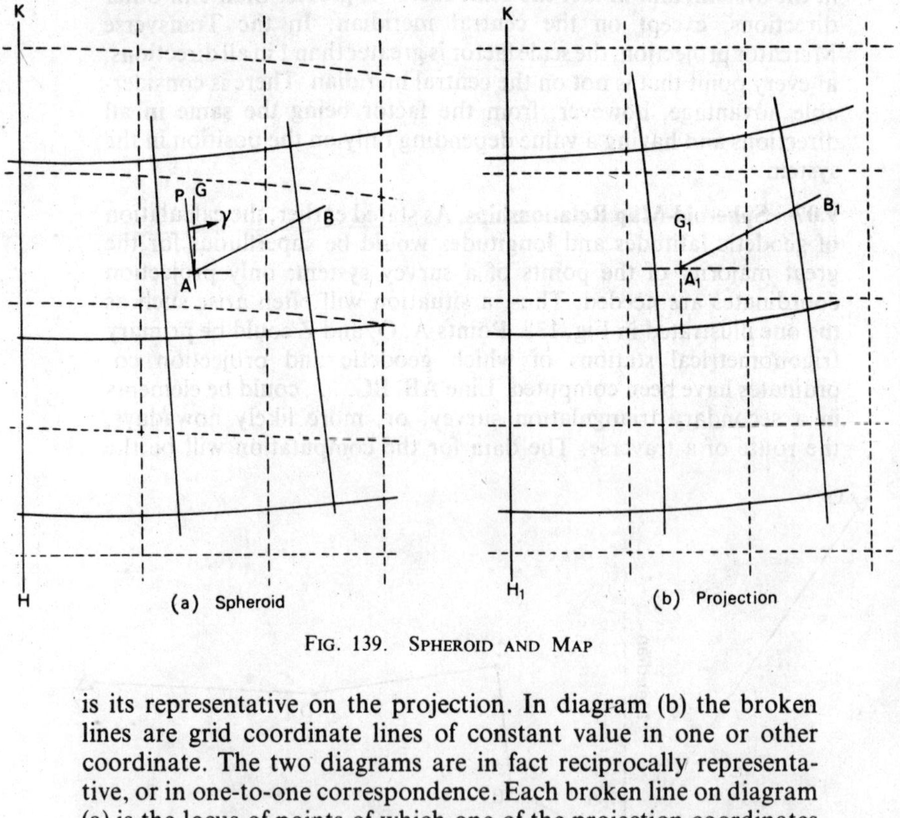

FIG. 139. SPHEROID AND MAP

is its representative on the projection. In diagram (b) the broken lines are grid coordinate lines of constant value in one or other coordinate. The two diagrams are in fact reciprocally representative, or in one-to-one correspondence. Each broken line on diagram (a) is the locus of points of which one of the projection coordinates has the value of the corresponding line on (b). That is, the projection framework has been shown on the spheroid just as the geodetic framework of meridians and parallels has been shown on the projection. Incidentally, if the projection is orthomorphic, all corresponding angles on the two diagrams will be equal: for instance, the grid lines on the spheroid will intersect exactly at right angles.

The broken lines through A are spheroid grid lines having co-ordinate values equal to N_A and E_A: the full line through A is the spheroid meridian.

Now the angle PAG labelled γ, between the meridian and the spheroid grid line of A, is called *convergence*. The angle GAB is the bearing of line AB referred to the spheroid grid line; we write GAB $= A-\gamma = \alpha$. There seems to be no accepted name for the angle α; of course if the projection is orthomorphic the angle $G_1A_1B_1$ will also be α and this is certainly to be called a grid bearing. At present we are concerned with angle GAB on the spheroid, and to be quite clear it will be called spheroid grid bearing where necessary.

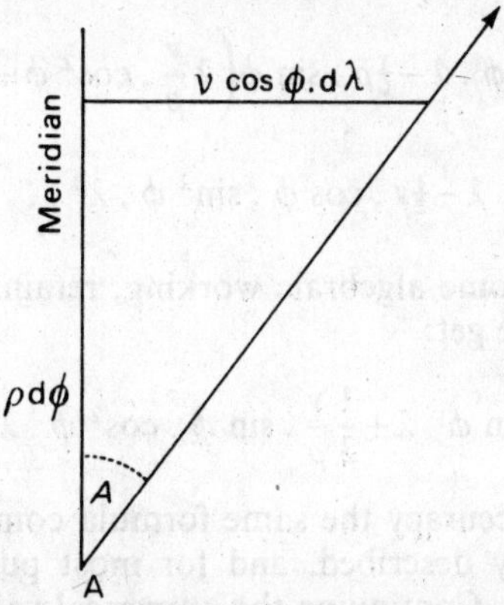

FIG. 140. DIFFERENTIAL TRIANGLE ON SPHEROID.

It turns out to be simpler mathematically and practically to use α instead of A; the reason for this is easily seen to be the fact that if α is zero the line takes off in the direction of the grid line along which the change of E is also zero. One can thus expect to have simple formulas by which the changes of projection coordinates will have values very close to $(s \,.\, \cos\alpha)$ and $(s \,.\, \sin\alpha)$.

To start the computation, knowing true bearing A, we need a formula for the convergence γ. Now if ϕ, λ are the latitude and longitude of A and smail changes $d\phi$ and $d\lambda$ are made, the corresponding linear shifts are as shown on Fig. 140, and we have

$$\tan A = \frac{\nu \,.\, \cos\phi \,.\, d\lambda}{\rho \,.\, d\phi}.$$

The E coordinate being a function of ϕ and λ, we can write the general differential

$$dE = \frac{\partial E}{\partial \phi} \,.\, d\phi + \frac{\partial E}{\partial \lambda} \,.\, d\lambda$$

and to keep the shifted point on the grid line AG we require $\mathrm{d}E = 0$, that is

$$\frac{\mathrm{d}\lambda}{\mathrm{d}\phi} = -\left(\frac{\partial E}{\partial \phi}\bigg/\frac{\partial E}{\partial \lambda}\right).$$

But γ is the bearing along which there is no change of E coordinate, so we have

$$\tan \gamma = -\left(\frac{\nu \,.\, \cos \phi}{\rho}\right)\left(\frac{\partial E}{\partial \phi}\bigg/\frac{\partial E}{\partial \lambda}\right).$$

Now refer to the formulas for the projection coordinates in terms of ϕ and λ. For the Polyconic projection we find that:

$$\frac{\partial E}{\partial \phi} = -\rho \,.\, \sin \phi \,.\, \lambda - \tfrac{1}{6}\rho \,.\, \sin \phi \left(2\frac{\nu}{\rho} \,.\, \cos^2 \phi - \sin^2 \phi\right)\lambda^3 \ldots$$

$$\frac{\partial E}{\partial \lambda} = \nu \,.\, \cos \phi \,.\, \lambda - \tfrac{1}{2}\nu \,.\, \cos \phi \,.\, \sin^2 \phi \,.\, \lambda^3 \ldots$$

By division, and some algebraic working, retaining only terms to third order in λ, we get:

$$\gamma = \sin \phi \,.\, \lambda + \frac{1}{3}\frac{\nu}{\rho} \,.\, \sin \phi \,.\, \cos^2 \phi \,.\, \lambda^3 \ldots$$

To this order of accuracy the same formula comes out for all the projections already described, and for most purposes the factor ν/ρ can be ignored. Continuing the numerical values already used, we find that:

at point A, $\gamma = 5515{\cdot}52'' + 0{\cdot}93'' = 1^\circ\ 31'\ 56{\cdot}4''$

at point B, $\gamma = 6663{\cdot}92'' + 1{\cdot}66'' = 1^\circ\ 51'\ 05{\cdot}6''$

9.08 Spheroid Grid Bearing Convergence. Before the formulas for the changes of coordinates between A and B are quoted, the problem of carrying forward the bearings must be dealt with. Figure 141 shows the line AB and the spheroid northing grid lines through A and B. Since the spheroid grid bearing α is used in the coordinate formulas, it will be desirable to have a formula to calculate the reverse spheroid grid bearing shown as β on the diagram, because the latitude and longitude of B would not be available for calculating the convergence there, by the formula given above, and one object of the present methods is to avoid having to compute these quantities. When β is found, the included angle between BA and BC is applied and the forward grid bearing of BC is obtained.

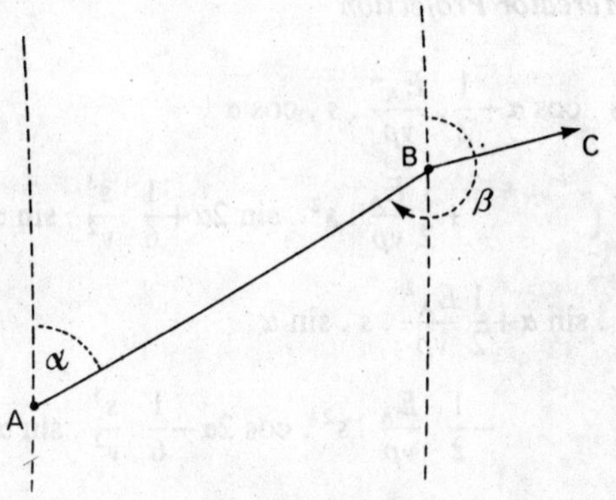

FIG. 141. GRID CONVERGENCE ON SPHEROID.

Thus, as with the geodetic coordinate formulas, a third formula giving the reverse spheroid bearing is required in practice, and it must be expressed in terms of the quantities used in the other formulas.

The reader will now see the point in using spheroid grid bearings instead of projection grid bearings. On the spheroid, the geometry, for example that of Fig. 138, is independent of the projection, particularly as to whether it is orthomorphic or not. Using a non-orthomorphic projection, there would otherwise be an additional complication of converting spheroid angles and bearings to their projection equivalents.

9.09 Point to Point Computation. To obtain the formulas for differences of coordinates in terms of s and α, the finite differences of latitude and longitude as given by Puissant formulas (Section 6.09) must be combined with the formulas for the projection, and bearing A must be replaced by $(\alpha+\gamma)$. The ensuing algebraic work is extremely tedious, and only the results will be quoted here. For the coordinate differences, the following formulas apply:

Polyconic and Cassini Projections

$$N_B = N_A + s \,.\, \cos\alpha + \frac{1}{2}\frac{E_A{}^2}{\nu\rho} \,.\, s \,.\, \cos\alpha + \frac{E_A}{\nu\rho} \,.\, s^2 \,.\, \cos\alpha \,.\, \sin\alpha + \frac{1}{3}\frac{s^3}{\nu^2} \,.\, \cos\alpha \,.\, \sin^2\alpha \ldots$$

$$E_B = E_A + s \,.\, \sin\alpha - \frac{1}{2} \,.\, \frac{E_A}{\nu\rho} \,.\, s^2 \,.\, \cos^2\alpha - \frac{1}{6} \,.\, \frac{s^3}{\nu^2} \,.\, \cos^2\alpha \,.\, \sin\alpha \ldots$$

Transverse Mercator Projection

$$N_B = N_A + s \,.\, \cos\alpha + \frac{1}{2} \,.\, \frac{E_A{}^2}{\nu\rho} \,.\, s \,.\, \cos\alpha$$

$$+\frac{1}{2}\frac{E_A}{\nu\rho} \,.\, s^2 \,.\, \sin 2\alpha + \frac{1}{6} \,.\, \frac{s^3}{\nu^2} \,.\, \sin\alpha \,.\, \sin 2\alpha \ldots$$

$$E_B = E_A + s \,.\, \sin\alpha + \frac{1}{2}\frac{E_A{}^2}{\nu\rho} \,.\, s \,.\, \sin\alpha$$

$$-\frac{1}{2} \,.\, \frac{E_A}{\nu\rho} \,.\, s^2 \,.\, \cos 2\alpha - \frac{1}{6} \,.\, \frac{s^3}{\nu^2} \,.\, \sin\alpha \,.\, \cos 2\alpha \ldots$$

For all three projections the reverse bearing formula is:

$$\beta = \alpha \pm 180° - \frac{E_A}{\nu\rho} \,.\, s \,.\, \cos\alpha - \frac{1}{2} \,.\, \frac{s^2}{\nu\rho} \,.\, \cos\alpha \,.\, \sin\alpha \ldots$$

$$= \alpha \pm 180° - \frac{s \,.\, \cos\alpha}{\nu\rho}(E_A + \tfrac{1}{2}s \,.\, \sin\alpha) \ldots$$

where the expression in brackets is for all practical computing purposes the easting of the middle point of the line AB. In this formula the terms in s will give values in radian: multiply by 206 265 to get angle seconds.

The direct computation of plane coordinates in the Transverse Mercator projection, using the points A and B of previous examples, goes as follows:

At A, convergence 1° 31′ 56·4″ given in Section 9·07 is subtracted from the true bearing 128° 44′ 17·2″ to give α = 127° 12′ 20·8″. So with s = 39 178·29 m we get $s \,.\, \cos\alpha$ = −23 690·306 m, $s \,.\, \sin\alpha$ = +31 204·286 m. The terms in the series for the T.M. projection are:

N_A	113 137·25	E_A	143 387·00
	−23 690·306		+31 204·286
	− 5·980		+ 7·876
	− 2·603		+ 0·726
	− 0·188		+ 0·053
N_B	89 438·17 m	E_B	174 599·94 m

The reverse bearing is

$$\beta = 127° \; 12' \; 20{\cdot}8'' + 180° + 19{\cdot}1'' = 307° \; 12' \; 39{\cdot}9''$$

As a check, the convergence at B makes the reverse true bearing BA to be 307° 12′ 39·9″ + 1° 51′ 05·6″ = 309° 03′ 45·5″ agreeing with the reverse bearing found by use of Clarke's formula (Section 6.08).

9.10 Scale Factor. The scale factor ds_1/ds (Section 9.06), is easily obtained from the formulas of the previous Section. For a short length, the value of $(ds_1)^2$ is $(N_B - N_A)^2 + (E_B - E_A)^2$. Referring to the formulas for the Polyconic and Cassini projections, we have:

$$(ds_1)^2 = (ds)^2 \,.\, \cos^2\alpha + \frac{E_A{}^2}{\nu\rho}(ds)^2\cos^2\alpha + \quad \ldots$$

$$+ (ds)^2 \,.\, \sin^2\alpha \quad \ldots$$

$$= (ds)^2 \quad + \quad (ds)^2 \,.\, \frac{E_A{}^2}{\nu\rho}\cos^2\alpha \ldots$$

Hence $(ds_1)^2/(ds)^2$ is $1 + \frac{E_A{}^2}{\nu\rho} \,.\, \cos^2\alpha$ plus terms in ds which will disappear as ds approaches zero. Taking the square root by the Binomial Theorem, we see that:

$$\text{Scale Factor} = ds_1/ds = 1 + \frac{1}{2}\frac{E_A{}^2}{\nu\rho} \,.\, \cos^2\alpha \ldots$$

for the Polyconic and Cassini projections. A similar operation on the formulas for the Transverse Mercator projection gives

$$\text{Scale Factor} = 1 + \frac{1}{2} \,.\, \frac{E_A{}^2}{\nu\rho} \ldots$$

thus illustrating that in this orthomorphic projection the scale is independent of direction at a point. At the points A and B the factors are respectively 1·000 252 and 1·000 374.

9.11 Chord Method. Another method widely used for computing projection coordinates directly from survey data is illustrated in Fig. 142. The curve A_1B_1, of length s_1, is the line on the projection system representing the line AB of length s on the spheroid model. The length of the straight chord A_1B_1 is denoted by c. Evidently, if the grid bearing G of this chord can be found, the differences of coordinates will be simply $c \,.\, \cos G$ and $c \,.\, \sin G$.

We suppose, for the present, that the projection in use is orthomorphic. In such case, the angle indicated by the curved arrow, that is the grid bearing at which the curve A_1B_1 leaves the point A_1, is the angle α described in previous Sections. Thus, to get the grid bearing G it is necessary to find the angle δ between curve and chord.

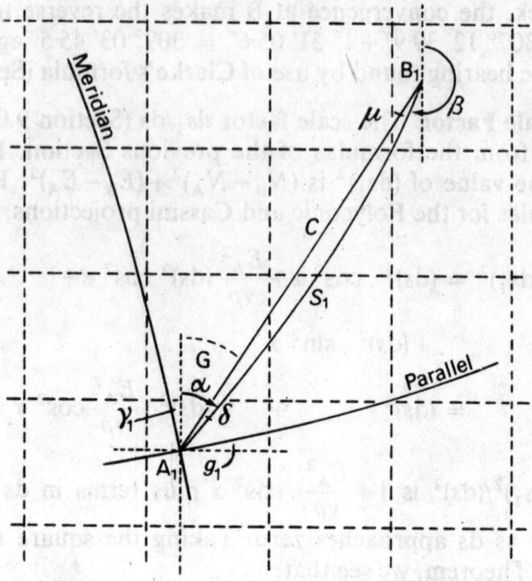

FIG. 142. CHORD ON PROJECTION.

This angle is obviously very small; in the literature it is usually called the '(t – T) correction', or 'arc-chord angle'.

The bearing G is given by

$$\tan G = \frac{E_B - E_A}{N_B - N_A},$$

and $\delta = \alpha - G$. By standard trigonometrical identity we get

$$\tan \delta = \frac{\sin \alpha \,.\, (N_B - N_A) - \cos \alpha \,.\, (E_B - E_A)}{\cos \alpha \,.\, (N_B - N_A) + \sin \alpha \,.\, (E_B - E_A)}$$

and using the formulas for the Transverse Mercator projection in Section 9·05, with some algebraic operation, we find that

$$\delta = \frac{s \,.\, \cos \alpha}{2\rho v}\left(E_A + \tfrac{1}{3}s \,.\, \sin \alpha\right) \quad \text{(radian)}.$$

The expression in brackets is the easting coordinate of the point one third of the way from A_1 to B_1.

Assuming, as before, that the computation is starting at a point where the spheroid latitude and longitude are known, the bearing α can be obtained from the true bearing by using the convergence. The next step is to calculate approximate values of $s \,.\, \cos \alpha$ and $s \,.\, \sin \alpha$ in order to obtain δ. With the data of the numerical example already used, we find $s \cdot \cos\alpha = -23\,690$ m and $s \cdot \sin\alpha = +31\,204$ m: the easting of the 'one third' point is thus $143\,387+10\,401 = 153\,788$ m, and δ comes out as $-9{\cdot}3''$. Hence the grid bearing of the chord is $(\alpha-\delta) = 127°\ 12'\ 20{\cdot}8''+9{\cdot}3'' = 127°\ 12'\ 30{\cdot}1'' = G$.

The length s_1 differs from s by the accumulated effect of the scale factor along the whole line. However, what we now want is the chord length c, and this is simply the square root of $(E_B-E_A)^2+(N_B-N_A)^2$. Using the formulas for the Transverse Mercator projection in Section 9.05, we find that

$$c^2 = s^2\left(1+\frac{E_A{}^2}{\rho\nu}+\frac{E_A \cdot s}{\rho\nu} \cdot \sin\alpha+\frac{1}{3}\frac{s^2}{\nu^2} \cdot \sin^2\alpha \ldots\right)$$

to fourth-order terms. After taking the square root and a little algebra, we get

$$c = s\left(1+\frac{E_A E_B}{2\rho\nu}+\frac{s^2}{6\nu^2} \cdot \sin^2\alpha \ldots\right)$$

The expression in brackets is the scale factor for the finite length. The value of s_1 is not actually required, but in practice it differs from c by an amount that is quite negligible.

Using the data of the worked example, we have:

$$c = 39\,178{\cdot}29+12{\cdot}04+0{\cdot}16 = 39\,190{\cdot}49 \text{ m}$$

comparing with 39 190·48 which is the square root of the sum of the squares of the coordinate differences already computed. The term in s^3 will be negligible for sufficiently short lines; it is less than 0·005 m when s is less than 10 km.

In order to proceed from point B_1 to get the differences of coordinates of the next line BC, the reverse spheroid bearing β must again be found. Of course the reverse chord bearing is simply $G \pm 180°$—in our example 307° 12′ 30·1″. The arc-chord angle μ must then be applied, to get the curve bearing β. Strictly, the formula for μ will be $\frac{s \cdot \cos\beta}{2\rho\nu}(E_B+\frac{1}{3}s \cdot \sin\beta)$, by analogy with the formula for δ, and we shall then have $\beta = G \pm 180°+\mu$, but obviously the same result will be obtained by using the formula for δ with the easting of the point $\frac{2}{3}$ of the way from A_1 to B_1, and reverse sign. The easting mentioned is about 16 419 m, and we find that μ is $+9{\cdot}9''$. The reverse bearing β is therefore $G \pm 180°+\mu = 307°\ 12'\ 40{\cdot}0''$, comparing with 307° 12′ 39·9″ calculated by the other method.

Since the projection is assumed to be orthomorphic, the angle at

B_1 between the curves B_1A_1 and B_1C_1 will be taken directly from survey data, and then computation of the next line can proceed.

It is to be noted that even when the arc-chord angles are very small, their effects will accumulate along a series of lines that lie in a roughly constant direction, as is often the case with traverses.

9.12 Non-orthomorphic Projection. When working in a non-orthomorphic projection, there is the additional complication that angles and bearings of the survey system on the spheroid cannot be directly used in the projection. It is perhaps best to start by looking at the geometry of the situation. Figure 143 shows points in

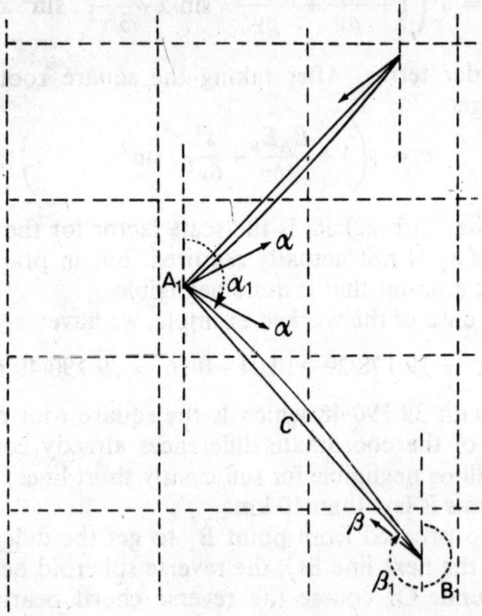

FIG. 143. CHORDS IN CASSINI PROJECTION.

the projection system with curves and chords as already described. On the Cassini projection the inevitable north–south 'stretching' of scale is not balanced by any east–west change of scale, and a little thought will indicate that the general effect will be that the projection grid bearings will be nearer to north–south than the corresponding spheroid grid bearings: in other words, if spheroid grid bearing α is represented on the projection by grid bearing α_1, then in the first and third quadrants $\alpha_1 < \alpha$, and in the second and fourth quadrants $\alpha_1 > \alpha$. This is shown on the diagram, where the spheroid grid bearings are indicated by the short lines with arrowheads.

In starting a computation of coordinate differences, the spheroid grid bearing α of the line must first be found by applying the convergence to the true bearing, then we require a formula to convert this to projection grid bearing α_1; the formula will depend on α and on the position of A in the projection system, but it will not depend on the length of the line to be computed since it is a property of the scale structure of the projection at point A. Next, the arc-chord angle must be applied, and of course this is a function of s.

Using the formulas for the non-orthomorphic projections in Section 9.09, we find that the chord bearing formula $\tan G = (E_B - E_A)/(N_B - N_A)$ can be expressed by the series:

$$\tan G = \tan\alpha - \frac{E_A{}^2}{2\nu\rho} \, . \tan\alpha - \frac{E_A}{2\nu\rho} \, . \, s \, . \sec\alpha(1+2\sin^2\alpha)$$

$$- \frac{1}{6}\frac{s^2}{\nu^2} \, . \tan\alpha(1+\sin^2\alpha) \ldots$$

As s approaches zero, only the first two terms will remain significant, and in the limit they are the expression of $\tan\alpha_1$. So, putting $\alpha_1 = \alpha - \theta$, $\tan\alpha_1 = \tan\alpha - \theta \, . \sec^2\alpha \ldots$, we find that

$$\alpha_1 - \alpha = \theta = \frac{E_A{}^2}{2\nu\rho} \, . \sin\alpha \, . \cos\alpha \text{ (in radian).}$$

Re-writing the series for $\tan G$ as

$$\tan\alpha_1 - \frac{s \, . \sec\alpha}{2\nu\rho}(E_A + \tfrac{1}{2}s \, . \sin\alpha)(1+\sin^2\alpha) \ldots,$$

putting $G = \alpha_1 - \delta$, $\tan G = \tan\alpha_1 - \delta \, . \sec^2\alpha_1 \ldots$ and ignoring the difference between α_1 and α in the small terms, we find that the arc-chord angle is:

$$\delta = \frac{s \, . \cos\alpha_1}{2\nu\rho}(E_A + \tfrac{1}{3}s \, . \sin\alpha_1)(1+\sin^2\alpha_1)$$

This may be compared with the formula in Section 9.11.

Referring to the numerical example, we have as before $\alpha = 127° \, 12' \, 20{\cdot}8''$ and θ works out as $-25{\cdot}1''$; hence $\alpha_1 = \alpha - \theta = 127° \, 12' \, 45{\cdot}9''$. The numerical values work out as follows:

$$s \, . \cos\alpha_1 \doteqdot -23\,694 \text{ m}, \qquad s \, . \sin\alpha_1 \doteqdot +31\,201 \text{ m},$$

$$E_A + \tfrac{1}{3}s \, . \sin\alpha_1 \doteqdot 153\,774 \text{ m}, \qquad \delta = -15{\cdot}0'',$$

$$G = 127° \, 13' \, 00{\cdot}9''.$$

From the formulas in Section 9.04 we find that

$$c = s + \frac{E_A E_B}{2\nu\rho} \,.\, s \,.\, \cos^2 \alpha + \frac{1}{6} \frac{s^3}{\nu^2} \,.\, \sin^2 \alpha \,.\, \cos^2 \alpha \ldots$$

Numerically, $c = 39\,178{\cdot}29 + 4{\cdot}40 + 0{\cdot}06 = 39\,182{\cdot}75$ m. The differences of coordinates, $c \,.\, \cos G$ and $c \,.\, \sin G$ are $-23\,699{\cdot}07$ and $+31\,203{\cdot}24$ respectively, giving for the Cassini projection $N_B = 89\,438{\cdot}18$ m and $E_B = 174\,578{\cdot}18$ m, which are to be compared with the values given in Section 9.04.

To continue, we find that the arc-chord angle at B_1 is $+16{\cdot}1''$ and the reverse grid bearing of the curve on the projection is $\beta_1 = 307° \; 13' \; 17{\cdot}0''$. To get back to the spheroid, we find that the projection bearing difference 'θ' is now $-37{\cdot}2''$, so the spheroid grid bearing is $\beta = \beta_1 - 37{\cdot}2'' = 307° \; 12' \; 39{\cdot}8''$. To this true bearing the true angle between BA and BC on the spheroid is applied to get the spheroid bearing of BC, and so the computation proceeds to the next line.

The various steps in the bearing calculations are briefly recapitulated below:

At Point A_1.	
True Bearing	128° 44′ 17·2″
Convergence γ	−1 31 56·4
Spheroid Grid Bearing	127 12 20·8
$\theta = \frac{E_A{}^2}{2\nu\rho} \,.\, \sin \alpha \,.\, \cos \alpha$	−25·1
Projection Grid Bearing $\alpha_1 = \alpha - \theta$	127 12 45·9
Arc-chord $\delta = \frac{s \,.\, \cos \alpha}{2\nu\rho} (E_{\frac{2}{3}})(1 + \sin^2 \alpha)$	−15·0
Chord Bearing $G = \alpha_1 - \delta$	127 13 00·9
At Point B_1.	
Chord Bearing to A_1: $G \pm 180°$	307 13 00·9
Arc-chord $\mu = \frac{s \,.\, \cos \beta}{2\nu\rho} (E_{\frac{2}{3}})(1 + \sin^2 \beta)$*	+16·1
Projection Grid Bearing $\beta_1 = G \pm 180° + \mu$	307 13 17·0
$\theta = \frac{E_B{}^2}{2\nu\rho} \,.\, \sin \beta_1 \,.\, \cos \beta_1$	−37·2
Reverse Spheroid Grid Bearing $\beta = \beta_1 + \theta$	307 12 39·8

* The chord bearing must be used in this formula since β is not yet known. $E_{\frac{2}{3}}$ means $(E_A + \frac{2}{3}s \,.\, \sin \alpha)$ or $(E_B + \frac{1}{3}s \,.\, \sin \beta)$, the easting of the point two-thirds of the way from A_1 to B_1.

The angle γ_1 shown on Fig. 142 is the meridian grid convergence on the projection and now of course it is not equal to γ. Coordinates N and E are functions of latitude and longitude and we have the general differential formulas

$$dE = \frac{\partial E}{\partial \phi}\,.\,d\phi + \frac{\partial E}{\partial \lambda}\,.\,d\lambda : \qquad dN = \frac{\partial N}{\partial \phi}\,.\,d\phi + \frac{\partial N}{\partial \lambda}\,.\,d\lambda.$$

Tan γ_1 is numerically equal to the special value of dE/dN when these two differentials are related so that there is no change of longitude, i.e. $d\lambda = 0$. Thus $\tan\gamma_1 = -\frac{\partial E}{\partial \phi}\Big/\frac{\partial N}{\partial \phi}$, the negative sign being appropriate because γ_1 will be positive when dE is negative. From the projection formulas:

$$\tan\gamma_1 = \frac{\lambda\,.\,\rho\,.\,\sin\phi + \frac{1}{3}\lambda^3\nu\,.\,\cos^2\phi\,.\,\sin\phi - \frac{1}{6}\lambda^3\rho\,.\,\sin^3\phi\,\ldots}{\rho - \frac{1}{2}\lambda^2\rho\,.\,\sin^2\phi + \frac{1}{2}\lambda^2\nu\,.\,\cos^2\phi\,\ldots}$$

which reduces to

$$\gamma_1 = \lambda\,.\,\sin\phi - \tfrac{1}{6}\lambda^3\,.\,\sin\phi\,.\,\cos^2\phi\,\ldots.$$

At point A_1, $\gamma_1 = 5515{\cdot}52'' - 0{\cdot}46'' \neq 1^\circ\,31'\,55{\cdot}1''$.

To complete the picture, the angle g_1 between the parallel and the grid line on the projection is given by $\tan g_1 = \frac{\partial N}{\partial \lambda}\Big/\frac{\partial E}{\partial \lambda}$ which reduces to $g_1 = \lambda\,.\,\sin\phi + \frac{5}{6}\lambda^3\,.\,\sin\phi\,.\,\cos^2\phi\,\ldots$ and at A_1 this is $1^\circ\,31'\,57{\cdot}8''$. Obviously the value of $(g_1 - \gamma_1)$ is the amount by which the intersection of meridian and parallel on the projection differs from exactly 90°: it is given by the formula $\lambda^3\,.\,\sin\phi\,.\,\cos^2\phi$, and at A_1 this is 2·8″, an angle far too small to be visible on any map.

9.13 Hotine Method. A method of dealing with non-orthomorphic projections, when computing a survey covering a rather small area, was suggested by Hotine. On the Cassini projection, for instance, at the point A used in numerical examples, the scale factor is 1·000 252 in the northing direction. We suppose that a surveyor has made some detail surveys in an area within 5 or 10 km of point A and has based his work on the national Cassini projection system.

Hotine's idea is to use temporary working coordinates referred to an origin in the area of the survey, a procedure which in any case helps to reduce the amount of arithmetic to be done. For instance, an origin at $N_0 = 110\,000$ m, $E_0 = 140\,000$ m would give the coordinates of A the temporary values 3137·25 and 3374·94. Now the procedure is to reduce all northing coordinates by removing the northing scale expansion of 252 parts per million. Thus the working coordinates of A will be 3136·46 and 3374·94. Any other control points used would be similarly treated. The effect is to make the working system locally orthomorphic. This simplifies computation, and when the surveyor has calculated his new points in the temporary system their national projection coordinates can be found by reversing the process. For example, a new point with temporary coordinates 1876·85 and 3527·51 would have the projection position

$$N = 110\,000 + 1877{\cdot}32 = 111\,877{\cdot}32$$

$$E = 140\,000 + 3527{\cdot}51 = 143\,527{\cdot}51.$$

The reader will observe that it is not really necessary to make any changes in the eastings, since the scale is true in that direction anyway.

9.14 Scale-Reduction Device. It will be seen that in all the projections described so far, the scale factor at all points and in all directions is either equal to or greater than 1: at the point B_1 in the numerical example, in the Transverse Mercator projection, the scale factor is 1·000 374, and at 200 km from the central meridian it is 1·000 491.

A practice commonly adopted nowadays is to apply an overall reduction factor, so that the scale is less than 1 in a central part of the system, equal to 1 along a particular easting grid line, and greater than 1 elsewhere. It is convenient to adopt a simple fraction, and in the case mentioned above a reduction of 1/4000 would make the scale factor 0·999 750 on the central meridian and 1·000 241 at 200 km on either side. The formula for actual scale factor would then be

$$1 + \frac{E^2}{2\nu\rho} - \frac{1}{4000}$$: this is equal to 1 when E is about 142 700 m.

Perhaps the simplest way to visualise and put into effect this overall reduction is to assume that one is working on a spheroid model of dimensions reduced, in this case by one part in 4000. In computations, the tabulated values of ρ and ν would be reduced to 3999/4000 of their true values, and any lengths measured or derived in survey operations would be reduced similarly before being used. The length 39 178·29 used in previous examples would come down to 39 168·50, then the projection scale effect of 12·19 m would be added to give the length 39 180·69 on the reduced projection system.

The reduction factor in the Ordnance Survey map system is 0·999 601 271: see Section 9.17.

9.15 Conical Projections. For mapping an area of considerable extent in longitude, a geometrical system based on the idea of an east–west centre line can be used, and in such a system the differences of latitude to be covered will be comparatively small.

Projections of the type described as *conical* are suitable. The geometrical relationships are illustrated in Fig. 144. The central

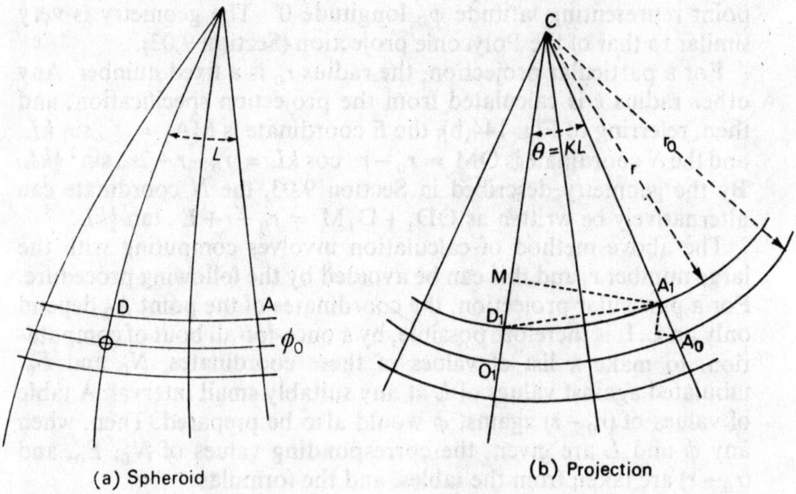

FIG. 144. CONICAL PROJECTION.

parallel of latitude ϕ_0 would be chosen to run through the middle of the area to be mapped. A reference meridian for longitude is also required and can conveniently be located in the middle of the area: longitude referred to this meridian is denoted by the capital *L* to emphasise that this angle cannot now be treated as small.

On the map system, all the meridians are represented by straight lines radiating from a common centre C. The angles between map meridians have a constant ratio with the true differences of longitude, the ratio *k* being usually a substantial fraction, of the order of 0·7.

The parallels on the map are parts of concentric circles centred at the point C, and their radii are functions of the latitude only. The lengths of these radii will be of the same order of magnitude as the dimensions of the spheroid, but the differences $(r_0 - r)$ will be limited to about the linear half-width of the strip covered by the system—say 250 km—corresponding with the east–west half-width in the transverse types of projection described previously.

The geometry of any particular conical projection is described by two specifications: (i) the value of the 'constant of the cone' *k*,

(ii) the functional relationship of the linear differences (r_0-r) to the latitude differences $(\phi-\phi_0)$. By suitable definition of these specifications, a conical projection can be given desired properties—such as being orthomorphic. Given the specification of a conical projection, rectangular coordinates of point A_1, representing point A of latitude ϕ longitude L on the spheroid, can be calculated as given below. An obvious point for the origin of coordinates is O_1, the point representing latitude ϕ_0 longitude 0°. The geometry is very similar to that of the Polyconic projection (Section 9.03).

For a particular projection, the radius r_0 is a fixed number. Any other radius r is calculated from the projection specification, and then, referring to Fig. 144(b), the E coordinate is $MA_1 = r \,.\, \sin kL$, and the N coordinate is $OM = r_0 - r \,.\, \cos kL = r_0 - r + 2r \,.\, \sin^2 \tfrac{1}{2}kL$. By the geometry described in Section 9.03, the N coordinate can alternatively be written as $OD_1 + D_1M = r_0 - r + E \,.\, \tan \tfrac{1}{2}kL$.

The above method of calculation involves computing with the large number r, and this can be avoided by the following procedure. For a particular projection, the coordinates of the point A_0 depend only on L. It is therefore possible, by a once-for-all bout of computation, to make a list of values of these coordinates, N_0 and E_0, tabulated against values of L at any suitably small interval. A table of values of (r_0-r) against ϕ would also be prepared. Then, when any ϕ and L are given, the corresponding values of N_0, E_0, and (r_0-r) are taken from the tables, and the formulas:

$$N = N_0 + (r_0 - r)\cos kL; \qquad E = E_0 - (r_0 - r)\sin kL$$

give the rectangular coordinates by use of numbers requiring only seven-figure order of accuracy.

Inverse Computation. Given N and E, the simple computations:

$$\tan\theta = E/(r_0 - N); \qquad L = \theta/k; \qquad r_0 - r = N - E \,.\, \tan \tfrac{1}{2}\theta$$

need no further explanation, and the latitude ϕ is obtained from (r_0-r) by inverse interpolation or formula.

Scale in a Conical Projection. Because of the simple structure of these projections, all meridians are geometrically identical. The scale factor in the meridian direction is numerically equal to the ratio of an element dr of the meridian to the 'true' length $\rho \,.\, d\phi$ on the spheroid, but the factor should be written $-dr/\rho \,.\, d\phi$ because in the notation now being used the radius r decreases as ϕ increases. Along each parallel the scale factor is constant, being simply the ratio of any length of parallel on the map to the length of the corresponding piece of the parallel on the spheroid model. On the model, for a difference of longitude L along parallel ϕ the length is $v \cdot \cos\phi \cdot L$ and on the projection the length is $k \cdot r \cdot L$, so the scale factor is $r \cdot k/(v \cdot \cos\phi)$.

Central Parallel. In analogy with the transverse projections, the construction of a conical projection can be based on making a central parallel, latitude ϕ_0, true to scale. Referring to the formula given above, this means that $k \cdot r_0 = v_0 \cdot \cos\phi_0$. If a cone with its vertex on the polar axis touches the spheroid along parallel ϕ_0, as indicated in Fig. 145, then the length of QA is $v_0 \cdot \cot\phi_0$, and

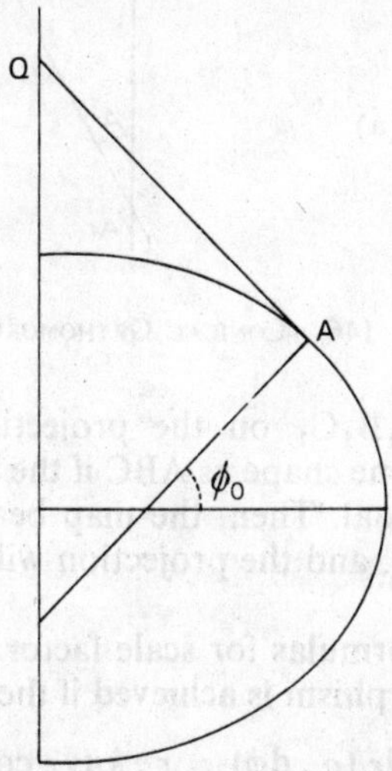

FIG. 145. STANDARD PARALLEL.

this will be the radius of the circle representing parallel ϕ_0 when the cone is slit along a meridian and laid out flat.

The tangent cone is analogous with the idea of associating cylindrical projections with a cylinder touching the spheroid along the equator.

Putting $r_0 = v_0 \cdot \cot\phi_0$ gives $k = \sin\phi_0$.

After adopting this or any other particular value for the constant of the cone, there remains the question of deciding on the spacing out of the parallels in relation to the latitude. For instance, if $(r_0 - r)$ is taken to be simply the linear distance along the meridian from latitude ϕ_0 to latitude ϕ, we get a Conical Equidistant projection: this will not be discussed further, because the only type of conical projection likely to be used for mapping is an orthomorphic one.

9.16 Conical Orthomorphic Projections. In any conical projection as defined above, the meridians and parallels necessarily cut at right-angles. Any projection having this feature will be orthomorphic if at any point the scale along the meridian is equal to the scale along the parallel. Triangle ABC on the spheroid, Fig. 146(a), is formed by infinitesimal bits of meridian AC and parallel CB, and

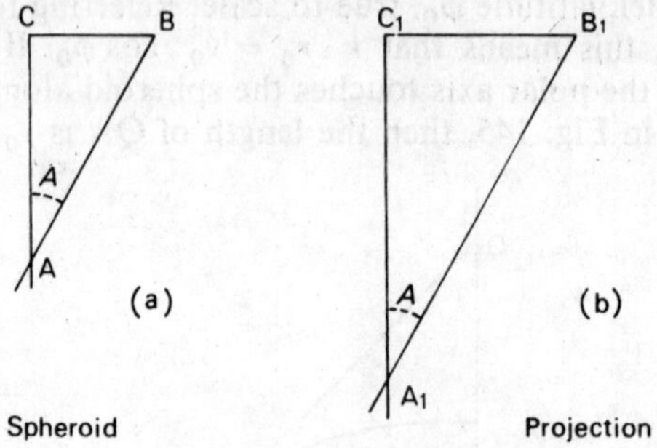

FIG. 146. CONICAL ORTHOMORPHIC.

its representative $A_1B_1C_1$ on the projection, being also right-angled, will be the same shape as ABC if the scale factors A_1C_1/AC and C_1B_1/CB are equal. Then, the map bearing A_1 will be equal to the true bearing A, and the projection will be angle-true, that is orthomorphic.

Referring to the formulas for scale factor given in Section 9.15, we see that orthomorphism is achieved if the relationship

$$-\mathrm{d}r/(\rho \,.\, \mathrm{d}\phi) = r \,.\, k/(\nu \,.\, \cos \phi)$$

is satisfied for all latitudes. This can also be written

$$-\frac{1}{k} \,.\, \frac{\mathrm{d}r}{r} = \frac{\rho}{\nu} \,.\, \sec \phi \,.\, \mathrm{d}\phi.$$

Radii of Parallels. For numerical computation of coordinates, using logarithms or a small calculator, a table of values of $(r_0 - r)$ against ϕ will be useful. Alternatively, there is a closed formula for r which can be used if sufficient computer power is available.

I. To make the table, one might think, in analogy with the transverse projections, of writing a series

$$r = r_0 - A(\phi - \phi_0) - B(\phi - \phi_0)^2 \ldots$$

and proceeding by the appropriate mathematical operations to find values of A B . . . that make the projection orthomorphic. This indeed can be done, but there are simpler methods.

It is obvious that in any reasonable projection the length $(r_0 - r)$

will be close to the meridian length $(S_\phi - S_{\phi_0})$ from latitude ϕ_0 to latitude ϕ. Denoting this length by m, one might try a series

$$r = r_0 - m - Pm^2 - Qm^3 \ldots$$

The condition for orthomorphism can then be written:

$$-\frac{dr}{dm} . v . \cos\phi = k . r . \rho . \frac{d\phi}{dm}$$

But of course $\rho . d\phi = dm$. Taking $k = \sin\phi_0$, $r_0 = v_0 . \cot\phi_0$, and writing in the series for r and dr/dm gives

$$(1+2Pm+3Qm^2+4Rm^3 \ldots)v . \cos\phi$$
$$= \sin\phi_0(v_0 \cot\phi_0 - m - Pm^2 - Qm^3 \ldots).$$

$v . \cos\phi$ must be replaced by its expansion as a difference from from $v_0 . \cos\phi_0$ in powers of m. By straightforward differentiation:

$$\frac{d(v . \cos\phi)}{d\phi} = -\rho . \sin\phi,$$

hence $$\frac{d(v . \cos\phi)}{dm} = \frac{d(v . \cos\phi)}{d\phi} . \frac{d\phi}{dm} = \frac{-\rho . \sin\phi}{\rho} = -\sin\phi.$$

The second differential coefficient is therefore:

$$-\cos\phi . \frac{d\phi}{dm} = \frac{-\cos\phi}{\rho}.$$

The next turns out to be $\dfrac{\sin\phi}{v\rho}\left(1+\dfrac{4e^2 . \cos^2\phi}{1-e^2}\right)$ and the next after that, omitting everything with e^2 as factor, is $\dfrac{\cos\phi}{v\rho^2}$. The coefficients in the series for r are then obtained by equating like powers of m on each side of the orthomorphism condition

$$(1+2Pm+3Qm^2 \ldots)\left(v_0 \cos\phi_0 - \sin\phi_0 . m - \frac{\cos\phi_0}{\rho_0} . \frac{m^2}{2} + \frac{\sin\phi_0}{v_0\rho_0}(1+4e^2 . \cos^2\phi_0)\frac{m^3}{6} + \frac{\cos\phi_0}{v_0\rho_0{}^2}\frac{m^4}{24} \ldots\right)$$
$$= v_0 \cos\phi_0 - m . \sin\phi_0 - Pm^2 . \sin\phi_0 - Qm^3 . \sin\phi_0 - Rm^4 . \sin\phi_0 \ldots$$

Equating coefficients of m gives $2Pv_0 . \cos\phi_0 - \sin\phi_0 = -\sin\phi_0$, so $P = 0$. Then we get $Q = 1/6v_0\rho_0$, and the full expansion for

$(r_0 - r)$ is found to be:

$$m + \frac{m^3}{6\nu_0\rho_0} + \frac{m^4 \,.\, \tan\phi_0}{24{\nu_0}^2\rho_0}(1 - 4e^2 \,.\, \cos^2\phi_0)$$
$$+ \frac{m^5}{120{\nu_0}^3\rho_0}(5 + 3\tan^2\phi_0)\ \ldots$$

in which terms up to fifth order only have been retained. From a basic table of meridian distances the values of $(r_0 - r)$ can be calculated: tabulation at intervals of 1′ of latitude is close enough for simple interpolation.

The actual scale factor at any point is $-\mathrm{d}r/\mathrm{d}m$, or

$$1 + \frac{m^2}{2\nu_0\rho_0} + \frac{m^3 \,.\, \tan\phi_0}{6{\nu_0}^2\rho_0}\ \ldots$$

II. The differential equation for orthomorphism is integrable in finite terms: the equation can be written:

$$-\frac{1}{k} \,.\, \frac{1}{r} \,.\, \frac{\mathrm{d}r}{\mathrm{d}\phi} = \frac{\rho}{\nu} \,.\, \sec\phi = \frac{(1 - e^2)\sec\phi}{1 - e^2 \,.\, \sin^2\phi}$$
$$= \left(1 - \frac{e^2\cos^2\phi}{1 - e^2\sin^2\phi}\right)\sec\phi$$
$$= \sec\phi - \frac{1}{2} \,.\, \frac{e^2\cos\phi}{1 - e\sin\phi} - \frac{1}{2}\,\frac{e^2\cos\phi}{1 + e \,.\, \sin\phi}$$

and the integral is

$$-\frac{1}{k}\ln r = \ln\tan(45° + \tfrac{1}{2}\phi) - \tfrac{1}{2}e \,.\, \ln\left(\frac{1 + e \,.\, \sin\phi}{1 - e \,.\, \sin\phi}\right) + \text{const.}$$

This equation is satisfied by the particular values r_0 and ϕ_0; on writing the equation with these values, subtracting, and taking antilogarithm, the result is:

$$\left(\frac{r}{r_0}\right)^{-\frac{1}{k}} = \frac{\tan(45° + \frac{1}{2}\phi)}{\tan(45° + \frac{1}{2}\phi_0)}\left[\frac{(1 + e \,.\, \sin\phi_0)(1 - e \,.\, \sin\phi)}{(1 - e \,.\, \sin\phi_0)(1 + e \,.\, \sin\phi)}\right]^{\frac{1}{2}e}$$

A computer of quite modest capacity can cope with this!

The scale factor at latitude ϕ is $-\dfrac{\mathrm{d}r}{\rho \,.\, \mathrm{d}\phi}$ or $\dfrac{kr}{\nu \,.\, \cos\phi}$.

A slight disadvantage in using a conical projection is that, if the map sheet lines are taken along grid coordinate lines the convergence angle between grid 'north' and meridian can be quite conspicuous, being of course equal to kL. This feature may be less obvious if meridians and parallels are used as sheet edges.

An overall scale reduction can be applied as already described (Section 9.14). This will have the effect of making the scale true along two different parallels, and between these the scale will be less than the nominal value.

Alternatively, it is possible to make the scale true along two chosen parallels ('standard parallels') by putting the two latitudes into the projection formula and solving for k and r_0.

9.17 Ordnance Survey Map Projection. The re-triangulation of Great Britain, completed in 1963, was an entirely new survey, even to the extent of including only eleven of the old primary stations in the new system. However, these eleven points played their parts in determining the structure of the new mapping framework, which has been computed on the same Airy spheroid as was used for the earlier surveys. The axes of this spheroid were given in feet and some researches had to be done in order to determine the relationship of those feet to the present accepted International Metre. The result is that the spheroid for mapping purposes has $a = 6\,377\,563{\cdot}39$ m, $1/f = 299{\cdot}324\,965$. The projection is Transverse Mercator with central meridian at 2°W longitude, and the origin for use in computation of coordinate formulas is at 49°N latitude. When coordinates have been calculated on this spheroid and origin from the standard Transverse Mercator formulas (Section 9.05), an overall reduction (Section 9.14) is applied by multiplying by the factor 0·999 601 271. Finally, in order to keep all coordinates within the map system positive, 400 000 m are added to all eastings since these would otherwise be negative for points west of 2°W: also, 100 000 m are subtracted from calculated northings.

REFERENCES

BOMFORD A. G. 'Transverse Mercator arc-to-chord and finite distance scale factor formulae'. *Empire Survey Review* No. 125, July 1962.

BRAZIER H. H. 'A skew orthomorphic projection with particular reference to Malaya'. *Commonwealth Survey Officers' Conference* 1947.

BURNS E. L. M. 'Point to point working for the conical orthomorphic projection'. *Empire Survey Review* No. 11, January 1934.

CHIAT B. 'Semigraphic adjustment and trilateration'. *Survey Review* No. 130, October 1963.

CLENDINNING J. 'Grid bearings and distances on the conical orthomorphic projection'. *Empire Survey Review* No. 48, April 1943.

DUFOUR H. M. 'Le calcul automatique des triangulations secondaires'. *Bulletin Géodésique* No. 84, 1967.

ERRITT W. A. 'The Transverse Mercator projection and its application to Tanganyika Territory'. *Empire Survey Review* No. 21, July 1936.

FANG J. T. 'Azimuth and distance corrections for the geodetic lines on the Lambert Conformal projection'. *Empire Survey Review* No. 75, January 1950.

HENDRICKZ D. R. 'The theory of Map Projections'. *Trigonometrical Survey of South Africa* 1949.

HOTINE M. 'The orthomorphic projection of the spheroid'. *Empire Survey Review* No. 62, October 1946.

KING C. W. B. 'Computation of ground control for air surveys at large scales'. *Empire Survey Review* No. 113, July 1959.

—— 'Survey computations on a small computer'. *Survey Review* No. 136, April 1965.

LEE L. P. 'The transverse Mercator projection of the spheroid'. *Empire Survey Review* No. 58, October 1945.

—— 'The Oblique Mercator projection'. *Empire Survey Review* No. 101, July 1956.

MAINWARING J. 'Map Projections'. *London* 1942.

MELLUISH R. K. 'Map Projections'. *Cambridge* 1931.

PAPWORTH K. M. 'The (t −T) correction for the Lambert No. 2 (conical orthomorphic) projection'. *Empire Survey Review* No. 56, April 1945.

REDFEARN J. C. B. 'Transverse Mercator formulae'. *Empire Survey Review* No. 69, July 1948.

ROYAL GEOGRAPHICAL SOCIETY *New Geodetic Tables* . . . 1927.

STEERS J. A. *An introduction to the study of map projections.* London 1962.

THOMPSON E. H. 'A direct method of transforming rectangular coordinates from the Cassini projection to the Gauss Conformal projection'. *Empire Survey Review* No. 18, October 1935.

WILLIAMS H. S. 'Nomographic treatment of the Cassini-Soldner projection correction terms'. *Empire Survey Review* No. 124, April 1962.

Universal Transverse Mercator Grid. U.S. Govt. Printing Office, 1971.

CONSTANTS, UNITS, ETC.

			logarithm
π =	3·141 592 653 6		0·497 149 872 7
1 radian =	57·295 779 513	degrees	1·758 122 632 4
=	3 437·746 770 8	minutes	3·536 273 882 8
=	206 264·806 25	seconds	5·314 425 133 2

The International System of Units (SI)

The units of length and time are now defined in terms of electromagnetic radiation:

The METRE, symbol m, is the length of 1 650 763·73 wavelengths of a certain transition in the krypton-86 atom, in vacuum.

The SECOND, symbol s, is the duration of 9 192 631 770 oscillations of the radiation associated with a certain transition in the caesium-133 atom. (this unit agrees with the mean solar second in the year 1900).

The KILOGRAM, symbol kg, is the mass of an internationally recognised piece of matter.

Some Spheroids

Clarke 1866	a = 6 378 206·40 m,	$1/f$ = 294·978 7
International (Hayford 1924)	6 378 388·00	297·000 0
Reference 1967	6 378 160·00	298·247 1

Velocity of Light in vacuum is 299 792·5 km/s.
(The *foot* is defined to be 304·8 millimetres exactly)

GREEK ALPHABET

A	α	Alpha	I	ι	Iota	P	ρ	Rho
B	β	Beta	K	κ	Kappa	Σ	σ	Sigma
Γ	γ	Gamma	Λ	λ	Lambda	T	τ	Tau
Δ	δ	Delta	M	μ	Mu	Υ	υ	Upsilon
E	ε	Epsilon	N	ν	Nu	Φ	ϕ	Phi
Z	ζ	Zeta	Ξ	ξ	Xi	X	χ	Chi
H	η	Eta	O	o	Omicron	Ψ	ψ	Psi
Θ	θ	Theta	Π	π	Pi	Ω	ω	Omega

ANSWERS TO EXAMPLES

Chapter II

1. 12° 20′ 06″ South
2. 21^h
3. Hour angle $4^h\ 01^m\ 52^s$
   Azimuth 43° 46′ or 316° 14′
   Altitude 45° 27′
5. LST $0^h\ 54^m\ 00{\cdot}3^s$
   Altitude 29° 42′ 21″
   Chronometer $21^h\ 20^m\ 40{\cdot}3^s$
6. 74° 25′ 46″ East
7. Position line cuts on parallel 6° 53′N at longitude 44° 51′ 30″W, at bearing 160° 38′.
8. 197° 14′ 06″
   (i) −7·5″: (ii) +2·0″

Chapter V

1. 306·798
2. 306·796
3. P 109·533
   Q 52·487
4. Adjustments:

−0·1″	+1·1″
+1·3	−0·9
−3·0	−1·2
−1·6	+0·1
−2·5	+0·9
−0·1	
−1·4	
+0·5	
−0·2	
+0·4	

5. Adjustments:

−4″	0″
−2	+2
−2	−2
0	0

7. $n = -0{\cdot}7$ cm
   $e = -1{\cdot}9$ cm

Chapter VI

1. 89° 53′ 34·1″
   270 06 25·9
2. 60° 01′ 29·5″
   72 57 40·2
   47 01 05·0
4. 19′ 03·570″
   24′ 57·431″

INDEX

The numbers refer to Sections

The numbers refer to Sections

The numbers refer to Sections

The numbers refer to Sections